Handbook of Technological Sustainability

In the context of an ever-changing technological world, it is crucial to prioritize the cultivation of sustainability and environmental consciousness. Adopting an interdisciplinary perspective, this handbook facilitates the integration of technology and sustainability by leveraging insights from many fields such as engineering, environmental science, economics, and sociology. The primary objective of this handbook is to provide a comprehensive analysis of the complex relationship between technology and the environment.

Handbook of Technological Sustainability: Innovation and Environmental Awareness includes recent and diverse case studies from a global perspective and demonstrates the utilization of technology to achieve sustainable development across several sectors, such as energy, agriculture, transportation, and urban planning. It explores innovative technologies emphasizing state-of-the-art and developing technologies, including renewable energy, circular economy practices, smart cities, and artificial intelligence-driven sustainability solutions. The handbook also examines the impact of laws, regulations, and international agreements on the advancement or impediment of technological sustainability. Written to be used as a reference, this handbook highlights the ethical and moral quandaries linked to technological sustainability analyzing various topics of significance, including environmental justice, privacy implications in smart technology, and the ramifications of artificial intelligence on employment and society. It provides practical methodologies and recommendations for individuals, enterprises, and governmental entities to adopt and integrate sustainable technologies effectively and furnish readers with a comprehensive guide for transitioning towards a more sustainable future.

As it showcases perspectives from technologists, sustainability professionals, and policymakers that have the potential to provide a range of opinions and practical insights derived from real-world experiences, this handbook is a book for all individuals, corporations, governments, researchers, and anyone seeking to harness technology to advance sustainability and effectively tackle the multifaceted issues posed by our dynamic global landscape.

Handbook of Technological Sustainability

Innovation and Environmental Awareness

Edited by

Pankaj Bhambri and Paula Bajdor

CRC Press is an imprint of the
Taylor & Francis Group, an **informa** business

Designed cover image: Shutterstock - PeachShutterStock

First edition published 2025
by CRC Press
2385 NW Executive Center Drive, Suite 320, Boca Raton FL 33431

and by CRC Press
4 Park Square, Milton Park, Abingdon, Oxon, OX14 4RN

CRC Press is an imprint of Taylor & Francis Group, LLC

ISBN: 978-1-032-75801-5 (hbk)
ISBN: 978-1-032-75864-0 (pbk)
ISBN: 978-1-003-47598-9 (ebk)

DOI: 10.1201/9781003475989

Typeset in Times
by Deanta Global Publishing Services, Chennai, India

Contents

PART II Technological Solutions for Sustainable Living

PART III Challenges, Risks, and Ethical Considerations

Preface

This handbook serves as a thorough guide that delves into the complex connection between technology and sustainability during a time of significant technical progress and increasing environmental issues. At the crossroads of advancement and duty, it is crucial to have a comprehensive grasp of how technology can influence and safeguard our world.

This handbook covers a wide range of topics, from the basic principles of sustainability to the ethical issues related to artificial intelligence in environmental solutions. The chapters explore the economic, social, and environmental aspects of sustainable technology, providing insights on renewable energy, waste management, agriculture, transportation, and other related topics. Our distinguished writers, specialists in their professions, explain the intricacies of technological sustainability, offering readers a guide to tackle obstacles and seize possibilities in the future.

This handbook's chapters assess the existing scene and predict future trends, innovations, and potential hazards. This handbook showcases the diverse aspects of the connection between technology and the environment, from analyzing South Africa's renewable energy sector to investigating the effects of deep fakes on media sustainability.

We see this handbook as a valuable resource for researchers, academics, industry experts, and anyone dedicated to promoting a sustainable future. We seek to stimulate discussion, encourage analytical thinking, and enable readers to participate in the continuous discourse on technological sustainability. We thank the contributors for their commitment to addressing important issues and anticipate that this handbook will stimulate valuable conversations and bring about beneficial transformations.

Let's work together to utilize technology to improve our planet and the welfare of future generations.

Editors

Pankaj Bhambri is affiliated with the Department of Information Technology at Guru Nanak Dev Engineering College in Ludhiana. Additionally, he fulfills the role of convener for his departmental board of studies. He possesses nearly two decades of teaching experience. His research work has been published in esteemed worldwide and national journals, as well as in conference proceedings. Dr. Bhambri has garnered extensive experience in the realm of academic publishing, having served as an editor for a multitude of books in collaboration with esteemed publishing houses including CRC Press. In addition to his editorial roles, he has demonstrated his scholarly prowess by authoring numerous books and contributing chapters to distinguished publishers within the academic community. Dr. Bhambri has been honored with several prestigious accolades, including the ISTE Best Teacher Award in 2023 and 2022, the I2OR National Award in 2020, the Green ThinkerZ Top 100 International Distinguished Educators Award in 2020, the I2OR Outstanding Educator Award in 2019, the SAA Distinguished Alumni Award in 2012, the CIPS Rashtriya Rattan Award in 2008, the LCHC Best Teacher Award in 2007, and numerous other commendations from various government and non-profit organizations. He has provided guidance and oversight for numerous research projects and dissertations at the postgraduate and PhD levels. He successfully organized a diverse range of educational programmes, securing financial backing from esteemed institutions such as the AICTE, the TEQIP, among others. Dr. Bhambri's areas of interest encompass machine learning, bioinformatics, wireless sensor networks, and network security.

Paula Bajdor is affiliated with the Faculty of Management at Częstochowa University of Technology and Faculty of Social Science at Calisia University. She has more than 10 years of teaching experience and 5 years of working experience in a Logistics company. She acquired Master of Management, Work and Safety and Management and Production Engineering degrees from the Częstochowa University of Technology, and a degree of post-graduate studies in financial and accounting from AGH University in Cracov. She obtained her doctorate in Social Science from Częstochowa University of Technology in 2012, by conducting research on Sustainable Development in Companies' Supply Chain Management. In 2022, she was habilitated based on her research about sustainable entrepreneurship. Her scientific interests include issues related to sustainable entrepreneurship, sustainable development, information and communication technologies, logistics and supply chain management, and cloud computing. For more than 10 years, Prof. Bajdor fulfilled many responsibilities such as organizational and scientific committee member for more than 40 international conferences; member of international associations – WSEAS, IETI, EBES, PTE, and NTIE, member of editor boards of four scientific journals, reviewers of scientific articles, conference's proceedings, monographs and academic and project's achievements, coordinator of SCAN scientific network within the CEEPUS framework, students' supervisor and expert in research conducted in cooperation with companies. She is an author of more than 120 scientific publications, published in scientific journals, conference proceedings, books, and monographs. She is an author of two books, chapters in academic textbooks, and co-author of one monograph. Prof. Bajdor was also the manager of two research projects (Improving Competencies in Data Analysis and Technology and Entrepreneurship Education – Bridging the Gap for Smart Product Development – TecHub 4.0) and a contractor in nine projects. She cooperates with many foreign research centers (Universitatea Europei de Sud Est Lumina, Slovak University of Technology in Bratislava, Valahia University of Targoviste, O.M. Beketov University in Kharkiv, Ukraine, University of Novi Sad, University of Cadiz, Spain, University of Fort Hare, South Africa) and domestic research centers from Warsaw, Cracov, Wroclaw, and Szczecin. The results of this cooperation are joint scientific research, lectures, classes with students, conferences, internships and publications. She has provided guidance for many dissertations for undergraduate, postgraduate, and PhD students. So far, she has received many awards and diplomas in recognition of her achievements, commitment, and scientific work.

Contributors

Paula Bajdor
Czestochowa University of Technology
Czestochowa, Poland

Pankaj Bhambri
Guru Nanak Dev Engineering College
Ludhiana, Punjab, India

T. Biju
University of Kerala
Thiruvananthapuram, Kerala, India

Priviledge Cheteni
University of Fort Hare
East London 5200, South Africa

N. Chithra
Anna Adarsh College for Women
Chennai, Tamil Nadu, India

Grzegorz Chmielarz
Czestochowa University of Technology
Czestochowa, Poland

Helena Fidlerová
Slovak University of Technology
Trnava, Bratislava, Slovakia

Andrii Galkin
University of Antwerp
Antwerp, Belgian

M. Ganeshkumar
SRM Institute of Science and Technology
Kattankulatur, Tamil Nadu, India

Katarzyna Grondys
Czestochowa University of Technology
Czestochowa, Poland

Moshood Issah
University of Ilorin
Ilorin, Kwara, Nigeria

Anandhi Kandhaswamy
Dhanalakshmi Srinivasan College of Arts and Science for Women (Autonomous)
Perambalur, Tamil Nadu, India

Dorota Klimecka-Tatar
Czestochowa University of Technology
Czestochowa, Poland

Robert Kuceba
Czestochowa University of Technology
Czestochowa, Poland

Tomasz Lis
Czestochowa University of Technology
Czestochowa, Poland

Izabela Małecka
University of Kalisz
Kalisz, Poland

Dejan Mirčetić
University of Novi Sad
Novi Sad, Republic of Serbia

P.F. Mishel
Bharathidasan University
Tiruchirappalli, Tamil Nadu, India

Paweł Patyk
ZF Passive Safety Systems Poland Sp. z o.o.
Czestochowa, Poland

Ilona Pawełoszek
Czestochowa University of Technology
Czestochowa, Poland

Umadevi Pongia
DhanalakshmiSrinivasan College of Arts and Science for Women (Autonomous)
Perambalur, Tamil Nadu, India

Janusz Przybył
Calisia Uniwersity
Kalisz, Poland

Aleksandra Ptak
Czestochowa University of Technology
Czestochowa, Poland

Izabela Rącka
University of Kalisz
Kalisz, Poland

Rachna Rana
Ludhiana Group of Colleges
Ludhiana, Punjab, India

S. Ruby
BJM Govt. College
Kollam, Kerala, India

P. Senthamizh Pavai
Dr. M.G.R. Educational and Research Institute
Chennai, Tamil Nadu, India

Rajneesh Sharma
Institute of Language Studies and Applied Social Sciences
Anand, Gujarat, India

Marta Starostka-Patyk
Czestochowa University of Technology
Czestochowa, Poland

P.F. Steffi
Cauvery College for Women (Autonomous)
Thiruchirappalli, Tamilnadu, India

L. Maria Suganthi
Dr. M.G.R. Educational and Research Institute
Chennai, Tamil Nadu, India

V. Mohana Sundari
SRM Institute of Science and Technology
Kattankulatur, Tamil Nadu, India

Leszek Szczupak
Calisia Uniwersity
Kalisz, Poland

A. Thangam
Dr. M.G.R. Educational and Research Institute
Chennai, Tamil Nadu, India

B. Thirumalaiyammal
Cauvery College for Women (Autonomous)
Thiruchirappalli, Tamilnadu, India

Ikechukwu Umejesi
University of Fort Hare
East London 5200, South Africa

J. Vigneshwari
Dr. M.G.R. Educational and Research Institute
Chennai, Tamil Nadu, India

Part I

Fundamentals of Technological Sustainability

1 The Principles of Sustainability

Paula Bajdor and Pankaj Bhambri

1.1 INTRODUCTION

This chapter is devoted to the issue of "sustainability" and, above all, the principles of it. The concept of "sustainability" comes from the Latin word "sustinere", where "sus" means up and "tenere" means hold, which indicates that "sustainability" expresses the ability to support specific causes. According to the definitions contained in dictionaries, "sustainability" means the ability to maintain activities related to, for example, economic development or the process of using natural environmental resources at a constant level. According to the *Cambridge Dictionary*, "sustainability" means the quality that allows an action to continue for a certain period, or about environmental issues, "sustainability" means a characteristic that causes human activities to cause little or no damage to the natural environment.

But in the literature, you can also find several definitions of "sustainability" as a long-term goal, assuming continuous economic growth with the lowest possible impact on the environment or ensuring continued economic growth while taking into account the harmonious cooperation of societies in such a way that future generations can also use natural resources necessary for survival.

This means that a concept closely related to "sustainability" is the concept of "sustainable development". This closeness results, for example, from the fact that very often, "sustainability" in English is translated as "sustainable development" in another language (e.g., this is the case in Polish). Therefore, both of these concepts are often treated as synonyms. However, sticking strictly to dictionary terminology, "sustainability" can be seen as a general phenomenon to achieve a long-term goal. At the same time, "sustainable development" refers to many processes and paths leading to its achievement.

The notion of "sustainable development" is presently regarded as a framework for contemplating the future, wherein factors related to the environment, society, and economy are harmonized in the pursuit of enhancing the standard of living, as articulated in the concept of sustainable development outlined in the Our Future as a Community report (1987). "Sustainable development refers to the possibility of achieving development at the current level of civilization while ensuring that the needs of the present generation are met without hindering the ability of future generations to meet their own needs" (Nagatsu et al., 2020; Shahadu, 2016).

DOI: 10.1201/9781003475989-2

The term "sustainability" or "sustainable development" is commonly understood as a development approach that not only fulfils the requirements of present generations but also guarantees that future generations will have the same opportunity to fulfil their demands (Singh et al., 2004). Both notions adopt a holistic approach, taking into account the impacts of current economic, ecological, and social actions. Moreover, the fundamental values and beliefs that form the basis of these concepts are rooted in overarching ideas such as fairness between generations, equal treatment of genders, acceptance of diverse social groups, alleviation of poverty, safeguarding the environment, preservation of biodiversity, and the establishment of equitable and peaceful communities (Jeronen, 2013).

1.2 THE HISTORICAL ROOTS OF SUSTAINABILITY

The concept of "sustainability" comes directly from a constant pursuit of discovering, learning, improving, or giving a physical character to ideals and ideas (Ashby et al., 2012). The motives for development are diverse – it may be the desire to discover or create new solutions, make life easier, provide help, or achieve financial success.

In turn, "sustainability", the emerging concept of sustainable development, suggested a more responsible use of forest resources (Malthus, 1926). Since then, it has been the subject of discourse and research, which constitute characteristic stages in the evolution of this concept (Table 1.1).

The term "sustainability" first appeared in a work written by J.S. Mill (1883), in which he assumed an unchanged state of capital and population but did not consider improving the conditions of human life. He believed that for the good of future generations, the world population should remain unchanged. He argued that improving people's lives, better living conditions, and even development, resulting in the

TABLE 1.1
Author's Characteristic Stages in the Evolution of Sustainable Development

Year	Author	Publication
1713	H.C. von Carlovitz	*Sylvicultura oeconomica, oder haußwirthliche Nachricht und Naturmäßige Anweisung zur wilden Baum-Zucht*
1798	T.R. Malthus	*Essay on the Principle of Population as it Affects the Future Improvement of Society*
1848	J.S. Mill	*Principles of Political Economy*
1864	G.P. Marsh	*Man and Nature*
1898	A.R. Wallace	*Our Wonderful Century*
1962	R. Carson	*Milcząca wiosna (Silent Spring)*
1968	P. Ehrlich	*The Population Bomb*
1972	E. Goldsmith	*A Blueprint for Survival*
1973	F. Schumacher	*Small Is Beautiful*
1987	Brundtland Commission	*Our Common Future*

extension of human life or a reduction in mortality among children and infants, may pose a threat because they may result in overpopulation, which is unfavourable for future generations. Another author, G.P. Marsh, described how the natural environment had been destroyed by human activity, claiming that over time, the Earth could become uninhabitable, leading to the extinction of all humanity. It should be noted, however, that G.P. Marsh (1965) did not mean protecting the environment for its own sake. Still, for the good of humanity, his approach is identical to the contemporary approach to sustainability. These three primary ideas formed a basic framework for the topics explored in Brundtland's Our Future Together report, which was the concluding document of the Earth's Assembly in Rio de Janeiro in 1992, almost a century after this work was published. Nevertheless, two decades prior, during the early 1970s, a significant advancement occurred in the understanding of the natural environment for its connection to socio-economic progress, sparking a discussion on ecology and the safeguarding of the environment. The impetus for this was the publication of R. Carson's book *Silent Spring* (1962), in which the author strongly emphasized the relationship between the excessive use of seeds enriched with poisonous substances (e.g., mercury) in agriculture and their impact on the extinction of local species. In another publication devoted to the natural environment and food security issues, P. Ehrlich (1972) expressed his belief that population development should be stopped or even lead to negative growth while improving food production to eliminate world hunger. In turn, E. Goldsmith (1972) recommended a solution according to which people should create small, decentralized, and deindustrialized communities. According to him, the arguments in favour of this solution are easier maintenance of morality in smaller groups, the possibility of more ecological agriculture, a higher level of satisfaction among people, and a reduced impact of a small community on the natural environment. It also assumed birth control, sustainable resource management, high social cohesion, and improvement of physical condition and mental health. In turn, F. Schumacher (1993) claimed that modern markets are unsustainable, and natural resources are treated as revenues and should be treated as capital. He also believed that nature's resistance to pollution is limited and may be exceeded at some point. The same issue was raised in the publication *The Limits to Growth*, in which the authors warn that the Earth's resources are limited and their irresponsible exploitation may lead to disaster. However, in the last decade of the 20th century, ecologists from highly industrialized and rich northern countries noted that global life support systems were undergoing progressive degradation resulting from various projects aimed at economic development (Guha, 2000). Aras and Crowther (2009) point out that sustainability is primarily about focusing on the consequences of actions regarding the exploitation of resources so that future generations can live in an equally diverse world as today regarding nature, climate, species, and diversity. Otherwise, these resources will be exhausted, significantly worsening future generations' quality of life. F. Ceschin and I. Gaziulusoy (2016) believe that sustainability is a system, and its elements do not influence the system's sustainability. The sustainability of the system's separate components can only be achieved when the structure as its entirety is in a state of equilibrium.

The 1970s were when the concept of sustainable development was fully established because, from then on, it became an issue widely discussed by scientists, which manifested itself in many publications, discussions, and research. Also, during this period, sustainable development began to be systematically mentioned in numerous programmes, declarations, and reports containing issues arising from three pillars – economic, ecological, and social, which constitute the foundations of this concept. Sustainable development is a concept known to almost everyone, at least because it is firmly documented in numerous publications and articles (Singh et al., 2005). Entering "sustainability" into an Internet search engine results in 1,990 million views. Such a high frequency of mentioning sustainable development in scientific literature has made it possible to identify and highlight three main stages that can be perceived as the main phases of the development of this concept:

1. The initial stage encompasses the time when several economic theories were devised; the authors cautioned about the detrimental consequences of ongoing economic progress, particularly its adverse influence on the environment. The period came to an end in 1972 with the convening of the Initial United Nations Convention on the Environment and Humanity in Stockholm. This meeting served as the debut of the notion of sustainable development. While it did not explicitly establish a direct correlation between environmental issues and development, it underscored the imperative for modifications in economic development policies (Drexhage & Murphy, 2010). This stage was marked by theoretical deliberations that failed to materialize in practical implementation.
2. The second phase witnessed a growing usage of phrases such as "development", "environment", "growth without damage", and "ecologically compatible development" in various publications. In 1980, the International Union for Conservation of Nature (IUCN) proposed the concept of integrating the economics and the environment through sustainable development. The United Nations World Commission on Environment and Development (WCED) was created in 1983 with the purpose of formulating a worldwide change agenda. The purpose of this programme is to increase awareness and concern regarding the adverse effects of economic and social growth on the environment and its natural resources. It aims to offer opportunities for sustainable growth in the long run that align with environmental protection and conservation (Drexhage & Murphy, 2010). In 1987, a commission led by Gro Harlem Brundtland, the Prime Minister of Norway at the time, published the Our Future as a Community report, also known as the Brundtland Report. This report, which involved 19 delegates from 18 countries, is notable for providing one of the most widely recognised descriptions of sustainable development. The study catalysed the establishment of a novel worldwide socio-economic strategy, wherein sustainable development emerged as a pivotal component in the management of the environment and other domains of human endeavour (Mebratu, 1998). This

level is characterized by theoretical considerations, while also providing practical advice and guidelines.

3. The third stage is said to be ongoing at present. Sustainability is presently being used across various domains through the utilization of sustainability concepts, such as biodiversity assessment models (Baxter et al., 2002; Bebbington & Frame, 2003), sustainability rating (Sustainability ratings, 2024), sustainability reporting guidelines (Idowu et al., 2013), sustainability risk management (Massey, 2024), sustainability-oriented innovation (Hansen & Grosse-Dunker, 2013), and technological sustainability, to which this book is devoted.

The pursuit of implementing sustainability is currently one of the most critical challenges facing national authorities, enterprises, various types of organizations, and the entire society.

1.3 PILLARS (DIMENSIONS) AND DOMAINS OF SUSTAINABILITY

1.3.1 Pillars (Dimensions) of Sustainability

Sustainable development is currently regarded as a solution to the apprehension that the natural systems of the planet would someday become incapable of managing the consequences of human activities on the economic, social, and environmental aspects (Poskrobko, 2009). It is worth mentioning, however, that the emphasis on its fundamental issues has changed over the years. In the 1970s, issues related to the natural environment played the most crucial role, and the actions taken and guidelines formulated focused mainly on ways to protect it. However, over time, the other two pillars also began to be considered – the social and economic aspects, apart from the environmental aspect, and reducing poverty among the population became one of the main challenges. However, the matter of environmental conservation has been given less importance compared to the improvement of living conditions and immediate economic growth. The primary sustainability concern that persists is the issue of the Industrial Age, specifically the unrestricted utilization of people and environmental resources (Starostka-Patyk, 2017). Sustainability is based on the idea that the three pillars of the economy, the environment, and society are interconnected and equally important. These pillars are analysed together, as shown in Figure 1.1.

The sustainability pillars presented above are closely interconnected, and each action taken within each pillar affects the others. There is a strong relationship between the environmental and economic pillars, where the use of responsible environmental practices determines not only the stability of the company but also its future development and competitiveness. There are also strong relationships between all three pillars, and sustainability organizes the previously known values and strives to combine them into a coherent system, the advantage of which is, among others, reconciling conflicting economic, social, and ecological interests (Ciegis, 2010).

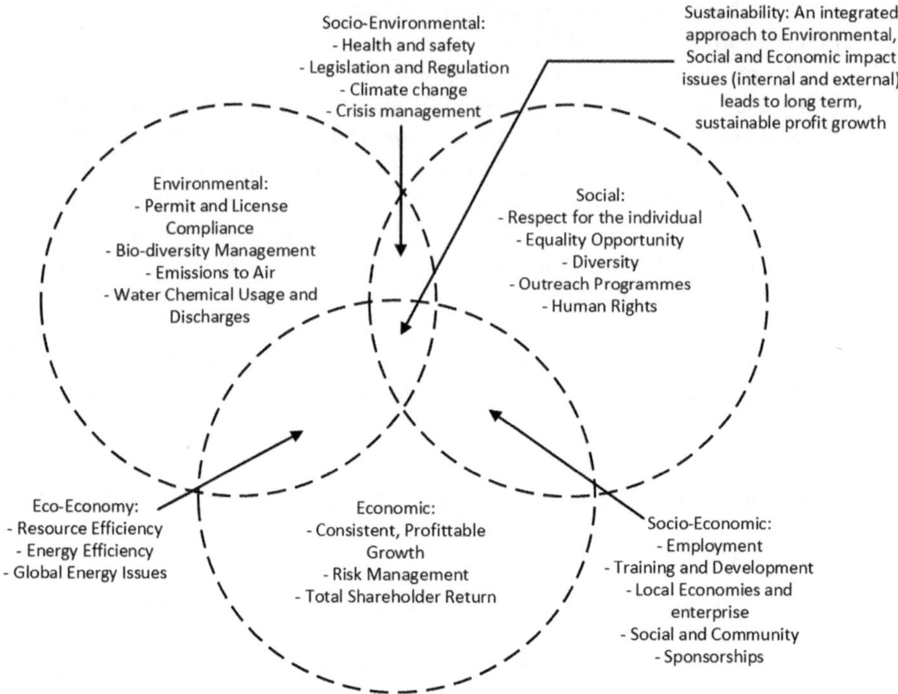

FIGURE 1.1 Pillars of sustainability. (Source: Ivković A., Ham M., & Mijoč J. (2014). Measuring Objective Well-Being and Sustainable Development Management. *Journal of Knowledge Management, Economics, and Information Technology*, 4, 1–8).

Environmental sustainability means continuously protecting and maintaining the appropriate natural environment conditions through proper practices and implementing policies aimed at this goal. In this way, you can ensure that future generations will be able to benefit from the natural environment to the same extent as current generations. However, it is essential to remember a number of factors that negatively affect not only the environmental balance but also the Earth's ability to support further life on the planet. These factors include primarily:

- ubiquitous climate changes cause violent weather phenomena (floods, hurricanes, tornadoes) in places where they have not occurred before;
- growing levels of air, water, and soil pollution resulting from constantly developing industry and human activities;
- release of excessive amounts of greenhouse gases into the atmosphere, which causes an increase in global temperature, resulting in rising ocean temperatures and melting glaciers;
- loss of biodiversity, resulting in the extinction of animal and plant species;
- overexploitation of available natural and non-renewable resources, resulting in their slow depletion.

The above-mentioned factors should determine the taking of further actions that would ultimately enable achieving environmental sustainability. These activities may take the form of the following objectives:

- reducing the level of greenhouse gas emissions;
- increasing the level of production and use of energy from renewable sources by emphasizing the possibilities of producing energy from wind, water, or solar;
- implementation of appropriate policies aimed at protecting biodiversity, but these policies should focus on eliminating the causes that negatively affect biodiversity, not the effects;
- introducing sustainable practices in a wide area, such as agriculture and food chains. These practices may take the form of appropriate precision farming strategies that are formulated and adopted. The use of proper technologies, regenerative agriculture, agrivoltaics, and hydroponic or aeroponic systems to continuously optimize cultivation and increase soil quality and productivity;
- implementing practices aimed at minimizing the amount of food waste, changing consumer attitudes, raising awareness of the amount of food wasted, understanding the causes of food waste, and presenting solutions that will reduce the level of food waste (food sharing, responsible grocery shopping).

Social sustainability aims to create inclusive societies and societies engaged in social matters while maintaining social cohesion and justice. In this aspect, the factors with a negative impact on social sustainability include:

- still high levels of poverty in some regions where people have to survive a day on one dollar. This poverty is mainly the result of socio-economic inequalities but also the result of policies implemented at national levels;
- discrimination and prejudice, leading to social exclusion. Discrimination does not only concern skin colour but may also concern age, gender, or religion;
- unequal access to natural resources; in this case, wealthier societies use the available resources to an unlimited extent, and poorer societies only to a limited extent, which only strengthens their weaker position;
- a sense of insecurity caused not only by the policies of a given country but also by conflicts at the local, regional, and global levels, which may develop into war over time;
- the use of economic models resulting in unsustainable consumption and unequal access to the products and services offered;
- the phenomenon of corruption, which is the result of irregularities in the implemented national policy.

In response to the above, in pursuit of social sustainability, the following actions should be implemented:

- ensuring equal access to resources for all people, which is possible by promoting appropriate practices and policies aimed at eliminating social and economic inequalities;
- respecting the fundamental human right to health and education by enabling them to have access to the education system and facilities offering appropriate medical services;
- developing and implementing practices that include people previously excluded from them as a result of discrimination based on skin colour, sexual orientation, or ethnic origin;
- increasing the sense of security in the place of residence by effectively fighting violence occurring in a given environment;
- effective administration of justice, which is possible with efficient and fast-acting justice bodies whose ranks are resistant to corruption;
- activities aimed at raising awareness and engaging people and entire communities in the issue of social sustainability.

However, economic sustainability is an approach which assumes that economic activities are carried out in a way that ensures the financial well-being of all participants in these activities, both now and in the future. This means striving to achieve a balance between economic growth, responsible management of resources, social equality, and financial stability. To achieve economic sustainability, the following actions should be taken in the form of:

- responsible management of natural (non-renewable) resources so that future generations can also benefit from them;
- creating compelling and innovative economic systems that ensure not only economic but also social development while respecting the rights of the natural environment;
- striving to achieve financial stability for all participants in a given economic system;
- increasing the level of involvement of each country in social issues, resulting in the creation of civil societies that are engaged and inclusive of all previously excluded groups;
- establishing and maintaining international cooperation between public administration and private business in individual countries, conducting common economic, social, and environmental policies, and applying collective responsibility.

The result of these activities is economic sustainability characterized by minimal impact on the natural environment, social and economic equality, a stable economy resistant to any turmoil, and organizations that base their activities on the principles of responsibility and ethics.

Awareness of the coexistence and interdependence between the pillars of sustainability makes it possible to consider a broader range of phenomena and activities, simultaneously identifying those that lead to beneficial effects in all three pillars. It also determines the activity of enterprises aimed not only at making profits but also at protecting the natural environment or improving the condition of the immediate social environment. However, harmonizing the connections and relations between these areas requires taking many different actions, e.g., developing environmentally friendly and more efficient technologies, introducing gradual restrictions on the use of environmental resources, or changing the way of life of modern societies.

1.3.2 Domains of Sustainability

Sustainability domains have recently been used as an alternative to the sustainability pillars presented above. One of the most popular approaches to presenting sustainability domains is Circles of Sustainability (CoS). CoS is a model that aims to identify the dominance of specific, socially specific modalities of space, time, embodiment, and knowledge (Ginsberg, 2016), enabling the assessment of macro-factors that create the local environment while detailing existing structures operating at the micro-level (James, 2015). CoS is also perceived as a method enabling understanding and assessment of sustainability in relation to the management of projects aimed at sustainable results in social, ecological, and economic aspects (James et al., 2015). CoS is based on four domains: economics, ecology, politics, and culture.

There are four primary domains, each of which is divided into seven subdomains. The assessment is made on a nine-point scale, ranging from "critical" to "vibrant" (James et al., 2015). This model is used by public organizations or administrations in urban planning and non-governmental organizations that implement projects of great importance for urban agglomerations.

The literature also explains that sustainability domains are areas in which, despite differences, the phenomenon of sustainability occurs, whether in the form of a concept, process, or implementation. Sustainability has several different domains (Idowu et al., 2013):

- biological, which assumes interactions between society and the natural environment or striving to preserve biodiversity and environmental capital for future generations;
- economic, which assumes interactions between modern production systems and the natural environment, the exhaustion of which will result in the collapse of entire economies or the use of instruments for the internationalization of environmental costs resulting from industrial activities;
- sociological, which assumes interactions between people and the natural environment, assuming that everyone has the right to equal access to the natural environment. This domain also addresses the issue of how human interest groups make decisions about the use of natural resources and the impact this has on other people in their daily lives;

- urban, which assumes the integration of human settlements with the natural environment. In this domain, the idea is to design cities or areas in such a way that the people inhabiting them live in conditions of integrated urban planning with the natural environment;
- ethics, this domain considers whether the natural environment has certain rights and whether the extinction rate of plants and species due to human activity is ethical. There are also considerations on whether people are part of nature or outside it and what moral choices they should follow.

Sustainability also occurs in domains related to:

- consumption, sustainability in consumption means using products and services in such a way as to reduce the impact on the natural environment as much as possible (Tseng et al., 2013);
- transport, sustainability in transport means the use of low- or zero-emission and energy-efficient vehicles powered by electric engines or using alternative fuels (Gong et al., 2019);
- logistics, sustainability in logistics means striving to reduce the ecological footprint generated by all logistic activities, including by using alternative solutions that do not reduce the productivity and effectiveness of these activities (Helm et al., 2018);
- marketing, sustainability in marketing means undertaking pro-ecological and pro-social activities by enterprises. It means promoting brands and products that are environmentally and socially good. It also means initiating sustainable practices by enterprises that take into account the long-term impact of their activities on the social and environmental environment (Hall et al., 2019);
- construction, sustainability in construction primarily means activities aimed at minimizing the impact of the construction process on the social and environmental surroundings. This assumes, among other things, using materials that are suitable for recycling, using solutions and technologies that reduce energy consumption, and building buildings that are zero-emission and energy self-sufficient (Zuo & Zhao, 2014);
- tourism, sustainability in tourism means using sustainable practices to minimize the negative effects of tourism on the social and environmental environment and maximize the positive impact (Florek, 2014);
- entrepreneurship, sustainability in entrepreneurship means running a business in such a way as to minimize the negative impact of this activity on the social and environmental environment. In the environmental aspect, this means minimizing the use of non-renewable resources, minimizing the generated waste, and implementing solutions that increase energy efficiency. In the social aspect, it means employing people based on employment contracts, active participation in charity campaigns, and taking care of the well-being of employees (Bajdor, 2021);

- management, sustainability in management means enterprise management that not only focuses on increasing profits and revenues but also includes activities focused on environmental protection and the well-being of the social environment. It is also the implementation of activities that accelerate global transformation, the implementation of organizational practices that will transform the organization into a sustainable organization, and the implementation of economic production and consumption that mitigates adverse environmental effects and increases the protection of available resources.

It can also be assumed that with continuous research on sustainability, the number of sustainability domains will increase.

1.4 BASIC ASSUMPTIONS OF SUSTAINABILITY

The basic assumptions of sustainability refer to a broad spectrum of contemporary problems of an ecological, social, economic, and even political nature. They serve not only to provide guidance on how to proceed when solving specific problems but also to define the role of various entities in solving them. Above all, however, the responsibilities of both states and people themselves in implementing sustainability are emphasized.

Man is treated as a central entity in sustainability (Piątek, 2007); however, this anthropocentrism is not arrogant, but has a natural character, consistent with biocentrism. This is therefore a new paradigm of placing man in the world in which he does not exist despite being above or next to nature, but creates unity with it. This means that through sustainable development, humans define a new role and place in the global ecosystem. Thus, humans have many responsibilities, including taking responsibility for people who suffer from hunger, feel the effects of a predatory economy, or are forced to leave their homes as a result of armed conflicts. In addition, there is also the fight against poverty, which is still a common phenomenon (in 2023, almost 10% of the entire population lived in extreme poverty, having less than $2.15 per day). Sustainability also assumes the full sovereignty of individual countries in natural resource management and the right to conduct independent ecological policies. This internal independence of states is probably the most significant right of every nation, even if, at the same time, the need to conduct many activities in cooperation with two or more states is emphasized, taking the form not only of economic cooperation but also of scientific or technological exchange (Gawor, 2006).

Another sustainability assumption points to the existence of inseparable relations between economic development and environmental protection, indicating the need to adopt long-term management of available resources, considering future generations' needs and interests. An important issue here is the problem of predatory economic activity, which contributes to many negative phenomena affecting the natural environment and entire communities. The eradication of the predatory economy can only be achieved through the collaborative efforts of mutual support and collaboration in developing and implementing legal solutions for environmental standards, establishing a financial system based on novel principles, and actively involving entire societies in the process of making choices of creating appropriate laws. As a result, this will significantly improve the condition of the natural environment and the communities around it. Another one of the main assumptions of sustainability is to emphasize the direct relationship between sustainability and securing human rights, especially in the field of equality. In this case, the assumption is also to strive to create societies that are engaged and actively participate in the activities undertaken by the state. Moreover, social participation reduces social inequalities and eliminates the phenomenon of discrimination that affects women, young people, indigenous societies, and those living in areas affected by armed conflicts.

Another assumption is developing and implementing a policy regarding broadly understood environmental protection. This policy calls for not only the responsible use of natural resources or conducting business activities without harming the natural environment but also introducing the "polluter pays" principle in connection with compensation for victims of pollution. Another measure that has become widely used relates to the problem of cross-border pollution and preventing its movement. First, the need to concentrate efforts on preventive actions, or in general, on an active environmental protection policy, is emphasized here. Great emphasis is also placed on immediately providing information about threats resulting from, e.g., ecological disasters. Such a solution is perceived to reduce the adverse effects of such events. However, it should be remembered that one of the more direct and specific recommendations about the implemented policy is recognizing environmental impact assessment as a tool for eco-development.

The set of assumptions also included those with a solid direct political dimension. The debate on sustainability raises the issue of conflicts, which are primarily considered a threat to development's sustainability. This means that residents living in conflict areas should be given special care.

All these assumptions have one thing in common: it is only possible to implement some sustainability activities and goals in a time suitable for one or two generations. Therefore, it seems correct to assume that if implemented, sustainability will become a revolution comparable to breakthrough moments in the history of humanity, also often referred to as revolutions (Pawłowski, 2009). Therefore, sustainability can be treated on an equal footing with previous revolutions in the industrial (Industrial Revolution), social (Social Revolution), scientific (Scientific Revolution), and agricultural (Agrarian Revolution) areas.

1.5 CONCLUSIONS

To sum up, it can be said that the principles of sustainability cover a wide range of areas taking the form:

- sustainable use of resources, which involves minimizing excessive consumption of natural resources such as water, energy, wood, and minerals and ensuring their efficient use;

- environmental protection through activities aimed at preserving biodiversity, purifying air and water, as well as minimizing pollutant emissions;
- increasing use of energy from renewable sources, which involves switching to solar, wind, or geothermal energy, which helps reduce greenhouse gas emissions and dependence on non-renewable energy sources;
- the use of recycling and reuse by promoting the recycling of materials and products and reducing the amount of waste;
- striving for social justice, ensuring decent working conditions, eliminating discrimination, and supporting the development of local communities;
- corporate social responsibility, manifested in ethical activities, respect for employees' rights, protection of the natural environment, and contributing to the well-being of societies;
- educating societies not only in the field of sustainability but also in promoting appropriate attitudes in terms of conscious choices and behaviour;
- sustainable economic development, which should take into account economic, social, and environmental aspects;
- conducting sustainable agriculture, taking into account the protection of soil, water, and biodiversity, avoiding excessive exploitation of natural resources, and minimizing the use of harmful chemicals;
- establishing and maintaining partnerships and cooperation between government, business, civil society, and other stakeholders to achieve sustainability.
- transparency and honesty in business, it is the responsibility of businesses to provide reliable information and apply fair business practices;
- adopting a long-term perspective because sustainability requires looking into the future and making decisions that benefit both the present and for subsequent generations.

REFERENCES

Aras G. & Crowther D. (2009). Making sustainable development sustainable. *Management Decision*, 47, 975–988. https://doi.org/10.1108/00251740910966686.

Ashby A., Leat M. & Hudson-Smith M. (2012). Making connections: A review of supply chain management and sustainability literature. *Supply Chain Management*, 17(5), 497–516. https://doi.org/10.1108/13598541211258573

Bajdor P. (2021). *Zrównoważona przedsiębiorczoś ć. Analiza i model.* Wydawnictwo Politechniki Częstochowskiej, Częstochowa.

Baxter T., Bebbington J. & Cutteridge D. (2002). The sustainability assessment model (SAM). In: Proceedings of the SPE International Conference on Health, Safety and Environment in Oil and Gas Exploration and Production (pp. 697–701).

Bebbington J. & Frame R. (2003). Moving from SD reporting to evaluation: The sustainability assessment model. *Chartered Accounting Journal of New Zealand*, 82(7), 11–13.

Carson R. (1962). *The Silent Spring*. Fawcett Publications, Greenwich.

Ceschin F. & Gaziulusoy İ. (2016). Evolution of design for sustainability: From product design to design for system innovations and transitions. *Design Studies*, 47. https://doi.org/10.1016/j.destud.2016.09.002.

Ciegis R. (2010). The concept of sustainable economic development and indicators assessment. *Management Theory and Studies for Rural Business and Infrastructure Development*, 21(2), 34–42.

Drexhage J. & Murphy D. (2010). Sustainable development: From Brundtland to Rio 2012. Background Paper Prepared for Consideration by the High Level Panel on Global Sustainability at Its First Meeting, 19 September 2010, UN Headquarters, New York.

Ehrlich P.R. (1972). *The Population Bomb*. Ballantine Books, New York.

Florek I. (2014). Sustainable tourism development. *Regional Form Development Studies*, 3, 157–166.

Gawor L. (2006). Wizja nowej wspólnoty ludzkiej w idei zrównoważonego rozwoju. *Problemy Ekorozwoju*, 1(2), 59–66.

Ginsberg N. (2016). Determining the context of an international development project. *The Journal of Developing Areas*, 50 (5), 431–442. https://www.jstor.org/stable/26415607

Goldsmith E. (1972). A blueprint for survival. *The Ecologist*, 2(1), 172–198.

Gong M., Gao Y., Koh L., Sutcliffe C., & Cullen J. (2019). The role of customer awareness in promoting firm sustainability and sustainable supply chain management. *International Journal of Production Economics*, 217, 88–96,

Guha R. (2000). *Environmentalism: A Global History*. Longman, New York.

Hall C., Kemper J., & Ballantine P. (2019). Marketing and sustainability: Business as usual or changing worldviews? *Sustainability*, 11. https://doi.org/10.3390/su11030780.

Hansen E. & Grosse-Dunker F. (2013). Sustainability-oriented innovation. https://doi.org/10.1007/978-3-642-28036-8_552.

Helm S.V., Pollitt A., Barnett M.A., Curran M.A., & Craig Z.R. (2018). Differentiating environmental concern in the context of psychological adaption to climate change. *Global Environmental Change*, 48, 58–167.

Idowu S.O., Capaldi N., Zu L., & Das Gupta A. (2013). *Encyclopedia of Corporate Social Responsibility*. Springer-Verlag, Berlin Heidelberg.

James L. (2015). Sustainability footprints in SMEs. *European Business Review*, 43, 453–473.

James P., Magee L., Scerri A., & Steger M. (2015). Urban sustainability in theory and practice: Circles of sustainability. https://doi.org/10.4324/9781315765747.

Jeronen E. (2013). Sustainability and sustainable development. In: Idowu S.O., Capaldi N., Zu L., & Gupta A.D. (eds.), *Encyclopedia of Corporate Social Responsibility*. Springer, Berlin, Heidelberg. https://doi.org/10.1007/978-3-642-28036-8_662

Malthus T.R. (1926). *First Essay on Population*. Macmillan, London.

Marsh G.P. (1965). *Man and Nature*. Belknap Press, Cambridge.

Massey F. (2024). Sustainability risk management: How can leaders develop a successful program? https://www.linkedin.com/pulse/sustainability-risk-management-how-can-leaders-part-1-frank/, access on 10.01.2024.

Mebratu D. (1998) Sustainability and sustainable development: Historical and conceptual review. *Environmental Impact Assessment Review*, 18, 493–520. http://dx.doi.org/10.1016/S0195-9255(98)00019-5

Mill J.S. (1883). *Principles of Political Economy, with Some of Their Applications to Social Philosophy*. Longmans, Green & Co., London.

Nagatsu M., Davis T., des Roches T., Koskinen I., MacLeod M.A.J., Stojanovic M., & Thoren H. (2020). Philosophy of science for sustainability science. *Sustainability Science*, 15(6), 1807–1817. https://doi.org/10.1007/s11625-020-00832-8

Pawłowski A. (2009). Rewolucja rozwoju zrównoważonego. *Problemy Ekorozwoju*, 4(1), 65–76.

Piątek Z. (2007). Przyrodnicze i społeczno–historyczne warunki równoważenia ładu ludzkiego świata. *Problemy Ekorozwoju*, 2(2), 5–18.

Poskrobko B. (2009). Współczesne trendy cywilizacyjne a idea zrównoważonego *rozwoju. Przegląd Komunalny*, 6, 25–29.
Schumacher E.F. (1993). *Small Is Beautiful*. Vintage Books, London.
Shahadu H. (2016). Towards an umbrella science of sustainability. *Sustainability Science*, 11. https://doi.org/10.1007/s11625-016-0375-3.
Singh, P., Singh, M., & Bhambri, P. (2004, November). Interoperability: A problem of component reusability. In: International Conference on Emerging Technologies in IT Industry (p. 60).
Singh, P., Singh, M., & Bhambri, P. (2005, January). Embedded systems. In: Seminar on Embedded Systems (pp. 10–15).
Starostka-Patyk M. (2017). *Reverse Logistics of Defective Products in Management of Manufacturing Enterprises*. Sophia, Katowice.
Sustainability ratings. https://www.swissre.com/investors/solvency-ratings/sustainability-ratings.html, access on 10.01.2024.
Tseng M.L., Fung Chiu S., Tan R.R., & Siriban-Manalang A.B. (2013). Sustainable consumption and production for Asia: Sustainability through green design and practice. *Journal of Cleaner Production*, 40, 1–5.
Zuo J. & Zhao Z-Y. (2014). Green building research – current status and future agenda: A review. *Renewable and Sustainable Energy Reviews*, 30, 271–281.

2 Fundamental Aspects of Sustainable Technological Development

Marta Starostka-Patyk

2.1 INTRODUCTION

Technology is developing rapidly, and everything indicates that this dynamic trend will be maintained in the coming years. Rapid technical and technological changes allow the production of more efficient and practical machines, devices, tools, and methods used in the economy and everyday life (Rüßmann et al., 2015). Thus, technological development improves work efficiency and better use of available resources.

The general definition of technological development frames it as a continuous technological progress and innovation process that includes the design, development, improvement, and implementation of new devices, methods, systems, and processes (Hakansson, 2015). This process involves systematically using scientific and technical knowledge to design and create new products or systems that meet requirements and solve problems. At the same time, the entire technological development process is characterized by innovation, where new ideas, concepts, and inventions are transformed into practical and commercially feasible technical solutions (Camisón and Villar-López, 2014).

In the characteristics of technological development, several elements are crucial for the correct implementation of this process (Baker, 2012):

- Innovation: It introduces new ideas, products, or methods that improve existing technologies or create new categories.
- Research and development (R&D): Investments in scientific research and technological development are the basis of technological progress.
- Diffusion of technology: The way how new technologies are adopted and accepted by society and the market.
- Digitization and automation: Advances in digital technologies and automation are transforming traditional industries and creating new business models.

DOI: 10.1201/9781003475989-3

- Challenges and regulations: Technological progress brings challenges such as privacy issues, data security, ethics, and the need for legal regulations.

Additionally, it should be taken into account that technological development is a change that affects various aspects of the functioning of society (social), the economy (economic), and the surroundings and the environment (ecological). Therefore, these changes align with contemporary requirements for developing technologies, which focus on sustainability following sustainable development's assumptions (priorities). Thus, it is necessary to emphasize that striving for a harmonious combination of technological progress with environmental protection, social justice, and economic efficiency is a paradigm of sustainable technological development (Hák et al., 2016). Hence, it is clear that sustainable technological development is characterized by a wide multidimensionality, taking into account not only the minimization of the negative impact of modern technologies on the natural environment but also social aspects such as the availability of technologies and their impact on the quality of life, and economic aspects in the form of, e.g., sustainable growth economic or promoting innovation in the technological area (Beckerman, 2017).

2.2 SUSTAINABLE DEVELOPMENT CONCEPT

The concept of sustainable development emerged in the 1970s as a response to warnings about unsustainable economic growth and its potentially disastrous effects on the natural environment and, therefore, on society. In the late 1980s, the concept was defined as "development that meets the needs of the present without compromising the ability of future generations to meet their own needs" (Jeronen, 2020). In the 1990s, this definition format was popularized worldwide, resulting in comprehensive action plans for sustainable development at global, national, and local levels.

In the subsequent years of the 21st century, it is essential to recognize the need to integrate sustainable development with economic and technological policy, resulting in the creation of concepts such as the green economy, clean technologies, and, in recent years, the circular economy. At the same time, the sustainable development goals were established, signaling the global commitment of countries to comply with them in three main pillars: environmental, social, and economic (Carley and Christie,2017). It is the technology that plays a crucial role in achieving these goals, achieving them through innovation and technological development in line with the principles of sustainability. Hence, developing sustainable technologies such as renewable energy sources, sustainable agriculture, energy efficiency, and smart cities has become an essential element of global strategies for sustainable development.

Thus, the paradigm of sustainable technological development was created in response to the challenges associated with traditional models of technological development, which often ignored their long-term effects on the environment, society, and the economy. It focuses on promoting technologies that are not only economically viable but also minimize the negative impact on the environment and, at the same time, improve society's quality of life, pointing to the deep integration of the concept of sustainable development with innovations and technological practices (Elliot, 2012).

2.3 THREE PILLARS OF SUSTAINABLE DEVELOPMENT

The basis for an integrated approach to progress that strives to balance humanity's needs and our planet's possibilities is three pillars (dimensions) of sustainable development: environmental, social, and economic (Purvis et al., 2019). In a technological context, they are closely interconnected, and technology plays a crucial role in supporting and balancing them.

In general, the environmental dimension of sustainable development focuses on protection of ecosystems, natural resources, and biodiversity; minimization of pollutant emissions; and effective and sustainable management of natural resources (Duić et al., 2015). In this area, sustainable technologies such as renewable energy sources (e.g., solar energy, wind energy), clean production technologies, recycling systems, and closed-loop economies are intended to reduce the negative impact of human activity on the environment and improve resource use efficiency.

The social pillar of sustainable development covers social equality, access to education and health care, social justice, decent working conditions, thus ensuring a high quality of life for all members of society (Murphy, 2012). Here, comprehensive and universal access to modern technologies, digitization of public services, e-education, telemedicine, and innovations aimed at improving living conditions in local communities are vital to promoting social inclusion (counteracting social and digital exclusion) and increasing access to essential services for society.

The economic dimension of sustainable development seeks to achieve sustainable economic growth, innovation, competitiveness, and economic efficiency while ensuring that profits are distributed fairly throughout society (Holden et al., 2017). From a technological point of view, the development and implementation of new technologies supporting sustainable development, such as intelligent energy management systems, sustainable technologies in production, and eco-innovations, are the engines of economic growth that do not harm the environment and promote social equality.

Integrating the three pillars mentioned above in a technological context is crucial to achieving sustainable development and its goals. Technologies that are economically viable while promoting social justice and minimizing negative environmental impacts are considered sustainable. For example, the development of renewable energy technologies supports the environmental pillar by reducing dependence on fossil fuels, the economic pillar by creating new jobs and financial sectors, and the social pillar by providing access to clean energy for marginalized communities (Mensah, 2019). This holistic integration of the pillars shows that technological progress can and should be carried out in a way that supports sustainable development in all three dimensions.

2.4 ENVIRONMENTAL, SOCIAL, AND ECONOMIC PROBLEMS WITH TECHNOLOGY SUPPORT

Considering the equal value of each of the three dimensions of sustainable development, it is possible to identify their main difficulties and interconnections in the context of technological development aimed at sustainability.

2.4.1 Environmental Dimension

Sustainable technological development is intended to create innovative solutions to many emerging environmental problems. These problems are complex and interconnected, making their solution necessary not only from the point of view of environmental protection but also for the global development of societies and economies. The most severe environmental challenges today are ongoing climate change; the degradation of natural resources; and various types of air, water, and soil pollution (Duić et al., 2015). In solving these problems, technological progress can bring many positive forms of support.

Analyzing the issue of climate change, the causes of which should be seen in the emission of greenhouse gases originating from human activities (combustion of fossil fuels, deforestation, intensive agricultural production), we feel their effects in the form of increasingly frequent extreme weather phenomena (droughts, floods, hurricanes) that threaten biodiversity, food security, and places of human existence (Biesbroek et al., 2010). This area can support developing and implementing low-emission, energy-efficient technologies that support adapting industrial processes to a changing climate.

The degradation and depletion of natural resources (water, soil, minerals, forests) result from their unsustainable use, which is exploited so intensively due to global population growth and increased consumption. As a result, the availability of these resources for future generations will be severely limited. It will additionally cause disruptions in ecosystems and reduce the ability of the environment to function and rebuild them (Reilly, 2012). Here, technological solutions are needed in the form of innovations aimed at sustainable management of natural resources, such as recycling, reusing, and circulating materials and raw materials; saving water and soil protection; and sustainable agriculture.

The problem of air, soil, and water pollution is very complex and multidimensional. It concerns various levels and branches of industry (production, burning of fossil fuels, poor waste management, and use of pesticides and fertilizers in agriculture). These pollutants translate into severe consequences for the health of members of society (diseases) and the natural environment (degradation of ecosystems and loss of biological balance) (Keeler et al., 2013). This area requires support in developing purification technologies that minimize pollutant emissions already at the production stage or even earlier, as well as innovative waste management systems, along with implementing the principles of circularity.

2.4.2 Social Dimension

In addition to counteracting environmental problems, sustainable technological development also faces challenges and emerging issues in the social area. Actions under this priority should strive to provide comprehensive technical benefits to all segments of society in line with the principles of social justice. In the social dimension of sustainable development, the most frequently emerging problematic issues are equal access to technology, the impact of technology on employment and the labor market, technological skills of the future, social exclusion and ethical challenges, and changes in the political and legislative framework (Murphy, 2012).

The main problem in society is the diversification of technological levels in different regions and social groups, further deepening inequality and the sense of social injustice. Access to technology, especially the latest developments, is often severely limited in less developed countries and among low-income social classes. The lack of this availability makes it very difficult for these communities to benefit from the advantages created by innovation (Benjamin, 2023). Therefore, technological progress to make this area sustainable must focus on initiatives democratizing access to technology by developing digital infrastructure, educational programs, and subsidies for innovative technologies for poorer communities.

Regarding the impact of technology on employment and the labor market, the biggest problem seems to be the replacement of human labor with machines, thanks to the achievements of increasingly widespread automation and digitalization. This leads to a growing level of unemployment and causes economic instability for workers, especially in groups with low professional qualifications (Sadovaya, 2019). The solution to problems in this area can be significantly supported by investments in training and professional retraining and by strengthening the development of industrial sectors that intensively use human labor. Creating social policy programs to keep employees during the transition period will also be helpful.

Another problem, similar to the above but of a slightly different nature, is the requirement for the workforce to have new digital skills, critical thinking skills, and adaptation skills to rapid changes resulting from the dynamic development of technology. These requirements may carry the risk that part of society will be unable to keep up with these changes, which will deepen competence gaps and social inequalities (Demir, 2019). Again, training and various educational programs will be helpful, as they should focus on the required skills of the future, including science, technology, engineering, and mathematics, but also on lifelong learning and flexible career paths.

The development of technological progress may also cause exclusion, mainly digital exclusion, of certain social groups from regular participation in the activities of citizens of this society as a result of, among other factors, unfavorable economic conditions (material poverty). Digital exclusion primarily concerns inequality in access to the Internet, digital technologies, and the skills necessary to use them effectively (Macdonald and Clayton, 2017). Effective counteraction should focus on the expansion and modernization of digital infrastructure, especially in rural areas, on facilitating access to technological equipment by subsidizing computers and mobile devices for poorer individuals, schools, and public institutions, and on the development of digital skills through the comprehensive promotion of educational programs and training courses improving digital competencies, with particular emphasis on social groups most at risk of digital exclusion (older people, people with disabilities, and residents of rural areas). It is also crucial to support innovative technological solutions, especially low-cost, easy-to-use ones, such as simpler user interfaces, language translations, and the availability of services to a broader group of recipients, including people with various types of disabilities (Macdonald and Clayton, 2017). Multiple types of activities should also be undertaken to integrate strategies to counteract digital exclusion with development, educational, and social policies, promoting technologies designed with the assumption of wide social accessibility, in accordance with international standards, as well as activities that build awareness and understanding of technological development and new digital skills. All these activities should adopt the digital inclusion of society as their primary goal.

2.4.3 Economical Dimension

Sustainable technological development must also address economic challenges that concern both the macroeconomic aspects of the global financial system and the microeconomic realities of individual businesses and consumers. The most severe and complex difficulties include high investment costs, risk, and uncertainty, the need to transform various economic sectors, global divergences, adjustment of markets, legal and political regulations, and the integration of new technologies with existing ones (Duić et al., 2015). Finding solutions to the problems mentioned is necessary for sustainable technological development to bring economic benefits while respecting the natural environment and supporting society.

Developing and implementing sustainable new technologies require high investment outlays, including research and development (R&D) and appropriate infrastructure. The amount of such costs, either for the economies of less developed countries or for individual enterprises, may constitute a barrier preventing the implementation of such challenges. Additionally, the risks associated with innovative ideas can hinder progress (Purvis et al., 2019). To eliminate this problem, state or EU financing programs, subsidies, various types of financial incentives, support mechanisms such as green bonds, funds or tax breaks for investors in R&D, and international scientific and research cooperation can help.

New technological solutions are usually burdened with high investment risk, and sustainable technological development often concerns new, previously untested markets. This level of risk and uncertainty regarding investment returns is not an incentive for potential investors (Baker, 2012). Therefore, in this area, it is beneficial to create a transparent and predictable regulatory and legal-political framework that will support these investments in the long term while minimizing risk and uncertainty.

Implementing innovations resulting from sustainable technological development very often requires the transformation of many traditional economic sectors (energy, transport, agriculture), which may result in economic and social disruptions

(Holden et al., 2017). Programs supporting such sectoral transformation, programs for retraining employees, or state support for the transformed sectors and regions may prove helpful here.

Significant discrepancies exist between developed and developing countries in their ability to invest in sustainable development. Planning additional investments in innovative technologies may deepen and create further inequalities between nations (Mensah, 2019). This challenge can be minimized by promoting international cooperation in the area of R&D, supporting technology transfer, international financing of various technological implementations, and providing additional support for developing countries.

Implementing new achievements in technological development, especially in the concept of sustainable development, requires a lot of effort to adapt markets, standards, legal and political aspects, as well as consumers and their preferences to these changes (Mensah, 2019). Here, it is necessary for the governments of interested countries to actively participate in shaping these tools, promoting sustainable consumerism, and introducing green standards and certifications.

The final major economic challenge for sustainable technological development is the integration of new sustainable technologies into these already existing systems. Given the need to develop compatibility of sustainable technologies with existing infrastructures and economic systems, these activities may be challenging and expensive at the same time (Sadovaya, 2019). The remedy here may be planning and implementing integration strategies to enable a smooth transition and compatibility of new technologies with existing systems.

Overcoming the economic challenges requires an integrated approach combining technological innovation, policy support, market flexibility, and international cooperation.

2.5 PRIORITY PRINCIPLES OF SUSTAINABLE TECHNOLOGICAL DEVELOPMENT

Environmental, social, and economic priorities are not separate or contradictory in sustainable development. They seek to integrate them (holistic approach) by assuming that sustainable development requires balance and synergy between these three dimensions. This approach highlights the interactions between the natural environment, social structure, and economic dynamics, indicating that these three areas are closely interconnected and influence each other (Purvis et al., 2019). Therefore, the fundamental principles of sustainable technological development aim to harmoniously combine technological progress with environmental protection, social justice, and economic stability, thereby helping to shape technological innovations to support future generations' sustainable development and well-being. The priority principles of sustainable technological development are as follows (Weaver et al., 2017):

- Environmental sustainability means that new technologies should be designed and implemented in such a way as to minimize and reduce their negative impact on the natural environment and should protect natural resources and biodiversity. This principle concerns primarily the reduction of greenhouse gas emissions, the prevention of air, water, and soil pollution, and the efficient use of resources.

- Social justice refers to enabling equal access to the benefits of technological development for all sectors of society, regardless of origin, economic, or geographical status. It also requires considering issues related to the affordability and usability of a given technology for various social groups. Priority should be given to rural communities, people with disabilities, and those with low incomes.
- Economic efficiency stems from sustainable technological development, contributing to economic growth and innovation while ensuring a fair distribution of monetary benefits. Additionally, new technologies should be cost-, material-, and energy-efficient, translating into widescale implications in industrial production and economic activity.
- Responsibility and transparency concern responsibility for the social, economic, and environmental effects resulting from the processes of creating and implementing new technologies. Implementing this principle requires the involvement of stakeholders and local communities in decision-making processes and monitoring and reporting the impact of a specific technology on each of the three dimensions of sustainable development.
- Adaptability and flexibility result from dynamic changes in each of the three dimensions of sustainable development and the related uncertainty regarding new technologies and the possibilities of their use. This principle is characterized by the desire to design technological innovations to make their long-term use possible while minimizing the risk of obsolescence. Modular and easy-to-update technologies, flexible in adapting to changing needs and conditions, are promoted.
- Orienting innovation to social needs emphasizes that technological development should focus on solving the real problems of people and societies rather than being driven solely by the pursuit of profit, economic growth, consumerism, and industrial domination.

The principles discussed indicate the direction in which sustainable technologies should be developed to serve both current and future generations.

2.6 PRIORITY AREAS OF SUSTAINABLE TECHNOLOGICAL DEVELOPMENT

The key to achieving sustainable development goals is to focus on priority areas of sustainable technological development, i.e., those in which innovation and technological progress can have the most significant impact on promoting sustainable development, covering environmental, social, and economic aspects.

2.6.1 Renewable Energy

One such area is renewable energy. In this area, technologies using renewable energy sources are developed and implemented to reduce dependence on fossil fuels and greenhouse gas emissions and ensure energy security. Sustainable technology development in this area focuses on renewable sources such as (Patel et al., 2023):

- Solar energy: Solar energy technologies convert sunlight into electricity or heat using various methods and materials. Most often, it is obtained from photovoltaic panels with high-efficiency solar cells and from flexible solar panels, the advantage of which is the possibility of installation on irregular or movable surfaces. Solar energy can also come from thermal solar collectors, thermodynamic solar systems, or hybrid systems, most often photovoltaic thermal systems.
- Wind energy: Wind energy technologies use the power of the wind to generate electricity. This energy is obtained from wind turbines, which, when constructed with appropriate technology, have higher efficiency in variable wind conditions and are easier to install in various environments, such as offshore turbines, which can take advantage of more robust and stable winds at sea, thereby increasing energy production. The wind energy sector is the subject of continuous research and development, focusing on increasing efficiency, reducing costs, and minimizing environmental and local community' impact. Innovations in this area include developing more extensive and efficient turbines, intelligent energy management and storage systems, and new materials and construction technologies that make wind turbines more durable and less environmentally invasive.
- Hydropower: Hydropower technologies use various methods of converting the energy contained in water into electricity. They cover many applications, from large hydropower plants to smaller systems used in local conditions. Most often, hydropower comes from traditional or run-of-river power plants and from small hydropower plants, which can be installed without the need to build large dams and reservoirs, thus minimizing the impact on the environment and local communities. It is more innovative to use the energy of tides and sea currents, as well as waves and thermal energy of the oceans, which have the potential to be a stable and predictable source of renewable energy.
- Geothermal energy: Technologies in this area use heat collected under the Earth's surface to produce electricity and for heating purposes. This form of renewable energy is characterized by high reliability and low variability, distinguishing it from other renewable sources. Energy is obtained from geothermal power plants, heat pumps, or direct use of geothermal water. The geothermal energy sector is constantly evolving, introducing innovations that aim to increase efficiency, reduce costs, and expand the possibilities of using this form of energy. New research directions focus on methods to increase the flow rate of geothermal reservoirs through

hydraulic stimulation, develop technologies using low-temperature geothermal sources, and integrate geothermal systems with other renewable technologies to maximize efficiency and energy sustainability.
- Energy from biomass: Technologies for obtaining energy from biomass use organic materials of plant and animal origin to produce heat, electricity, or fuel. Biomass energy is considered renewable because the materials from which it is made can be regenerated relatively quickly. Biomass can come from many sources, including agricultural, forestry, urban waste, and specially cultivated energy plants. Technologically, energy from biomass is produced through direct combustion, gassing, pyrolysis, and methane fermentation processes. Biomass can also be processed into liquid biofuels as substitutes or additives for gasoline and diesel oil.

2.6.2 Energy Efficiency

The next priority area of sustainable technological development, next to renewable energy, is energy efficiency. This area is characterized by innovations aimed at increasing the efficiency of energy use in industry, construction, transport, and households, which translates into reduced energy demand and emissions of harmful substances. Sustainable technological development in energy efficiency requires cooperation between governments, the private sector, research institutions, and consumers. It is an ongoing process that, in addition to introducing new technologies, also involves modernizing existing systems and practices to achieve a sustainable energy future. Technological innovations in the area of energy efficiency concern (Worrell et al., 2018):

- Modernization of energy infrastructure through the implementation of intelligent energy networks (smart grids) allows for better management of energy flow, flexible response to changes in demand, integration of renewable energy, and modernization and insulation of heating and cooling systems in buildings to reduce energy losses.
- Optimization of production processes using technologies enables more efficient use of energy in industrial processes, e.g., waste heat recovery or process automation, to minimize energy loss.
- Designing and using new and technologically advanced construction insulation materials significantly reduce the need for heating and cooling interiors. Implementing materials with better heat conduction parameters in electronic components and machines increases their energy efficiency.
- Use energy management systems in commercial, industrial, and residential buildings that monitor and control energy consumption, adapting it to actual needs.
- Development of lighting technologies by promoting LED lighting and other energy-saving lighting technologies that offer long life and low energy consumption compared to traditional light sources.

- Innovation in electromobility, i.e., the development of electric vehicles and charging infrastructure, is critical to reducing the consumption of fossil fuels and emissions from transport.
- Promote a culture of energy saving by implementing educational programs and social campaigns to raise awareness of the importance of energy efficiency and ways to save energy.

2.6.3 Water Resources Management

Water resources management is the next important area among the priorities for sustainable technological development (Rattan et al., 2005). Technology in water resources management focuses on developing and implementing innovative technical solutions and management systems that optimize water use, protect its quality, and ensure access to clean water for all communities while minimizing negative impacts on the natural environment. Sustainable technological development in water resources management requires an interdisciplinary approach, i.e., a combination of technological innovations with environmental policy, spatial planning, and the involvement of local communities. All this is to manage this valuable resource effectively in a way that benefits society, the economy, and the environment. Technological support in this area includes (Cosgrove and Loucks, 2015):

- Efficient use of water, i.e., technologies that reduce water consumption in agriculture, industry, and households.
- Advanced wastewater treatment systems, i.e., modern technologies that effectively remove contaminants and pathogens, enable water to be reused for various purposes.
- Water resources management uses digital technologies, i.e., the Internet of Things (IoT), artificial intelligence (AI), and extensive data analysis to monitor and manage water resources in real time. Thanks to these technologies, optimizing water distribution, detecting leaks early, and preventing the overexploitation of water resources are possible.
- Developing energy-efficient desalination technologies, i.e., transforming seawater or salt water from another source into fresh water, is suitable for drinking and other purposes. These technologies are essential for regions suffering from water shortages.
- Effective management of water resources in agriculture, i.e., advanced and precise irrigation systems supplying water directly to plant roots. This form of irrigation reduces losses and increases water use efficiency.
- Protection of aquatic ecosystems through technologies for protecting and renovating marine ecosystems (natural methods of water purification, systems for protecting riverbanks against erosion, technologies supporting the renovation of degraded wetlands).
- Education and public awareness are enabled by using digital platforms and social media to educate and raise awareness about sustainable water

management, which is vital for changing behavior and promoting a water-saving culture.

2.6.4 Sustainable Agriculture

The priority areas of sustainable technological development also include the area of sustainable agriculture. Sustainable technological development in this area involves introducing innovations and practices that increase agricultural productivity and efficiency (Jain and Bhambri, 2005). This must be done while minimizing negative impacts on the natural environment, improving the well-being of communities, including rural ones, and ensuring the long-term stability of ecosystems. These technologies aim to create efficient, sustainable agricultural systems from ecological, economic, and social points of view. Sustainable technological development in agriculture requires continuous evaluation of the impact of new technologies on the environment, economy, and communities, as well as adaptation of practices to local conditions. It needs to ensure long-term sustainability and food security. The development of technology in the area of sustainable agriculture is based on the following (Altieri, 2018):

- Organic farming methods are agricultural practices that promote sustainable management of natural resources, biodiversity conservation, environmental protection, and human health. This crop type does not use synthetic fertilizers, pesticides, herbicides, fungicides, or genetically modified organisms (GMOs). The goal of organic farming is to produce food sustainably.
- Precision agriculture uses advanced technologies such as GPS, drones, field sensors, and data management systems to monitor and optimize agricultural processes accurately. This makes it possible to apply fertilizers, water, and plant protection products precisely, which increases efficiency and minimizes waste and environmental damage.
- Digital agritechnologies use information and communication technologies (ICT), such as mobile applications, online platforms, and big data, to support agricultural decisions, resource management, and market access. These activities increase productivity and contribute to sustainable resource management.
- Biotechnology and genetic engineering involve the development of plant varieties resistant to pests, diseases, and extreme climatic conditions. Although these technologies are controversial, they can help increase yields and reduce the need for chemical plant protection products.
- Integrated Pest Management (IPM) focuses on developing pest control methods that combine various agricultural, biological, mechanical, and chemical practices to reduce their impact on crops while limiting negative environmental impacts.
- Regenerative agriculture and agroforestry promote agricultural practices that restore and strengthen soil health, increase biodiversity, and sequester carbon.

2.6.5 Food Supply Chain Management

Right next to the area of sustainable agriculture, there is a priority area of managing food supply chains. Sustainable technological development in this area focuses on using technological innovations to create more efficient, transparent, and resilient supply systems that minimize negative environmental impacts, improve food safety, and support social justice. The aim is to optimize production, transport, and distribution processes to reduce food loss and waste, reduce greenhouse gas emissions, and ensure access to healthy and sustainable food for the wider population. Sustainable technological development in the management of food supply chains requires a holistic approach that combines technological innovation with social and environmental responsibility, involving all participants in the supply chain – from farmers through processors and distributors to consumers. Among the activities undertaken in this area, it stands out (Beske et al., 2014):

- Data-driven supply chains use advanced data analytics and the Internet of Things (IoT) to monitor and optimize logistics processes. These systems enable traceability of food products from farm to fork, providing better control over food quality and safety and minimizing loss and waste.
- Green food packaging technologies, i.e., the use of innovative packaging materials that are biodegradable, compostable, or reusable, reduce waste and plastic pollution.
- Digital distribution platforms in the form of online platforms and mobile applications optimize distribution processes, allowing for better matching of supply and demand, reducing food losses, and facilitating access to markets for small producers.
- Social initiatives and cooperation networks based on the development of community food banks and applications supporting the exchange or distribution of unused food among those in need contribute to reducing food waste and promoting social solidarity.
- Certification and international sustainability standards for agriculture, manufacturing, and trade support environmentally-friendly and fair practices for producers.

2.6.6 Circular Economy

Sustainable technological development in the following priority area, the circular economy, involves creating and implementing innovative technologies that efficiently use resources, minimize waste, and promote recycling and reuse of materials. Here, efforts focus on creating production and consumption systems that work harmoniously with natural ecological cycles, reducing environmental impact, and ensuring sustainable socio-economic development. To realize its full potential, the circular economy requires cross-sector collaboration, policy support, and consumer engagement. New and innovative solutions in this area concern (Winans et al., 2017):

- Minimizing waste by creating technologies that enable the design of products and processes in such a way as to reduce the amount of waste generated from the very beginning. These technologies also concern the production of materials that are easier to recycle, dismantling technologies, and production methods that reduce the waste of raw materials.
- Designing recyclable products, i.e., products designed so that they can be easily broken down into components and raw materials at the end of their life. These, in turn, will be suitable for reuse in producing new products, thus closing the material cycle.
- Reuse and remanufacturing technologies supporting reuse systems for products and their components. The possibility of regenerating materials is also essential, as they can be refurbished and put back into economic circulation, reducing the need to use new resources.
- Technologies for designing and producing ecological and biodegradable materials or those from renewable sources to replace those that are harmful to the environment or difficult to recycle. Such materials are primarily biopolymers, natural composites, and innovative plant-based materials.
- Obtaining energy from waste, i.e., using waste as an energy source. This is possible, for example, by producing biogas from organic waste or by burning non-recyclable waste to produce heat and electricity, with appropriate pollution control measures.
- Possibilities of using digital technologies in the circular economy, i.e., the use of digital technologies such as the Internet of Things (IoT), artificial intelligence (AI), and blockchain, to monitor and optimize the flow of materials, manage supply chains, as well as to promote transparency and traceability of products.
- Eco-innovation and product life cycle modeling, i.e., technologies and methods for assessing the impact of products on the environment throughout their entire life cycle, from production to disposal. Eco-innovation also includes developing business models that promote services instead of ownership (e.g., sharing economy), reducing the total amount of goods produced and consumed.

2.6.7 Sustainable Mobility

Sustainable mobility is also a priority area of sustainable technological development. It focuses on creating, implementing, and promoting transport technologies that minimize negative environmental impacts, improve energy efficiency, reduce emissions of harmful substances, and support social and economic inclusion. Its main goal is to create transport systems that are ecologically sustainable, economically efficient, and accessible to all segments of society. Sustainable mobility is achieved through (Silva et al., 2018; Weaver et al., 2017):

- Electrification of transport, or the shift from fossil fuel-powered vehicles to electric vehicles (EVs), including cars, buses, trucks, and two wheelers. Electrification of transport is vital to reducing CO_2 emissions and other air pollutants. However, investments are needed in the development of charging infrastructure, the introduction of electric public transport systems, and innovations in batteries with higher energy density and faster charging.
- Mobility as a Service (MaaS) integrates various public and private transport forms into a unified, digitally accessible service that allows you to plan, book, and pay for trips using a single application platform. The apps enable users to take advantage of car-sharing, bike-sharing, taxis, public transport, and more to optimize routes and reduce the need for a private vehicle.
- Autonomous transport, i.e., the development of autonomous vehicles (AVs) that can drive themselves without direct human intervention, potentially reducing accidents caused by human errors, improving road capacity, and reducing congestion.
- Green transport infrastructure means designing and constructing infrastructure that supports sustainable transport, including cycle paths, sidewalks, green bus stops, and intelligent traffic management systems.
- Sustainable development of aviation and shipping in the form of innovations aimed at reducing their impact on the environment through improving fuel efficiency, using alternative fuels, and developing low-emission technologies.

2.6.8 Smart Cities and Societies

The final priority area is intelligent cities and communities. Sustainable technological development uses advanced technologies and innovations to create more efficient, safe, and environmentally friendly urban ecosystems. The priority here is to improve residents' quality of life, minimize the impact on the natural environment, and promote social and economic sustainability. The following technological possibilities are used in this area (Silva et al., 2018):

- Smart grids are for efficient energy distribution, integration of renewable energy, and consumption optimization so that cities can reduce emissions and energy costs.
- Intelligent transport systems (ITS), such as on-demand public transport, traffic management systems, vehicle-sharing applications, and electric vehicle infrastructure, aim to reduce congestion, emissions, and improve transport accessibility.
- Intelligent construction, i.e., automated building management systems (BMS), is used to optimize the consumption of energy, water, and other resources, improving the comfort and safety of residents.
- Information and communication technologies (ICT), such as devices and sensors based on the Internet of Things (IoT), are integrated with urban

infrastructure, which collects real-time data and enables their analysis for better city management. Additionally, data management platforms, i.e., central systems for analyzing and managing data from various sources (transport, energy, public safety), support decision-making and improve city services.

- Intelligent security systems that increase public safety include cameras with image recognition, lighting that responds to the presence of people, and rapid response systems in the event of failures or disasters. Additionally, technologies monitoring environmental parameters provide information about air quality and can contribute to preventive actions in public health protection.
- Social participation is the digital platform for residents where appropriate tools and applications enable residents to participate in decision-making processes, report problems, and participate in community life.

The priority areas of sustainable technological development characterized above constitute the foundations of this concept. Also, the technologies and solutions mentioned in this area are only the most important ones from the entire pool, which is much more significant because innovations are created continuously. Some of the presented possibilities are already implemented and functioning in economies and societies. Others are still being refined for implementation soon. Regardless of the level of advancement, technological development continues. Since it is intended to represent positive outcomes and bring benefits, it is crucial to ensure that it develops sustainably.

2.7 FUTURE PERSPECTIVES AND CONCLUSIONS

Sustainable technological development has the potential not only to minimize adverse environmental impacts but also to improve the quality of social life and ensure sustainable economic growth. Therefore, in addition to the technological achievements that already exist and are in operation, future activities should be monitored, considering the dynamics of change. Development prospects in this area focus on using innovation and technological progress to solve critical environmental, social, and economic challenges.

Therefore, further investments and innovations in renewable energy technologies are necessary to integrate diverse sources and thus increase their share in the energy mix while reducing dependence on fossil fuels. We should continue to develop innovative city technologies that support efficient resource management, mobility, infrastructure, and public services, contribute to creating sustainable and resident-friendly cities, and maintain the improvements that have already been achieved (Silva et al., 2018). In the era of digitization and automation, it seems obligatory to use advanced technologies such as artificial intelligence (AI), the Internet of Things (IoT), and robotics to optimize production, resource management, and waste reduction in industry and agriculture(Heeks, 2010). There is also a need to continuously improve circular economy processes and promote business models and technologies that enable material reuse, recycling, and waste minimization, thus supporting the transition from a linear to a circular economy. Additionally, there should be a focus on developing technologies in the area of sustainable agriculture that will improve efficiency and support the production of sustainable food. The hydrological crisis in some parts of the world highlights the need to implement innovations in water purification and sanitation technologies that enable sustainable management of water resources and access to clean water for all (Cosgrove and Loucks, 2015). It is also essential to develop educational technologies to promote access to knowledge and skills that are key to sustainable development, as well as to combat digital exclusion and promote social equality. New solutions to support adaptation to the impacts of climate change, such as extreme weather events and rising sea levels, are essential to protect communities and ecosystems. Finally, an overarching necessity is strengthening international cooperation in the research, development, and diffusion of sustainable technologies to support global sustainable development efforts (Weaver et al., 2017).

Both the current state and prospects indicate the need for a holistic approach to sustainable development and technology implementation, considering environmental, social, and economic needs, and emphasizing the role of innovation as a tool to achieve lasting, sustainable development on a global scale.

REFERENCES

Altieri, M. A. (2018). *Agroecology: The science of sustainable agriculture*. CRC Press.

Baker, J. (2012). The technology–organization–environment framework. *Information Systems Theory: Explaining and Predicting Our Digital Society*, 1, 231–245.

Beckerman, W. (2017). 'Sustainable development': Is it a useful concept?. In *The economics of sustainability* (pp. 161–179). Routledge.

Benjamin, R. (2023). Race after technology. In *Social theory re-wired* (pp. 405–415). Routledge.

Beske, P., Land, A., & Seuring, S. (2014). Sustainable supply chain management practices and dynamic capabilities in the food industry: A critical analysis of the literature. *International Journal of Production Economics*, 152, 131–143.

Biesbroek, G. R., Swart, R. J., Carter, T. R., Cowan, C., Henrichs, T., Mela, H., ... Rey, D. (2010). Europe adapts to climate change: Comparing national adaptation strategies. *Global Environmental Change*, 20(3), 440–450.

Camisón, C., & Villar-López, A. (2014). Organizational innovation as an enabler of technological innovation capabilities and firm performance. *Journal of Business Research*, 67(1), 2891–2902.

Carley, M., & Christie, I. (2017). *Managing sustainable development*. Routledge.

Cosgrove, W. J., & Loucks, D. P. (2015). Water management: Current and future challenges and research directions. *Water Resources Research*, 51(6), 4823–4839.

Demir, A. O. (2019). Digital skills, organizational behavior and transformation of human resources: A review. *Ecoforum*, 8(1), 92–113.

Duić, N., Urbaniec, K., & Huisingh, D. (2015). Components and structures of the pillars of sustainability. *Journal of Cleaner Production*, 88, 1–12.

Elliott, J. (2012). *An introduction to sustainable development*. Routledge.

Hák, T., Janoušková, S., & Moldan, B. (2016). Sustainable development goals: A need for relevant indicators. *Ecological Indicators*, 60, 565–573.

Hakansson, H. (Ed.). (2015). *Industrial technological development (routledge revivals): A network approach*. Routledge.

Heeks, R. (2010). Do information and communication technologies (ICTs) contribute to development?. *Journal of International Development*, 22(5), 625–640.

Holden, E., Linnerud, K., & Banister, D. (2017). The imperatives of sustainable development. *Sustainable Development*, 25(3), 213–226.

Jain, V. K., & Bhambri, P. (2005). *Fundamentals of information technology & computer programming*. KATSONS.

Jeronen, E. (2020). Sustainable development. In *Encyclopedia of sustainable management* (pp. 1–7). Springer International Publishing.

Keeler, E., Michael, S., & Richard, Z. (2013). The optimal control of pollution. In *Economics of natural & environmental resources* (Routledge Revivals) (pp. 409–439). Routledge.

Macdonald, S. J., & Clayton, J. (2017). Back to the future, disability and the digital divide. In *Disability and technology* (pp. 128–144). Routledge.

Mensah, J. (2019). Sustainable development: Meaning, history, principles, pillars, and implications for human action: Literature review. *Cogent Social Sciences*, 5(1), 1653531.

Murphy, K. (2012). The social pillar of sustainable development: A literature review and framework for policy analysis. *Sustainability: Science, Practice and Policy*, 8(1), 15–29.

Patel, A. K., & Sharma, A. K. (Eds.). (2023). *Renewable energy innovations: Biofuels, solar, and other technologies*. John Wiley & Sons.

Purvis, B., Mao, Y., & Robinson, D. (2019). Three pillars of sustainability: In search of conceptual origins. *Sustainability Science*, 14, 681–695.

Rattan, M., Bhambri, P., & Shaifali, M. (2005). Information retrieval using soft computing techniques. In National Conference on Bio-informatics Computing (p. 7).

Reilly, J. M. (2012). Green growth and the efficient use of natural resources. *Energy Economics*, 34, S85–S93.

Rüßmann, M., Lorenz, M., Gerbert, P., Waldner, M., Justus, J., Engel, P., & Harnisch, M. (2015). Industry 4.0: The future of productivity and growth in manufacturing industries. *Boston Consulting Group*, 9(1), 54–89.

Sadovaya, E. (2019). Digital economy and a new paradigm of the labor market. *Mirovaiaek onomikaimezhdunarodnyeotnosheniia*, 62(12), 35–45.

Silva, B. N., Khan, M., & Han, K. (2018). Towards sustainable smart cities: A review of trends, architectures, components, and open challenges in smart cities. *Sustainable Cities and Society*, 38, 697–713.

Weaver, P., Jansen, L., Van Grootveld, G., Van Spiegel, E., & Vergragt, P. (2017). *Sustainable technology development*. Routledge.

Winans, K., Kendall, A., & Deng, H. (2017). The history and current applications of the circular economy concept. *Renewable and Sustainable Energy Reviews*, 68, 825–833.

Worrell, E., Bernstein, L., Roy, J., Price, L., & Harnisch, J. (2018). Industrial energy efficiency and climate change mitigation. In *Renewable energy* (pp. Vol1_548–Vol1_568). Routledge.

3 Technology and the Environment

Janusz Przybył

3.1 INTRODUCTION

In an era of global climate change and increasing environmental degradation, understanding the role that technology plays in these processes is becoming extremely important. Human history shows how technological innovations, from the Industrial Revolution to the modern digital era, have had an undeniable impact on the natural environment. This study aims to explore and understand this complex relationship, considering both the negative and positive aspects of technology's environmental impact.

3.2 CHAPTER 1: HISTORICAL CONTEXT OF TECHNOLOGICAL DEVELOPMENT

At the beginning of the Industrial Revolution in the 18th century, mass production and the use of steam engines initiated an era of intensive use of natural resources and emissions of pollutants. This era provides the foundation for contemporary environmental problems such as climate change and air pollution. Historical studies, such as the work of John U. Nef in *The Rise of the British Coal Industry*, shed light on these early connections between technological progress and environmental degradation.

3.2.1 Changes in the 20th and 21st Centuries

In the 20th and 21st centuries, technological development has accelerated, resulting in both positive and negative consequences for the environment. On the one hand, progress in the field of technology has contributed to improving the quality of life; on the other, it has led to increasing ecological problems. The work of authors such as Vaclav Smil in *Energy and Civilization: A History* (Smil, 2017) provides a cross-sectional view of this evolution. In the 20th and 21st centuries, dynamic technological development not only brought enormous benefits to humanity in the form of medical, communication, and technological progress but also posed new environmental challenges. Progress in the fields of electronics, computer science, and robotics, although contributing to increased efficiency and making everyday life easier, also resulted in an increase in the demand for raw materials and energy, which had a direct impact on the natural environment.

DOI: 10.1201/9781003475989-4

3.2.2 Digital Revolution and the Environment

The digital revolution, beginning in the second half of the 20th century, brought with it new challenges, such as the production and disposal of electronic equipment, as well as increased greenhouse gas emissions resulting from energy consumption by data centers and telecommunications networks. On the other hand, the development of digital technologies has contributed to the optimization of many production processes and resource management, which could have a positive impact on the environment.

3.2.3 Paradigm Shift in the 21st Century

In the 21st century, more and more attention has been paid to sustainable development and renewable energy sources. The development of technologies such as photovoltaics, wind turbines, and electric vehicles has begun to be perceived not only as an alternative to traditional energy sources but as a key element in the fight against climate change. Authors such as Jeremy Rifkin in his book *The Third Industrial Revolution* emphasize the importance of these changes in the context of the new economic and social model.

3.2.4 Conclusions from the History of Technological Development

By analyzing the historical context of technological development, we can see that each technological era has brought both benefits and challenges to the natural environment. The conclusions of this analysis are clear: technological progress is inevitable and necessary, but its direction and implementation must be carefully considered in the context of long-term environmental effects. Constant monitoring of the environmental impact of technology and the adaptation and development of sustainable technologies should be key elements of any future technology strategy.

3.3 CHAPTER 2: CONTEMPORARY ENVIRONMENTAL CHALLENGES

3.3.1 Climate Change and CO_2 Emissions

The greatest challenge of our times is climate change, driven mainly by greenhouse gas emissions from human activities. In this context, technology is both part of the problem and part of the solution. Reports from the Intergovernmental Panel on Climate Change (IPCC) analyze these dynamics in detail.

3.3.2 Environmental Pollution and Electronic Waste

Another important problem is environmental pollution, especially in the form of electronic waste. The growing consumption of electronic devices leads to the generation of huge amounts of waste, which is often not properly treated. Studies such as Baldé et al.'s (2017), "The Global E-waste Monitor", provide alarming data on the scale of this problem.

3.3.3 Impact of Electronic Waste on the Environment

The problem of electronic waste, also known as e-waste, has become one of the most pressing environmental challenges of our time. The pollution associated with this includes not only physical waste but also the emissions of harmful chemicals during the production and disposal of electronic equipment. These substances, such as lead, mercury, and cadmium, can have long-term effects on human health and ecosystems.

3.3.4 Recycling and Electronic Waste Management

In response to the growing problem of e-waste, many countries and organizations have begun to introduce recycling and e-waste management programs. It is important that these programs are effective and accessible to prevent the illegal dumping and export of e-waste to developing countries. An example of such activities is the European Union Directive on Waste Electrical and Electronic Equipment (WEEE), which imposes responsibility on producers for collecting and processing waste from their products.

3.3.5 Innovations in Electronic Waste Processing

New technologies such as advanced recycling methods, robotics in waste separation, and improved processing techniques can significantly contribute to reducing the environmental impact of e-waste. Innovative approaches to product design, such as design for recycling (DfR) and product life extension, also play a key role in minimizing electronic waste.

3.3.6 Education and Social Awareness

A key element in the fight against the problem of e-waste is education and raising public awareness about the responsible use and disposal of electronic devices. Information campaigns, educational programs in schools, and cooperation with technology companies can significantly contribute to changing society's attitudes toward the consumption and disposal of electronics.

3.4 CHAPTER 3: TECHNOLOGICAL INNOVATION AND ENVIRONMENTAL PROTECTION

3.4.1 Technology for Sustainable Development

In the face of global environmental challenges, technology can play a key role in promoting sustainable development. Renewable energy innovations such as solar, wind, and geothermal energy are becoming more available and efficient, offering alternatives to fossil fuels. Authors such as REN21 in their report "Renewables 2021 Global Status Report" emphasize the importance of these technologies in counteracting climate change.

3.4.2 Green Technologies and Waste

Green technologies, including innovative methods of recycling and waste management, are key to minimizing the negative impact of waste on the environment. Research in biotechnology, nanotechnology, and materials science opens up new opportunities for the efficient use of raw materials and waste reduction. An example is the work "Waste to Wealth" by Peter Lacy and Jakub Rutkowski, which analyzes the potential of the circular economy.

3.4.3 Development of Digital Technologies and the Environment

Digitization and the development of information technologies also have their place in the discussion about the environment (Rattan et al., 2005). While they generate some environmental impact in their own right, they also offer tools to monitor climate change, manage resources more efficiently, and promote environmental awareness. Works such as "Digitalization for a Sustainable Society", edited by Melchiorre and others, shed light on these aspects.

3.4.4 The Use of Big Data and Artificial Intelligence in Environmental Protection

Modern digital technologies such as Big Data and artificial intelligence (AI) play an important role in monitoring the environment and predicting climate change. Using satellite data and advanced algorithms, it is possible to precisely track changes in ecosystems, pollution, and weather patterns. As a result, researchers and decision-makers are better equipped with the tools necessary to make informed decisions about environmental protection.

3.4.5 Digitalization in Natural Resources Management

Information technologies also contribute to more effective management of natural resources. AI-based systems can help optimize water use, manage forests, and protect biodiversity by analyzing huge amounts of data and drawing practical conclusions from them. For example, smart irrigation systems in agriculture allow you to reduce water consumption while maintaining high crop productivity.

3.4.6 Digital Technologies in Education and Environmental Awareness

Digital educational technologies such as mobile applications, educational games, and e-learning platforms play a key role in raising environmental awareness. They provide information about sustainable development, climate, and environmental protection in an accessible and engaging way, reaching a wide audience. Thanks to this, it is possible to develop pro-ecological attitudes from an early age.

3.4.7 Challenges Related to Digital Technologies

Despite many benefits, digital technologies also pose some environmental challenges. The energy consumption of data centers, the production of electronic devices, and the generation of e-waste are just some of them. Therefore, it is important that the development of digital technologies takes place in accordance with the principles of sustainable development while minimizing their negative impact on the environment.

3.5 CHAPTER 4: SOCIAL AND ECONOMIC DIMENSIONS OF TECHNOLOGY

3.5.1 Technology and Social Change

The use of technology also has a social dimension. The development of sustainable technologies can contribute to the creation of new jobs, changing consumption patterns and promoting a more conscious lifestyle. However, there is also a risk of deepening inequality when access to modern, environmentally friendly technologies is limited. Authors such as Rifkin in *The Third Industrial Revolution* analyze these social aspects of the technological revolution.

3.5.2 Economy and Sustainable Development

From the economic point of view, the development of sustainable technologies is becoming not only an ecological but also an economic necessity. A challenge like climate change can have far-reaching economic implications, and investments in green technologies can include economic benefits. Economic reports, such as those issued by the World Bank, often discuss these issues.

3.5.3 Economics of Sustainable Development

In an economic context, investments in sustainable technologies not only respond to environmental challenges but also open up new market opportunities. The transformation toward a sustainable economy can stimulate innovation, create new markets and jobs, and support long-term economic stability.

3.5.4 Investments in Green Technologies

Investments in green technologies, such as renewable energy sources, energy efficiency, and sustainable transport, have the potential not only to reduce harmful gas emissions but also to generate economic growth. For example, developing the renewable energy sector is often seen as a key element in stimulating local economies and creating jobs.

3.5.5 Sustainable Business Models

There is also growing interest in sustainable business models that take into account the environmental and social aspects of their operations. Concepts such as the circular economy, responsible sourcing of raw materials, and corporate social responsibility (CSR) are becoming more and more important for companies around the world.

3.5.6 Challenges for Developing Economies

However, the transformation toward a sustainable economy also brings challenges, especially for developing economies. Access to modern technologies, financing of transformation, and the need to retrain the workforce are key challenges in these regions. International cooperation and support from international organizations such as the World Bank can play an important role in mitigating these challenges.

3.6 CHAPTER 5: FUTURE DIRECTIONS IN TECHNOLOGICAL DEVELOPMENT

3.6.1 Innovation and a Sustainable Future

In the context of future technological developments, it is important to direct innovation towards sustainability. The technological potential of artificial intelligence, robotics, and the Internet of Things (IoT) can contribute to the creation of more effective and less harmful natural resource management systems. Works such as *The Fourth Industrial Revolution* by Klaus Schwab provide a vision of how new technologies can shape the future in line with sustainable development.

3.6.2 Ethical and Regulatory Challenges

With every new technology comes ethical and regulatory challenges. This requires a responsible approach to the design and implementation of technologies, taking into account potential environmental impacts. In the debate about ethics in technology, works such as *Technology and the Virtues* by Shannon Vallor emphasize the importance of responsible technological development.

3.6.3 Education and Environmental Awareness

Technological development must go hand in hand with education and raising environmental awareness. Educational programs and social campaigns can play a key role in shaping attitudes and behaviors that support sustainable development. Examples include initiatives such as the Global Environmental Education Partnership (GEEP), which promote environmental education around the world.

3.6.4 The Importance of Education in Promoting Sustainable Development

Education plays a key role in shaping environmental awareness and promoting sustainable development (Bhambri and Gupta, 2005). It is not only about imparting knowledge, but also about developing critical thinking skills, understanding environmental complexities, and the ability to make sustainable decisions. Educational programs that integrate sustainability knowledge can prepare the next generations to create and maintain a more sustainable world.

3.6.5 The Role of Technology in Environmental Education

Digital technologies such as e-learning platforms, mobile applications, and virtual reality can significantly enrich the environmental education process. They enable interactive and engaging learning experiences that can help understand complex environmental issues and promote active engagement in environmental action.

3.6.6 International Cooperation in Environmental Education

International cooperation in the field of environmental education is crucial in building global awareness and responsibility. Organizations such as the Global Environmental Education Partnership (GEEP) play an important role in sharing best practices and educational resources and supporting global education initiatives. These actions can help build a coherent global vision of sustainable development.

3.6.7 Preparing for a Sustainable Future

Environmental education and awareness are essential to prepare society to deal with the challenges of sustainable development. Technological development, although offering tools and solutions, must be supplemented with solid knowledge and ecological awareness. Only in this way can we ensure that future generations have the necessary competencies to create a sustainable world.

3.7 CHAPTER 6: GREEN TECHNOLOGY SUCCESS CASES

3.7.1 Renewable Energy Sources – Breakthrough Projects

One of the most promising areas in ecotechnology is renewable energy sources. Projects such as wind farms in the North Sea or developing solar panel technologies in various regions of the world show how technology can contribute to sustainable energy production. Case studies such as "Renewable Energy Sources and Climate Change Mitigation" (IPCC, 2012) offer in-depth analysis of these initiatives.

3.7.2 Clean Water Technologies

Another key area is innovations related to the purification and management of water resources. Desalinization and wastewater treatment technologies open up new opportunities for access to clean water, which is crucial in the context of global climate change and the growing demand for water. Examples of such technologies are described in detail in the work "Water 4.0" by David Sedlak.

3.7.3 Innovations in Recycling and Waste Management

Recycling and waste management technologies also play an important role in reducing the human impact on the environment. Innovative approaches to recycling, such as recycling composite materials and electronics, are changing the way societies deal with waste. Reports such as Global Recycling Markets: Plastic Waste (ISWA) provide detailed information on these trends.

3.8 CHAPTER 7: CHALLENGES AND BARRIERS TO ECOTECHNOLOGY IMPLEMENTATION

3.8.1 Social and Political Resistance

One of the main challenges in implementing ecotechnologies is social and political resistance. It often results from a lack of knowledge, understanding, or economic threats. An example would be resistance to windmills due to concerns about the landscape or noise. Publications such as *The Politics of Green Transformations* (Scoones et al., 2015) analyze these phenomena.

3.8.2 Costs and Financing

Another barrier is the cost of implementing new technologies and access to financing. Although the long-term ecological and economic benefits are significant, the initial investment can be a barrier. Reports such as "Green Finance for Developing Countries" (UN Environment Program) present strategies for financing green technologies.

3.8.3 Technological and Infrastructure Challenges

Finally, there are technological and infrastructure challenges, such as the need to modernize energy networks so that they can support renewable energy sources. Research in this area, such as "Grid Integration of Renewable Energy Sources" (IRENA), highlights these challenges and proposes solutions.

3.9 SUMMARY

3.9.1 Integration and Innovation as the Key to Sustainable Development

In summary, technology plays a complex role in the context of environmental protection. It is both part of the problem and a potential solution. Success in achieving sustainable development will depend on the ability to integrate technological innovations with environmental concerns. This requires a combination of scientific, technological, social, and ethical approaches to development.

This chapter reviews key aspects of the relationship between technology and the environment, presenting both the challenges and opportunities available. Sustainable technological development requires not only innovation but also education, social awareness, and responsible political decisions as well as sources of financing and financial incentives for entrepreneurs and governments. The existing relationship between technology and the natural environment is deeply complex and two-sided. On the one hand, history shows that technological development has often led to the overuse of natural resources and the deterioration of the environment. On the other hand, the current technological era offers unprecedented opportunities to combat ecological problems and promote sustainable development. The future of technology in the context of environmental protection is a challenge full of possibilities, real, possible to apply and implement, and necessary. To exploit this potential, investments in research and development of sustainable technologies, as well as the adaptation of existing solutions to new ecological standards, will be crucial. It is also necessary to increase public awareness of the impact of technology on the environment and to promote responsible consumption and waste management. In this context, governments, businesses, and international organizations will play an important role and should cooperate to create and implement policies that support sustainable development. The active role of consumers will also be necessary, as they can shape the market toward more sustainable products and services through conscious choices. In an era where climate change is one of the greatest challenges for humanity, technology is no longer just a tool for progress but also the key to the survival of our planet. Technological innovations, if developed and applied with attention to their impact on the environment, can contribute to creating a sustainable future in which harmony between progress and the protection of our environment is not only possible but necessary. Ultimately, responsibility for the future of our planet lies in the hands of each of us – both as creators and users of technology and as citizens of the Earth. Sustainable technological development, combining innovation with care for the environment, is not only a choice but a necessity to ensure a healthy and prosperous future for subsequent generations.

REFERENCES

Baldé, C. P., Wang, F., Kuehr, R., & Huisman, J. (2017). *The Global E-waste Monitor 2017: Quantities, Flows, and Resources.* United Nations University (UNU), International Telecommunication Union (ITU) & International Solid Waste Association (ISWA). – Provides data and analysis on the global electronic waste problem.

Bhambri, P., & Gupta, S. (2005, March). A survey & comparison of permutation possibility of fault tolerant multistage interconnection networks. In National Conference on Application of Mathematics in Engineering & Technology (p. 13).

IPCC. (2012). *Reports of the Intergovernmental Panel on Climate Change.* – Detailed analyzes of the impact of technology on climate change.

Rattan, M., Bhambri, P., & Shaifali. (2005). Institution for a sustainable civilization: Negotiating change in a technological culture. In National Conference on Technical Education in Globalized Environment Knowledge, Technology & The Teacher (p. 45).

Scoones, I., Leach, M., & Newell, P. (2015). *The Politics of Green Transformations.* Routledge. – Researches the social and political aspects of implementing green technologies.

Smil, V. (2017). *Energy and Civilization: A History.* MIT Press. – Discusses the historical impact of technological development on energy use and its consequences for the environment.

4 Sustainable Solutions for Global Waste Challenges

Integrating Technology in Disposal and Treatment Methods

K. Geetha, J. Vigneshwari, Pankaj Bhambri, and A. Thangam

4.1 INTRODUCTION

The escalating challenges of waste generation on a global, regional, and local scale are a consequence of rapid industrialization, urbanization, and improved living standards. As society consumes energy and raw materials from the environment, it inevitably produces solid waste as an output. This sustained imbalance in input and output over the long term results in environmental degradation. To ensure the enduring sustainability of any community, it becomes imperative to consistently remove and safely dispose of these wastes, whether they are solid, liquid, or gaseous. Employing technology is key to addressing this issue. Implementing smart waste management systems equipped with sensors and IoT devices allows for real-time monitoring and the optimization of collection routes. This technological approach is essential in the extraction of waste from society, fostering a more efficient and sustainable waste management system. Waste management systems can become more efficient, environmentally friendly, and economically viable. However, it is essential to balance technological solutions with proper waste reduction strategies and recycling initiatives with public awareness.

4.1.1 Concept of Waste Management

The concept of waste management encompasses the systematic handling, disposal, and treatment of various types of waste materials to minimize environmental impact and promote resource conservation (Singh, Singh, Kaur et al., 2005). It involves strategies such as recycling, composting, incineration, and landfilling, as well as efforts to reduce waste generation through sustainable practices and public education. The

 DOI: 10.1201/9781003475989-5

penal procedure is also essential to check violations of civic rules. Public awareness and community engagement are integral components of effective waste management. Educational programs and campaigns promote responsible waste disposal, recycling practices, and the importance of reducing, reusing, and recycling materials. Moreover, extended producer responsibility (EPR) promotes sustainable production and consumption by urging manufacturers to oversee their products' entire lifecycle. Effective waste management requires citizens to receive adequate training and education on public issues such as sanitation and hygiene.

4.1.2 Revolution of Technology in Waste Management

Technology is revolutionizing waste management, offering innovative solutions to enhance efficiency and environmental sustainability. Radio frequency identification (RFID) and GPS technologies enable real-time tracking of waste containers and vehicles, optimizing collection routes and reducing fuel consumption. Smart bins equipped with sensors and Internet of Things (IoT) capabilities monitor fill levels, facilitating timely and efficient waste collection. Advanced data analytics and predictive modeling harness big data to discern waste generation patterns, enabling informed decision-making and resource allocation.

Technologies like optical sensors and robotic arms simplify the process of separating recyclables from mixed waste streams, making automation an essential component of waste sorting. In e-waste management, recycling technologies and extended producer responsibility (EPR) systems ensure the responsible disposal and recovery of valuable materials from electronic products. Blockchain technology enhances transparency in waste management supply chains, reducing the risk of illegal dumping.

Mobile applications engage the public, providing information on recycling practices, collection schedules, and mechanisms for reporting issues. While technology offers promising solutions, a holistic approach that combines technological advancements with waste reduction strategies and public awareness is crucial for building a comprehensive and sustainable waste management framework.

The waste management sector is in dire need of innovation since it not only benefits the environment but also opens up new avenues for advancement (Singh, Bhambri et al., 2005). Benefits of the smart waste management are depicted in Figure 4.1. Demand and restrictions are driving up the cost of the waste management

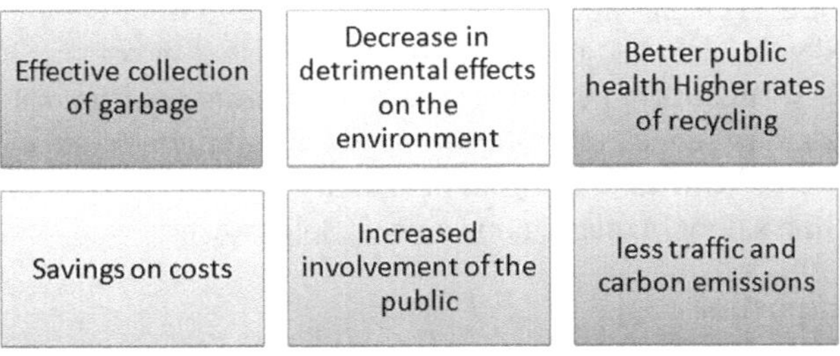

FIGURE 4.1 Benefits of smart waste management.

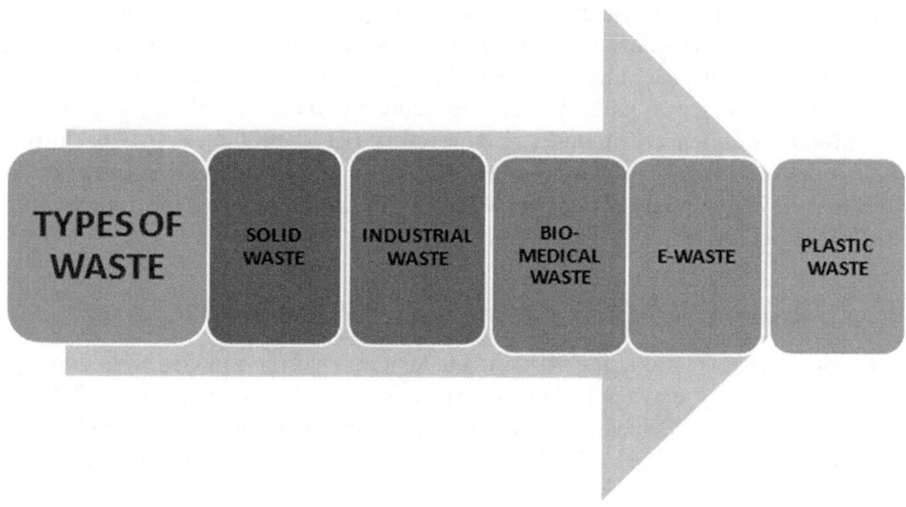

FIGURE 4.2 Types of waste.

practices used today. This is where organizations' and consumers' workloads and expenses may be reduced by using smart waste technology and embracing smart waste management. The following are the ways that it helps the world:

4.2 METHODS OF MANAGING DIFFERENT TYPES OF WASTE

Effectively managing various types of waste requires a multifaceted approach. Types of waste is shown in Figure 4.2. Technology aids in sorting and recycling processes, utilizing automated systems with optical sensors and robotics.

4.2.1 Solid Waste

Municipal waste, industrial waste, and dangerous waste are categorized as solid waste. Municipal waste arises from man's domestic activities.

4.2.1.1 Sanitary Landfill

A sanitary landfill is a garbage removal site designed to limit ecological effects and well-being risks. It includes a progression of defensive measures, including liners and leachate assortment frameworks, to keep contaminants from saturating the soil and groundwater. To reduce odor, vermin, and air pollution, waste is regularly compacted and covered with soil. The goal of sanitary landfills is to reduce environmental dangers while safely managing and storing solid waste.

4.2.1.2 Composting

Aerobic and anaerobic microorganisms decompose organic matter into compost, which is valuable manure. This is a biochemical process (Singh, Singh, Bhambri et al., 2005). It contains a variety of nutrients that are needed for the healthy growth of plants. Compost also increases the quality of the soil and hence it is called soil conditioner.

4.2.1.3 Vermicomposting

Worm composting, or vermicomposting, is a sustainable method of recycling organic waste using worms. In this process, specific species of earthworms are introduced to a controlled environment, such as a compost bin or vermiculture system. The worms consume organic matter like kitchen scraps and garden waste, breaking it down into nutrient-rich compost. This eco-friendly approach not only reduces the volume of waste sent to landfills but also produces high-quality compost that enhances soil fertility. Vermicomposting is a small-scale, efficient solution for households and businesses looking to minimize their environmental impact and contribute to nutrient cycling in a natural and sustainable way.

4.2.1.4 Biomethanation

One of the most cutting-edge methods for treating urban solid waste is biomethanation, which uses organic manure and biogas to recover resources. The biogas can be utilized for warming or power generation, while the ooze from the treatment plant is used as natural compost. As a result, it can be used in fruit processing plants and dairy farms where methane gas can be used immediately.

4.2.2 Industrial Waste

The unwanted byproduct of industrial processes is industrial waste. These processes include operations in mining and manufacturing.

4.2.2.1 Segregation

Glass, plastic, and paper are recyclable industrial wastes. As a result, this waste should be separated from biodegradable waste, hazardous waste, and solid non-hazardous waste by a business. To do this, the recyclable material ought to be distinguished by the plant or organization, and a different waste assortment framework can be laid out for them.

4.2.3 Biomedical Waste

In healthcare facilities, there are two main categories of biomedical waste: paper, packaging materials, non-infected plastic, and cardboard are examples of non-hazardous waste: (a) infectious waste, which includes sharps, non-sharps, disposable plastics, liquid waste, and so on. (b) Glass, chemical waste, cytotoxic waste, and other items are examples of non-infectious waste.

4.2.3.1 Incineration

It is the waste disposal method through burning. At sufficiently high temperatures, incineration destroys 99.99% of toxic organic compounds by decomposing them to carbon dioxide, water, and harmless gases. The other methods include steam sterilization or autoclaving, irradiation, thermal inactivation, chemical inactivation, and others.

4.2.4 E-Waste

The term "e-waste", also known as "electronic waste", refers to the discarded items and equipment of the electronic and telecom industry. E-squander contains toxic materials like lead, mercury, and cadmium, posing environmental and well-being chances if improperly discarded. Recycling e-squander is critical to recover valuable resources and forestall contamination (Singh et al., 2005a). However, its hazardous nature necessitates proper handling, highlighting the significance of environmentally and health-wise e-waste management practices.

4.2.4.1 Extended Producer Responsibility

The extended producer responsibility is viewed as a powerful apparatus for tracking down answers for the perplexing issue of item removal and contamination avoidance.

The E-Squander (the Board and Dealing with) Rules, 2011, presented by the Public Authority of India, aims to control the age, assortment, and removal of electronic waste to limit its natural and well-being influences. These standards command makers, buyers, and recyclers to comply with explicit rules for the safe handling, storage, and removal of electronic waste. Manufacturers are expected to take responsibility for the earth sound management of their products throughout their lifecycle. Furthermore, the principles lay out guidelines for the assortment, treatment, and recycling of e-waste to ensure sustainable waste management practices and moderate natural contamination.

4.2.5 Plastic Waste

Plastic waste refers to discarded plastic materials, including bottles, bags, packaging, and containers, which pose significant environmental and health challenges worldwide. Plastic waste is non-biodegradable and can persist in the environment for hundreds of years, leading to pollution of oceans, rivers, and landfills. It harms marine life, contaminates soil, and enters the food chain, posing risks to human health. Effective plastic waste management strategies include recycling, reducing single-use plastics, promoting alternative materials, and implementing policies to reduce plastic pollution and promote a circular economy.

4.3 CAUSES OF WASTE MANAGEMENT

The main causes of waste are practices and behavior, i.e., a capitalist or consumerist society that characterizes modern life. We are already purchasing and consuming massive quantities of consumer goods. Such goods may be plastic, metal, organic (such as food), etc., and may be functional objects or merely packaging contained in them for transport. One of the key problems is that if we no longer want an object, we

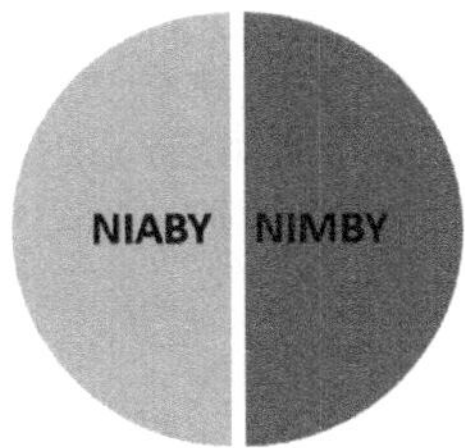

FIGURE 4.3 Strategies in waste management.

try to throw it away. Another big concern is that certain plastic products take almost forever to break down and can be hard to reuse or recycle. One thing that's clear is that there wasn't even a waste management issue before plastic. In fact, the vast majority of waste was reused, i.e., animals were given food, clothes were repaired, waste metal, etc., was sold to rag-and-bone men. In reality, landfills were originally invented as a way to fill holes left by mining operations in the earth.

4.4 NEW STRATEGY WITH RESPECT TO WASTE MANAGEMENT

There is an increasing emphasis on public involvement in contemporary times. Figure 4.3 is showing the strategies in waste management. The Polluter Pays principle is gradually finding implementation, encapsulated by two attitudes:

(i) NIMBY (Not In My Back Yard) and
(ii) NIABY (Not In Anybody's Back Yard).

The public is becoming more informed about environmental issues, expressing concerns about actual or perceived contamination and the potential hazards or nuisances associated with waste. The public may adopt a "Not In My Backyard" (NIMBY) mentality, especially when authorities propose the establishment of a waste or pollution disposal facility. This mindset is invoked when suggestions are made for locating a facility for the disposal of toxic waste or pollution. Environmental activists and NGOs often oppose such plans.

4.5 ALTERNATIVE WASTE MANAGEMENT SOLUTIONS

Active waste management should supplant passive methods. Addressing urban waste entails grappling with its multifaceted challenges, thus integrating socioeconomic, environmental, and societal factors into planning and management processes.

(a) Materials recovery:

For recovery and reuse, a range of materials found in urban and industrial solid waste are appropriate. It would seem that the most likely materials to be retrieved from the waste are paper, cardboard, plastics, glass, non-ferrous metals, and ferrous metals. It is frequently stated that reuse and recycling are not novel concepts; rather, they have been practiced formally and informally for many years. The reuse and recycling of products is also an important aspect of daily life in developing nations.

(b) Zero waste:

Future solid waste disposal plans: The concept of zero waste is emerging as a promising concept. It is essential for our long-term survival as well as the future prevention of uncontrolled explosive crises. Zero waste is a philosophical goal that aims to eliminate waste rather than manage it. Since no device is 100% effective, the goal of eliminating waste is unlikely. However, there may be ongoing progress toward the zero-waste goals during the process, which may result in significant shifts in how waste is actually viewed.

(c) Prevention of pollution:

Avoidance of contamination implies limiting waste materials created from all aspects of the manufacturing interaction to the creation cycle (Singh et al., 2005b). This is achieved through various techniques, including the adjustment of the products themselves, changing the strategies for creation to make them more viable, and reusing materials utilized in the process with the goal that there is less waste. Counteraction of discharges constitutes the "mixed media" way to deal with natural impact, with the core value being that it is easier to stop the development of waste in any case than to deal with or discard it later on.

4.5.1 Fresh Waste Management Approaches under a Trial in the World

- Organic waste is heated in anaerobic conditions to turn it into useful gases like methane, which can be used as fuel or a fuel-based substance. Organic waste is converted into sugar or protein.
- The pressure of waste into building blocks with additional solid materials. One of these plants is in operation in Japan.
- The use of degradable plastic bags for urban waste collection and transport instead of garbage cans.
- The transport of waste in pipelines as liquid slurry, a technique now used and is under review.

4.5.2 Effective Waste Management and Reduction Strategies Using Technology

Extracting waste from society involves using technology to reduce, recycle, and manage various types of waste efficiently.

4.5.2.1 Waste Sorting Technologies

Deploying advanced sorting technologies such as robotic sorting systems and AI-powered machines to separate recyclables from non-recyclables more efficiently. Thus utilizing computer vision and machine learning to enhance automated sorting in recycling facilities will reduce waste separation complications.

4.5.2.2 Blockchain for Waste Tracking

Implementing blockchain technology to create a transparent and traceable system for tracking waste from its source to its final destination. Also, it helps in identifying areas of inefficiency and ensuring proper disposal and recycling practices.

4.5.2.3 Data Analytics for Waste Prediction

Using data analytics to predict waste generation patterns based on historical data and seasonal trends. This can aid in proactive planning for waste management, helping to allocate resources more effectively.

4.5.2.4 3D Printing with Recycled Materials

3D printing innovation helps to make new items from reused materials. By reducing the need for new raw materials and reducing waste, this can encourage a circular economy.

4.5.2.5 Extended Producer Responsibility (EPR) Platforms

EPR stages are frameworks that consider makers responsible for the whole lifecycle of their items, including their end-of-life disposal. Producers are encouraged to take responsibility for managing their products after use and ensuring that they are recycled, reused, or disposed of appropriately by these platforms. To set up collection systems, recycling facilities, and take-back programs for EPR platforms, producers, recyclers, governments, and other stakeholders frequently work together. By moving the burden of waste management to makers, EPR stages advance sustainable production and utilization rehearses while reducing environmental impacts associated with product disposal.

4.5.2.6 Augmented Reality (AR)

Developing AR applications to educate individuals on proper waste disposal and recycling practices. AR can provide interactive and engaging experiences to raise awareness about the environmental impact of different types of waste. The integration of AR technology in waste management aligns with the broader trend of leveraging digital solutions for more sustainable and efficient practices. By providing real-time information, enhancing training, and optimizing processes, AR contributes to the overall effectiveness of waste management systems, reducing environmental impact and promoting a circular economy.

4.5.2.7 Drones for Monitoring Illegal Dumping

Use of drones equipped with cameras to monitor and detect illegal dumping activities. This technology can help authorities quickly identify and address instances of improper waste disposal.

4.5.2.8 Recycling Apps

Recycling facilities face a considerable challenge in separating recyclable and non-recyclable items from contaminated waste, particularly amid the pandemic. To tackle this issue, the organization introduced two recycling apps, namely

- Recycle Nation
- iRecycle.

These applications are designed to decrease the influx of non-recyclable materials at recycling centers. They empower individuals to enhance their recycling practices by offering information on recycling rates, center locations, and detailed lists of materials suitable for recycling. These waste management apps serve as convenient tools, assisting users in easily identifying recyclable items and promoting improved recycling habits.

4.5.3 Other Smart Ways in Waste Reduction

Some of the below-mentioned advanced treatment methods play a crucial role in minimizing environmental impact, optimizing resource recovery, and advancing the transition toward more environmentally friendly and energy-efficient waste management practices.

4.5.3.1 Pneumatic Waste Collection System

It revolutionizes urban waste management by utilizing an intricate network of underground tubes. Waste from various collection points is deposited into designated underground containers, and a centralized vacuum system propels it through the pneumatic tubes to a central collection station. Then, automated sorting and compaction processes enhance efficiency.

This innovative approach minimizes traditional garbage truck pickups, reducing traffic congestion, environmental impact, and operational costs. The closed system mitigates odors and noise, while the space-saving design is particularly advantageous in densely populated urban areas. Pneumatic waste collection systems represent a sustainable and efficient solution for modernizing waste disposal infrastructure.

4.5.3.2 Advanced Thermal Treatment Technologies

It encompasses innovative methods like gasification, pyrolysis, plasma gasification, and fluidized bed combustion, designed for efficient waste management. These processes utilize heat to break down organic and inorganic waste, generating valuable by-products such as syngas or bio-oil and recovering energy. Technologies like incineration with energy recovery, anaerobic digestion, and superheated steam treatment further contribute to sustainable waste disposal.

4.5.3.3 Plasma Gasification

It is an advanced waste treatment technology that employs extremely high temperatures generated by a plasma torch to convert organic and inorganic materials into synthesis gas, or syngas. In this process, waste is exposed to a plasma arc, typically reaching temperatures exceeding 5,000 degrees Celsius, causing the waste to break down into its elemental components. The syngas produced can be utilized for energy generation, and the process is highly efficient in minimizing the volume of waste while reducing environmental impact. Plasma gasification offers a sustainable alternative to traditional incineration, providing a cleaner and more resource-efficient approach to waste disposal.

The research affirms that plasma gasification can effectively eliminate CO_2 emissions. Despite being a relatively new technology, its limited adoption is attributed to a slow shift in preferences. However, its remarkable potential is evident, as it offers a promising solution for environmental sustainability by significantly reducing carbon dioxide emissions. Though not widely embraced yet, the impressive capabilities of plasma gasification underscore its potential to transform waste management practices and contribute to a cleaner and more sustainable future.

4.5.3.4 Circular Economy

It is an economic system that aims to make the most of resources and reduce waste. Not at all like the customary direct economy, which follows a "take, make, arrange" model, a roundabout economy centers on sustainability, longevity, and shut circle processes. The circular economy is viewed as a comprehensive strategy for addressing environmental issues, encouraging economic expansion, and ensuring resource stewardship. By closing the loop on production and consumption cycles, a circular economy aims to create a regenerative and sustainable system that benefits both businesses and the environment.

4.6 CONCLUSION

Improper waste management is a common practice that is unsafe and should be replaced with composting, which is a safer waste management method. Because our planet has already begun to face the consequences of dumping tons of garbage, waste management is emphasized a great deal. The legislatures and the neighborhood metro bodies should come up with more new procedures to reduce squander and ought to likewise make mindfulness among individuals about the advantages of using eco-friendly products. The government will be able to more effectively dispose of and recycle waste wherever possible if it incorporates the most recent and innovative technologies into its overall waste reduction and management efforts.

REFERENCES

Ahad, M. A., Paiva, S., Tripathi, G., & Feroz, N. (2020). Enabling technologies and sustainable smart cities. *Sustainable Cities and Society*, 61, 102301.

Akhil Jabbar Meerja, M. B. (2022). *Emerging Technologies and Applications for a Smart and Sustainable World*. Bentham Science Publishers.

Chadar, C. S. (2017). 'Solid Waste Pollution: A Hazard to Environment'. *Recent Advances in Petrochemical Science*, 2(3), 41–43.

Deepti Gupta, D. K. (2020). *Energy, Environment Ecology & Society*. Dhanpat Rai & Co. (P) Ltd .

Dr. Aradhana Salpekar, D. P. (2005). *Encyclopaedia of Ecology and Environment*. Jnanada Prakashan.

Heiman, M. K. (1990). From 'Not in My Backyard!' to 'Not in Anybody's Backyard!'. *Journal of the American Planning Association*, 23, 359–362.

Kumar U, A. M. (2018). *Biodiversity: Principles and Conservation*. M/s AGROBIOS (India).

Nižetić, S., Djilali, N., Papadopoulos, A., & Rodrigues, J. J. (2019). Smart technologies for promotion of energy efficiency, utilization of sustainable resources and waste management. *Journal of Cleaner Production*, 231, 565–591.

Patil, M., Boraste, S., & Minde, P. (2022). A comprehensive review on emerging trends in smart green building technologies and sustainable materials. *Materials Today: Proceedings*, 65, 1813–1822.

Rameshwar, R., Solanki, A., Nayyar, A., & Mahapatra, B. (2020). Green and smart buildings: A key to sustainable global solutions. In *Green Building Management and Smart Automation* (pp. 146–163). IGI Global.

Rani, N. (2022). *IoT-Based Smart Waste Management for Smart Cities*. CRC Press.

Riekki, J., & Mämmelä, A. (2021). Research and education towards smart and sustainable world. *IEEE Access*, 9, 53156–53177.

Singh, M., Bhambri, P., & Kaur, K. (2005, March). Network security. In National Conference on Future Trends in Information Technology (pp. 51–56). (10)

Singh, M., Singh, P., Bhambri, P., & Sachdeva, R. (2005). A comparative study: Security algorithms. In Seminar on Network Security and Its Implementations (p. 14). (10)

Singh, M., Singh, P., Kaur, K., & Bhambri, P. (2005, March). Database security. In National Conference on Future Trends in Information Technology (pp. 57–62). (8)

Singh, P., Singh, M., & Bhambri, P. (2005a). Internet security. In Seminar on Network Security and Its Implementations (p. 22). (9)

Singh, P., Singh, M., & Bhambri, P. (2005b). Security in virtual private networks. In Seminar on Network Security and Its Implementations (p. 11). (9)

Webliography

https://www.upperinc.com/blog/waste-management-technologies/

https://www.startus-insights.com/innovators-guide/waste-management-trends-innovation/

http://tumkuruniversity.ac.in/oc_pg/es/Recent%20trends%20in%20global%20solid%20waste%20processing%20technologies..pdf

5 Environmental Stewardship in the Digital Age

A Technological Blueprint

V. Mohana Sundari, M. Ganeshkumar, and Pankaj Bhambri

5.1 INTRODUCTION

As we enter the 21st century, we are living in a revolutionary period known as the "Digital Age", which is defined by the pervasiveness of digital technologies that have completely changed the way we interact with one another, live, and work. This epoch is characterized by the digitization of information, the interconnectedness of global networks, and the accelerated evolution of technology. Understanding the trajectory of the Digital Age necessitates a closer examination of its definition, key characteristics, and the rapid pace at which technological advancements have unfolded in recent years.

The Digital Age is marked by the convergence of various technologies, fundamentally altering the fabric of our societies. At its core, this age is distinguished by the digitization of data, enabling the seamless transfer and manipulation of information across borders and disciplines. This transformation has been fueled by the evolution of technology over time, tracing its roots from the Industrial Revolution to the present day. Key milestones, such as the invention of the internet, the proliferation of personal computing, and the widespread adoption of mobile devices, have played pivotal roles in shaping the contemporary landscape.

Recent years have seen a tremendous acceleration in technical breakthroughs. Technological advances in the fields of artificial intelligence, big data analytics, and the Internet of things have not only reshaped the landscape of digital possibilities but have also impacted every aspect of our everyday life. The enormous opportunities and challenges presented by the rapid pace of invention call for detailed knowledge of the ways in which the Digital Age is affecting all facets of human existence.

As we delve into the intricate interplay between digital technologies and environmental stewardship, it is imperative first to grasp the backdrop of the Digital Age. This exploration will pave the path for a thorough examination of the role these

DOI: 10.1201/9781003475989-6

technologies play in disseminating knowledge, steering economies toward sustainability, and addressing pressing environmental challenges. In this regard, our study attempts to balance the quest for economic prosperity with the necessity of environmental stewardship, all the while unlocking the transformational potential of digital technology.

5.2 IMPORTANCE OF ENVIRONMENTAL STEWARDSHIP

5.2.1 Recognition of Environmental Challenges

5.2.1.1 Climate Change

- Climate change stands as one of the most pressing global challenges, marked by shifts in temperature, precipitation patterns, and sea levels.
- Extreme weather, increasing sea levels, and ecological disturbances are among the effects.

5.2.1.2 Depletion of Resources

Resource depletion is a result of the unsustainable extraction and use of natural resources, including minerals, freshwater, and fossil fuels.

- Recognizing the finite nature of these resources is crucial for long-term environmental sustainability.

5.2.1.3 Biodiversity Loss

- The alarming decline in biodiversity, attributed to habitat destruction, pollution, and climate change, poses a threat to ecosystems and human well-being.
- The loss of species diversity can disrupt ecosystems, affecting food chains and ecosystem services.

5.2.2 Need for Sustainable Practices

Sustainable practices are imperative to address these challenges and ensure a harmonious coexistence between human activities and the environment.

Since ecological health and human well-being are intertwined, sustainable practices seek to satisfy current demands without jeopardizing the capacity of future generations to satisfy their own.

5.2.3 Role of Technology in Addressing Environmental Issues

Technological innovation plays a pivotal role in mitigating and adapting to environmental challenges.

Renewable energy: By lowering dependency on fossil fuels and reducing greenhouse gas emissions, technology helps switch to renewable energy sources.

Smart resource management: Technologies such as IoT (Internet of Things) enable efficient resource management, optimizing energy consumption and reducing waste.

Environmental monitoring: Advanced technologies allow for real-time monitoring of environmental indicators, aiding in the early detection and response to environmental threats.

Green infrastructure: The development and implementation of green technologies contribute to sustainable urban planning, promoting biodiversity, and reducing the environmental footprint of human settlements.

Innovative solutions: Technological advancements, including artificial intelligence and biotechnology, contribute to the development of innovative solutions for environmental restoration, waste management, and conservation efforts.

In essence, recognizing environmental challenges, adopting sustainable practices, and harnessing the power of technology are integral components of environmental stewardship. As we navigate the Digital Age, integrating these principles becomes paramount in ensuring a resilient and sustainable future for our planet.

5.3 REVIEW OF LITERATURE

In the last decade, artificial intelligence (AI) has undergone remarkable advancements, leading to the emergence of a multitude of practical applications that are reshaping the world [Ford, 2021]. Within academic literature, the definition of AI lacks unanimity, but artificial intelligence-driven results can broadly be characterized as structures possessing the capability to act wisely [Bhambri and Singh, 2005]. These systems are skilled at interpreting outside input and using it to carry out particular tasks in adaptable configurations; on occasion, they can replicate human behaviors with emotional, social, and cognitive intelligence [Di Vaio et al., 2020].

Artificial intelligence (AI) is essentially the term used to describe machines that do cognitive tasks like learning, interacting, and problem-solving that are normally performed by human minds [Nilsson, 1971]. The accessibility and ownership of organizational and customer data are recognized as fundamental advantages for firms seeking to innovate using AI and learn quickly [Hartmann, 2020] and recognize the pivotal role that data plays in navigating and thriving in the AI era.

Different definitions of AI systems are condensed into four categories by Russell and Norvig along two dimensions: the reasoning–behavior axis and the human performance–rationality dimension. These categories include systems that behave and think like people as well as those that act and think rationally.

For an AI system to be deemed effective, it should possess several key capabilities, including natural language processing for communication in a human-like manner, knowledge representation to store information, automated reasoning for utilizing stored information to answer questions and draw new conclusions, and machine learning for adapting to new circumstances and discerning and extrapolating patterns [Huang et al., 2019].

5.3.1 Digitalization and Sustainability

The relationship between sustainability and digitization reflects broad developments impacting the modern economy and society, calling for significant changes [Del Río,

2021]. First, it is necessary to clarify these two key ideas. Although there is no one definition for "sustainability", the UN Brundtland Commission defined it as "development that meets the needs of the present without compromising the ability of future generations to meet their own needs" in 1987. This definition is now widely accepted [Stuermer et al., 2017; Imperatives S. report, 1987]. The digital age has emerged at the same time as a result of technology's quick and unrelenting advancement, changing both the social and economic landscapes [Ji, Zhou et al., 2023].

3.2 Digital Innovation and Society

The range of issues facing the economy and society is diverse; the phrase "Great Challenges" refers to particular crucial obstacles whose removal could make a substantial contribution to resolving significant social issues with a high probability of having an international influence through widespread adoption [George et al., 2016]. According to Davidson et al. (2023), the United Nations has established 17 sustainable development goals, which include things like ending hunger, poverty, and promoting health and well-being as well as gender equality and high-quality education.

Within this framework, digital innovation shows promise as a means of addressing these issues and related criticalities, especially in conjunction with innovative scientific and technological organizational strategies. An example of this is how digital innovation can help with active aging by facilitating social activity.

5.3.3 Digital Technologies and Dissemination of Knowledge

The body of research indicates that digital technologies are essential to the spread of knowledge. Cross-border information exchange is made possible by digital content repositories, social media, and online platforms. A study conducted in 2011 by Anderson and Dron highlights the value of social and collaborative learning made possible by digital technologies. Furthermore, research like Siemens (2005) emphasizes the value of networked learning and the part that technology plays in building accessible and networked knowledge ecosystems.

5.3.4 Digital Technologies and Low-Carbon Economies

According to researchers like Schiederig, Tietze, and Herstatt (2012), digital technologies play a key role in promoting sustainability, which includes the shift to low-carbon economies. Energy-efficient technology, data analytics, and smart grids all help to maximize resource use and lower carbon footprints. The World Economic Forum (2018) highlights how digital technology will drive the Fourth Industrial Revolution, which will shape low-carbon and sustainable economies.

5.3.5 Digital Transformation Methods for Solving Environmental Problems

Studies indicate that digital transformation techniques, such as big data, artificial intelligence, and the Internet of Things, can effectively tackle environmental issues.

An analysis of the application of AI in environmental management and monitoring was conducted by Chen et al. (2018). Manyika et al. (2011) have emphasized how big data analytics facilitates thorough investigation of environmental trends, which helps with well-informed decision-making. Digital technology integration is thought to be essential to environmental conservation efforts.

5.3.6 Digital Technologies on Achieving Sustainable Development Goals 2030

Research shows that using digital technologies in citizen science projects helps significantly advance the sustainable development goals (SDGs). For example, Haklay (2013) talks about how citizen science may improve data gathering and play a part in environmental monitoring. Follett and Strezov's (2015) paper highlights how digital platforms democratize science and encourage public involvement in the achievement of the sustainable development goals. Bowser et al. (2014) highlight the use of online platforms and mobile applications in citizen scientific programs.

The foundation for comprehending the complex interplay between digital technologies and environmental stewardship is provided by these literary insights and makes room for additional study.

5.4 ENVIRONMENTAL STEWARDSHIP AND DIGITAL AGE: A DISCUSSION

In the dynamic landscape of the Digital Age, the intersection of digital technologies and environmental stewardship has emerged as a pivotal area of exploration. As we navigate a rapidly evolving technological landscape, it becomes imperative to understand how digital tools contribute to the dissemination of knowledge, drive the transition towards low-carbon economies, effectively address environmental problems through digital transformation methods, and empower citizen science initiatives toward achieving the sustainable development goals (SDGs) by the year 2030 (Bhambri et al., 2020). This inquiry delves into the multifaceted role of digital technologies in shaping our approach to environmental sustainability and knowledge dissemination. From online education platforms to smart environmental monitoring systems, the synergy between digital advancements and environmental stewardship presents unprecedented opportunities for positive change. This exploration aims to unravel the transformative potential of digital technologies in addressing pressing environmental challenges and advancing a sustainable future. In essence, this information sets the stage by acknowledging the overarching theme of the digital age's influence on environmental stewardship and then delineates specific questions that will guide the exploration of various dimensions of this intricate relationship as follows:

5.4.1 In What Ways Do Digital Technologies Play a Role in Spreading and Sharing Knowledge?

Digital technologies significantly contribute to the dissemination of knowledge through various channels. Online platforms, educational websites, and digital libraries provide accessible repositories of information (Sharma et al., 2020). Social media platforms facilitate real-time sharing and collaborative learning. E-learning platforms, webinars, and online courses empower individuals to remotely acquire new skills and knowledge. Moreover, digital technologies support interactive and multimedia formats, enhancing the effectiveness of knowledge transmission.

5.4.2 What Role Do Digital Technologies Play in the Transition Toward Low-Carbon Economies?

Digital technologies play a pivotal role in facilitating the shift toward low-carbon economies by enabling more efficient resource management and sustainable practices. Smart grids optimize energy distribution, while IoT devices facilitate real-time monitoring and control of energy consumption. Digital platforms contribute to the advancement and seamless integration of renewable energy sources. Big data analytics aids in identifying patterns for more sustainable resource use. Additionally, digital technologies contribute to the creation of eco-friendly solutions and the promotion of environmental awareness.

5.4.3 How Effective Are Digital Transformation Methods in Solving Environmental Problems?

Digital transformation methods, such as the Internet of Things, big data analysis, and artificial intelligence, are highly effective in addressing environmental problems. AI can process vast amounts of data to identify environmental trends and offer predictive insights. Big data analytics helps in monitoring and managing natural resources efficiently. IoT devices offer up-to-the-minute environmental data, enhancing the quality of decision-making processes. Digital transformation enhances the accuracy and speed of environmental monitoring and allows for the development of innovative solutions to ecological challenges.

5.4.4 What Is the Impact of Citizen Science Initiatives Utilizing Digital Technologies on Achieving Sustainable Development Goals 2030?

Citizen science initiatives leveraging digital technologies have a significant impact on achieving Sustainable Development Goals (SDGs) 2030. Through mobile apps, online platforms, and digital tools, citizens actively contribute to environmental monitoring, biodiversity conservation, and climate change research (Bhambri et al., 2019). Digital technologies enhance data collection, analysis, and dissemination, fostering collaboration between scientists and citizens. This collaborative approach empowers communities to participate in achieving SDGs, promoting transparency, inclusivity, and the democratization of scientific knowledge.

5.5 FINDINGS

Achieving the goals outlined in your research questions – leveraging digital technologies for knowledge dissemination, transitioning toward low-carbon economies, addressing environmental problems through digital transformation, and making an impact on sustainable development goals (SDGs) with citizen science initiatives – requires a comprehensive and strategic approach.

5.5.1 Promoting Knowledge Dissemination through Digital Technologies

1. Invest in E-learning platforms:
   Develop user-friendly and accessible e-learning platforms that cater to diverse learning styles.
   Incorporate multimedia content, interactive assessments, and real-world applications.
   Offer certifications and recognition for completing online courses to motivate learners.
2. Encourage Open Educational Resources (OER):
   Support the creation and curation of OER repositories.
   Establish collaborations with educational institutions and content creators to contribute to OER.
   Promote the use of creative commons licenses to facilitate sharing and adaptation.
3. Leverage social media and online communities:
   Establish official social media channels to share educational content and engage with learners.
   Create online communities or forums for students to discuss topics, ask questions, and collaborate.
   Use social media advertising to reach a broader audience and attract learners.

5.5.2 Transitioning towards Low-Carbon Economies

1. Smart infrastructure investments:
   Allocate funds for the development and implementation of smart infrastructure projects.
   Collaborate with technology companies and startups specializing in smart solutions.
   Implement policies that incentivize businesses to adopt energy-efficient technologies.
2. Encourage the adoption of renewable energy:
   Deliver tax or subsidy incentives for businesses investing in the adoption of renewable energy.
   Encourage and invest in research and development initiatives for the advancement of renewable energy technologies.
   Improve partnerships with renewable energy providers for widespread adoption.
3. Implement IoT for energy management:

Invest in IoT devices for monitoring energy consumption in industries, homes, and public spaces.

Develop a centralized platform for real-time data analytics and decision-making.

Educate businesses and consumers on the benefits of IoT-enabled energy management.

4. Advocate for sustainable practices:

Engage in public awareness campaigns promoting sustainable practices.

Collaborate with industry associations to develop and implement sustainability standards.

Establish green business certification programs to recognize and encourage sustainable practices.

5.5.3 Addressing Environmental Problems through Digital Transformation

1. Implement AI for environmental monitoring:

Collaborate with AI experts and research institutions for the development of environmental monitoring AI.

Deploy AI algorithms for analyzing satellite imagery, climate data, and biodiversity information.

Establish partnerships with environmental agencies for data sharing.

2. Leverage big data analytics:

Develop big data analytics platforms for processing large datasets related to environmental issues.

Collaborate with research institutions to access relevant datasets.

Use analytics to identify trends, correlations, and areas for intervention.

3. Deploy IoT devices:

Implement IoT sensors for real-time monitoring of air quality, water quality, and other environmental factors.

Create a network of connected devices to collect comprehensive data.

Integrate IoT data into decision-making processes for urban planning and environmental conservation.

4. Foster innovation through digital technologies:

Establish innovation hubs or accelerators focusing on environmental technologies.

Encourage startups and researchers to propose and develop digital solutions.

Provide funding and support for pilot projects that demonstrate the effectiveness of innovative technologies.

5.5.4 Making an Impact on SDGs through Citizen Science Initiatives

1. Facilitate digital citizen science platforms:

Develop user-friendly mobile applications and web platforms for citizen science projects.

Incorporate features for data collection, image recognition, and collaborative mapping.

Provide tutorials and resources to guide citizens in participating effectively.

2. Promote community engagement:

Organize community workshops and events to raise awareness of citizen science opportunities.

Establish partnerships with local community organizations to facilitate engagement.

Recognize and reward community contributions to citizen science projects.

3. Collaborate with NGOs and Governments:

Partner with non-governmental organizations (NGOs) specializing in environmental conservation.

Collaborate with government agencies for data validation, policy advocacy, and project scalability.

Seek grants and funding from international organizations supporting citizen science initiatives.

4. General considerations:

(a) Policy support:

Advocate for policies that promote the integration of digital technologies in education, energy, and environmental sectors.

Collaborate with policymakers to develop and update regulations supporting sustainability.

(b) Education and training:

Offer workshops and training programs to educators and professionals on the effective use of digital technologies.

Integrate digital literacy and environmental education into school curricula.

5. Data governance and privacy:

Establish clear data governance frameworks to confirm accountable and right use of data.

Prioritize user privacy along with data security in the development and deployment of digital technologies.

6. Monitoring and evaluation:

(a) Establish metrics:

Define specific metrics for each initiative, such as the number of learners reached, energy savings achieved, or environmental impact.

Regularly review and update metrics based on project goals and outcomes.

(b) Regular assessments:

Conduct periodic assessments to measure the effectiveness of each initiative.

Solicit feedback from stakeholders, users, and communities to identify areas for improvement.

(c) Feedback loops:

Institute processes for perpetual feedback loops to ensure continual enhancement. Foster a culture of ongoing learning and adaptation guided by real-time feedback.

5.6 DISCUSSION

In synthesizing the findings of this study, it becomes evident that digital technologies are intricately linked to environmental stewardship (Kaur et al., 2019). The examination of efficiency gains, innovation for sustainability, knowledge dissemination, and citizen engagement highlights the multifaceted contributions of digital tools. Recommendations for policymakers emphasize the importance of integrating digital literacy programs into education, updating regulatory frameworks, and introducing incentive structures to promote sustainable digital practices. For businesses and society, suggestions include prioritizing eco-friendly technologies, actively engaging with local communities through digital platforms, and supporting environmentally responsible products and services. However, the study acknowledges certain limitations, including temporal constraints due to the focus on recent literature and potential challenges in ensuring a diverse sample for qualitative data. Looking ahead, future research avenues involve longitudinal studies, cross-cultural analyses, exploration of ethical considerations, and investigations into the integration of digital tools in policy implementation. In summary, this research adds to the ongoing conversation about the impact of digital technologies on promoting a sustainable and environmentally conscious future, providing practical insights for diverse stakeholders.

5.7 CONCLUSION

This study has explored the transformative potential of digital technologies in advancing environmental stewardship. The study synthesized findings that unveiled profound connections between digital technologies and sustainability, emphasizing efficiency gains, innovation, knowledge dissemination, and citizen engagement. Recommendations for policymakers, businesses, and society were outlined, emphasizing the need for digital literacy, regulatory frameworks, incentives, sustainable practices, community engagement, and conscious consumer choices. While recognizing limitations such as temporal constraints and potential sample biases, the study sets the stage for future research exploring longitudinal trends, cross-cultural variations, ethical considerations, and the integration of digital tools in policy implementation. Overall, this research underwrites to the evolving treatise on the pivotal role of digital technologies in determining a more sustainable future, offering practical insights and recommendations for stakeholders navigating the intersection of technology and environmental stewardship.

REFERENCES

Anderson T, Dron J. Three generations of distance education pedagogy. *International Review of Research in Open and Distributed Learning*. 2011;12(3):80–97.

Bhambri P, Singh M. Artificial intelligence. In Seminar on E-Governance - Pathway to Progress (p. 14), 2005.

Bhambri P, Sinha VK, Dhanoa IS. Development of cost effective PMS with efficient utilization of resources. *Journal of Critical Reviews*. 2020;7(19):781–786.

Bhambri P, Sinha VK, Dhanoa IS, Kaur J. Genome DNA sequence matching using HBM algorithm. *International Journal of Control and Automation*. 2019;12(5):531–539.

Bowser A, Hansen D, He Y. The role of mobile technologies in promoting health for communities in crisis. *Journal of Health Communication*. 2014;19(1):89–103.
Chen J, Shyu ML, Shu LS. Smart city applications and case studies in environmental monitoring. *Sensors*. 2018;18(11):3768.
Davidson E, Wessel L, Winter JS, Winter S. Future directions for scholarship on data governance, digital innovation, and grand challenges. *Information and Organization*. 2023;33(1):100454.
Del Río CG, González Fernández MC, Uruburu CÁ. Unleashing the convergence amid digitalization and sustainability towards pursuing the sustainable development goals (SDGs): A holistic review. *Journal of Cleaner Production*. 2021;32:122204
Di Vaio A, Palladino R, Hassan R, Escobar O. Artificial intelligence and business models in the sustainable development goals perspective: A systematic literature review. *Journal of Business Research*. 2020;121:283–314. DOI: 10.1016/j.jbusres.2020.08.019
Follett R, Strezov V. An analysis of citizen science based research: Usage and publication patterns. *PloS One*. 2015;10(11):e0143687.
Ford M. *Rule of the Robots: How Artificial Intelligence Will Transform Everything. Basic Books*. London: Hachette UK Company; 2021.
George G, Howard-Grenville J, Joshi A, Tihanyi L. Understanding and tackling societal grand challenges through management research. *Academy of Management Journal*. 2016;59(6):1880–1895. DOI: 10.5465/amj.2016.4007
Haklay M. Citizen science and volunteered geographic information: Overview and typology of participation. In *Crowdsourcing Geographic Knowledge* (pp. 105–122). Springer; 2013.
Hartmann P, Henkel J. The rise of corporate science in AI: Data as a strategic resource. *Academy of Management Discoveries*. 2020;6(3):359–381. DOI: 10.5465/amd.2019.0043
Huang MH, Rust R, Maksimovic V. The feeling economy: Managing in the next generation of artificial intelligence (AI). *California Management Review*. 2019;61(4):43–65.
Imperatives S. Report of the world commission on environment and development: Our common future. *Report of the World Commission on Environment and Development*. 1987;10:1–300.
Ji Z, Zhou T, Zhang Q. The impact of digital transformation on corporate sustainability: Evidence from listed companies in China. *Sustainability*. 2023;15(3):2117. DOI: 10.3390/su15032117
Kaur J, Bhambri P, Kaur S. SVM classifier based method for software defect prediction. *International Journal of Analytical and Experimental Model Analysis*. 2019;11(10):2772–2776.
Manyika J, Chui M, Brown B, Bughin J, Dobbs R, Roxburgh C, Byers AH. *Big Data: The Next Frontier for Innovation, Competition, and Productivity*. McKinsey Global Institute; 2011.
Nilsson NJ. *Problem-Solving Methods in Artificial Intelligence*. McGraw-Hill; 1971
Siemens G. Connectivism: A learning theory for the digital age. *International Journal of Instructional Technology and Distance Learning*. 2005;2(1):3–10.
Schiederig T, Tietze F, Herstatt C. Green innovation in technology and innovation management—an exploratory literature review. *R&D Management*. 2012;42(2):180–192.
Sharma R, Bhambri P, Sohal AK. Energy bio-inspired for MANET. *International Journal of Recent Technology and Engineering*. 2020;8(6):5580–5585.
Stuermer M, Abu-Tayeh G, Myrach T. Digital sustainability: Basic conditions for sustainable digital artifacts and their ecosystems. *Sustainability Science*. 2017;12:247–262.
Winter SJ, Butler B. Creating bigger problems: Grand challenges as boundary objects and the legitimacy of the information systems field. *Journal of Information Technology*. 2011;26(2):99.

6 Economic, Social, and Environmental Dimensions of Sustainable Technology

Aleksandra Ptak and Tomasz Lis

6.1 INTRODUCTION

Sustainable technologies, also known as green technologies or low-impact technologies, are a key element in achieving sustainable development on a global scale. Their goal is to minimize negative impacts on the natural environment and promote the efficient use of resources to meet humanity's current and future needs (Williams, 2024).

In today's society, which is increasingly aware of limited natural resources and the effects of climate change, sustainable technologies are becoming an essential tool to achieve a balance between the requirements of economic development and environmental protection. Through innovations in areas such as renewable energy, energy efficiency, sustainable transport, circular economy, organic farming, and water conservation, sustainable technologies seek to transform the way one produces, consumes, and uses resources (Corbera et al., 2019).

Implementing sustainable technologies has the potential not only to reduce greenhouse gas emissions and other pollutants but also to stimulate economic growth, create new jobs, and improve the quality of life for communities around the world (Davos Agenda, 2022). It is therefore not only a matter of protecting the environment but also of public health and social justice and ensuring a better future for the next generations.

Additionally, sustainable technologies can help increase communities' resilience to climate change and other environmental threats. By building more flexible and efficient energy, transport, and production systems, one can reduce the risks associated with extreme weather events and provide better protection against environmental disasters (Argyroudis, 2022). Therefore, investing in sustainable technologies not only protects the natural environment but also constitutes a strategic step towards building a more resilient and sustainable society.

 DOI: 10.1201/9781003475989-7

However, for sustainable technologies to reach their full potential, the involvement of all stakeholders, including governments, businesses, local communities, and civil society, is necessary. Collaboration at global, regional, and local levels to promote innovation (Al-Emran & Griffy-Brown, 2023), exchange best practices, and support the development of sustainable technologies on a large scale is crucial.

6.2 ECONOMIC DIMENSION OF SUSTAINABLE TECHNOLOGIES

The economic dimension of sustainable technologies is a key factor in achieving sustainable economic development. Switching to sustainable technologies can deliver cost savings through the efficient use of natural resources and energy, which in turn leads to lower production and business costs. For example, using renewable energy as an alternative to fossil fuel energy can reduce electricity costs in the long run because the costs of operating renewable installations are often lower than those of fossil fuels. Investments in sustainable technologies can also stimulate innovation and the development of new economic sectors. For example, the development of photovoltaic technologies and wind energy creates new opportunities for the industry and service sectors related to renewable energy. The long-term economic benefits of using sustainable technologies also include reducing dependence on fossil raw materials, which can protect the economy against possible price shocks and raw material shortages (Prosumentklasteroze, 2024). Companies that invest in the research and development of sustainable technologies can gain a competitive advantage by creating innovative products and services that meet growing market demands for environmental protection.

Moreover, sustainable technologies can generate new jobs in sectors related to production and their implementation (Hrstandard, 2021). Creating jobs in the sustainable sector can help reduce unemployment and improve social conditions.

Sustainable technologies can also bring benefits to consumers in various areas of their everyday life and consumption. The development of products and services based on sustainable technologies can contribute to creating healthier, safer, and more comfortable options for consumers. For example, food and cosmetics produced using sustainable technologies can be free of toxic substances, resulting in better health and safety for consumers (Tarapata, 2015). In the opinion of the authors of the chapter, the development of sustainable technologies often goes hand in hand with an increase in the ecological and social awareness of consumers. More and more people prefer products and services that are environmentally friendly and socially responsible. By choosing such options, customers support companies that care about protecting the environment and promoting sustainable practices.

All in all, the economic dimension of sustainable technologies is extremely important for both enterprises and entire national economies, because it contributes to creating new development opportunities, increasing competitiveness, and improving the quality of life of society, while caring for the natural environment and the planet's resources.

6.3 ENVIRONMENTAL DIMENSION OF SUSTAINABLE TECHNOLOGIES

The environmental dimension of sustainable technologies is a key element in responding to the challenges of climate change, environmental degradation, and biodiversity loss. In the face of growing ecological problems, technological development is becoming not only a matter of progress but also an indispensable tool to achieve harmony between people and the planet (Chovancova et al., 2023).

Sustainable development-orientated technologies emphasize innovation and energy efficiency and minimize the negative impacts on the environment. Examples of such technologies include renewable energy sources such as solar, wind, and geothermal energy, which reduce greenhouse gas emissions and dependence on fossil fuels. Another area of interest is energy efficiency technologies in construction and transportation that reduce energy consumption and carbon dioxide emissions. These solutions may include intelligent energy management systems, green buildings, or electric vehicles powered by energy from renewable sources. Technologies supporting environmental monitoring and protection, such as satellite remote sensing systems, sensors monitoring air quality, and IT platforms for analyzing ecological data, are also important. These solutions enable a better understanding of processes occurring in nature and more effective actions to protect ecosystems.

In industry, clean production technologies and green innovations aim to minimize pollutant emissions, reduce waste, and optimize the use of natural resources. Closed-loop production processes, the use of biomaterials, and high-efficiency recycling technologies are necessary to reduce the negative impact of industry on the environment (de Mello et al., 2021).

6.4 SOCIAL DIMENSION OF SUSTAINABLE TECHNOLOGIES

The social dimension of sustainable technologies is the third important aspect of achieving sustainable development. Although sustainable technologies are often perceived through the prism of their environmental impact and economic benefits, their significant importance for local and global communities cannot be ignored.

First, sustainable technologies can contribute to improving the quality of life of residents by providing access to necessary resources, such as clean water, electricity, and transport infrastructure. For communities, especially those in rural regions and developing countries, new technologies can mean a breakthrough in eliminating fuel poverty and improving sanitation (Mouratidis, 2021).

Moreover, sustainable technologies often go hand in hand with the creation of high-quality jobs. The development of sectors such as renewable energy, energy efficiency, and public transport generates demand for qualified labor, which contributes to the reduction of unemployment and economic growth (The House Project, 2023).

Another important social aspect of sustainable technologies is strengthening local communities and their ability to develop independently. By engaging residents in decision-making processes, educating them about the benefits of sustainable practices, and supporting local initiatives, these technologies can help build more sustainable and resilient communities (Nogalski & Karpacz, 2007; Toniolo et al., 2023). It should also be noted that sustainable technologies can contribute to reducing social inequalities by ensuring equal access to resources and development opportunities for all social groups, regardless of economic status or geographical location. Finally, the social dimension of sustainable technologies also includes cultural and social aspects. Technological innovations can be more effective in integrating traditional practices with modern solutions, respecting and promoting the cultural diversity and heritage of local communities (The House Project, 2023).

6.5 OPPORTUNITIES AND CHALLENGES RELATED TO THE INTRODUCTION OF SUSTAINABLE TECHNOLOGIES

The introduction of sustainable technologies is not only an opportunity to create innovative solutions, improve economic efficiency, and reduce the negative impacts on the environment, but also constitutes a challenge related to the need to adapt to new trends, restructure economic sectors, and balance technological progress with the principles of sustainable development (Williams, 2024). In this context, it is important to identify potential benefits and at the same time consciously manage threats to ensure the harmonious development of society, the economy, and the natural environment.

One of the opportunities created by the introduction of sustainable technologies is the possibility of significantly reducing the ecological footprint. These activities cover a wide spectrum of innovations, starting with the use of renewable energy sources such as solar, wind, and geothermal energy. This not only reduces greenhouse gas emissions but also reduces dependence on conventional, often environmentally harmful, energy sources (Gielen et al., 2019).

More efficient use of resources is another aspect that helps minimize the negative impact of human activity on the environment. By introducing innovative technologies, such as advanced monitoring systems (Ullo & Sinha, 2020), precision agriculture (Zaman, 2023), or intelligent water management (Cherlinka, 2024), it is possible to reduce the waste of raw materials and minimize negative effects on ecosystems.

In the transport sector, the use of sustainable technologies translates into reduced exhaust emissions and improved energy efficiency of vehicles. Electric vehicles, developing autonomy technologies and vehicle-sharing systems, are examples of innovations that not only reduce the impact of transport on the environment but can also generate new jobs in the sectors related to the production, service, or development of these technologies (Sher et al., 2021).

The dynamic development of fields related to ecology, renewable energy sources, and environmental technology opens new professional prospects. Sustainability specialists, environmental engineers, and energy efficiency experts are becoming more and more sought after, which contributes to the diversification of the labor market and strengthens the economy based on sustainable values (Smith, 2021).

However, implementing these solutions also comes with challenges. We have dealt with some of them before, but on a smaller scale, others are completely new and require facing both companies and state legal regulators, as well as other economic market participants (Avelar et al., 2024). They open up prospects for reducing energy consumption, more efficient use of resources, and improving the overall quality of life.

The coronavirus pandemic brought accelerated digital transformation and thus highlighted the scale of social problems that had previously been hidden or unnoticed, such as digital exclusion. Digital exclusion, which involves the lack of access or limited access to digital technologies and the ability to use them, has become an even more pressing problem in the era of the pandemic (Jaumotte et al., 2023). People on the margins of society, including older people, people with disabilities, residents of rural areas, and those with low economic status, have faced additional difficulties in accessing remote work, remote learning, health services, and other basic needs that have become increasingly dependent on technology (Bhambri et al., 2019). This situation prompted many positive social reactions, including local and social initiatives aimed to help those who found themselves in difficult situations due to lack of access to digital technologies (Park & Humphry, 2019). Examples of such activities include organizing online training, providing computer equipment, and establishing social contacts to support people in need.

The development of intelligent systems, the use of the Internet of Things (IoT), and blockchain technology all constitute potential attack points for cybercriminals. Attacks on energy infrastructure, urban traffic management systems, and autonomous vehicles are becoming more and more real. Securing these systems against cyberattacks is therefore becoming a priority, while at the same time, the challenge is to ensure that sustainable development goes hand in hand with appropriate protection measures against cyber threats. The development of smart cities carries risks related to citizens' privacy. Monitoring systems that collect data to optimize urban functions require appropriate security measures to ensure the privacy and protection of residents' data (Netmetis, 2023).

Problems related to e-waste management constitute one of the significant social challenges resulting from technological development. With the advancement of electronics and the rapid pace of replacement of electronic equipment, the amount of e-waste continues to increase. Complex storage and disposal of this waste may generate threats to the environment and public health. Valuable raw materials contained in electronics, such as gold and silver, are often lost when e-waste is improperly managed. Therefore, it is crucial to develop effective recycling systems and educate society about the responsible handling of used electronic devices (Shahabuddin et al., 2023).

The loss of jobs in traditional sectors is another social challenge related to changes toward sustainable development. Economic transformation toward greener practices may lead to restructuring of the labor market, which in turn requires retraining of workers. Investment in training and education programs is essential to enable people to adapt to changing economic conditions and find new employment opportunities in sustainable sectors (World Economic Forum, 2023).

Additionally, the implementation of sustainable practices may be understood and implemented differently on a global scale. Developed countries often have greater financial resources, which allows them to adapt more quickly to modern sustainable technologies. At the same time, developing countries may encounter difficulties related to limited resources and access to modern technologies, which generates inequalities in access to the benefits of sustainable development. International cooperation and supporting countries with less economic potential are therefore becoming key elements in building global balance in the field of sustainable development (Greene-Dewasmes & Wilson, 2023).

6.6 INNOVATIVE SUSTAINABLE TECHNOLOGIES INTRODUCED BY ENTERPRISES – EXAMPLES

In the next part of the chapter, the authors presented examples of companies that actively invest in innovative sustainable technologies, striving to improve their business practices and reduce the negative impacts on the environment. These advanced solutions not only reduce emissions and resource consumption but also open the door to financial savings, improve operational efficiency, and build a positive brand image.

6.6.1 Tesla

Tesla has been a pioneer in the field of electric motoring for years. One of the key innovations introduced by Tesla is advanced lithium-ion batteries, which are the driving heart of their electric vehicles (Anand & Bhambri, 2018). These batteries provide high capacity, long life, and fast charging, which translates into increased range and efficiency of vehicles. The company is also distinguished by the continuous development of battery technology, which leads to better parameters and lower production costs. Tesla has invested heavily in expanding its charging infrastructure, which includes Supercharger stations. These stations allow vehicles to be quickly charged on the route, eliminating concerns about range and charging times. A developed network of Superchargers makes traveling by electric car more comfortable and attractive for customers. The company continues to work on autonomous driving, introducing increasingly advanced features and driver assistance systems. Tesla Autopilot offers features such as automatic lane keeping, adaptive cruise control, and emergency braking systems, which improve driving safety and comfort. The company constantly introduces innovations related to the safety and comfort of the use of its vehicles. New features such as advanced environmental monitoring systems, warnings about potential hazards, and improved infotainment systems that provide better service and communication for the driver and passengers are being introduced (Forbes, 2022).

6.6.2 Google

Google is seen as one of the largest players in the technology industry and is taking numerous actions to improve energy efficiency and reduce its carbon footprint. Google actively invests in projects related to renewable energy sources, such as solar and wind energy. The company purchases electricity from renewable installations around the world, which allows it to reduce its dependence on fossil fuels and lower greenhouse gas emissions. Google focuses on effective energy management in its data centers by using advanced cooling technologies and optimizing energy consumption. The company employs modern monitoring and control systems that allow it to accurately track energy consumption and identify areas where improvements can be made to save energy. Google is also involved in research into new technologies related to renewable energy. The company invests in research projects on geothermal energy, which uses the Earth's heat as an energy source, and solar technologies, which enable the efficient use of solar energy. Google actively partners with other companies, non-governmental organizations and academic institutions to promote sustainable development and fight climate change. The company is involved in various initiatives, such as the "Climate Savers Computing Initiative" and the "Renewable Energy Buyers Alliance", which aim to promote the use of renewable energy and reduce greenhouse gas emissions in the technology sector (Google Environmental Report, 2023).

6.6.3 Nestlé

Nestlé, a global giant in the food industry, attaches great importance to the development of sustainable practices in its operations. The company consistently invests in innovative technologies that support sustainable development goals and minimize negative impacts on the environment. One of the key areas in which Nestlé uses technology is sustainable agriculture. By using advanced irrigation systems, monitoring soil parameters, and using natural plant protection methods, the company strives to increase crop efficiency and minimize the use of water and chemicals. Nestlé is also actively working on sustainable supply chain management. It uses advanced information technology and monitoring systems to optimize transportation and logistics, minimizing greenhouse gas emissions associated with delivery and distribution. In the context of packaging, Nestlé is researching innovative solutions such as packaging from renewable materials and alternative recycling options to reduce plastic waste and promote recycling. The company also focuses on effective water management, using technologies to monitor water consumption, wastewater treatment, and rainwater recovery to reduce negative impacts on water resources and the natural environment. In the field of product innovation, Nestlé is introducing more and more vegan, organic, and locally produced product options to the market, responding to the growing consumer demand for environmentally friendly products (Nestle, 2023). These activities demonstrate Nestlé's commitment to developing sustainable business practices and using innovative technologies to achieve a balance between economic development and environmental protection.

6.6.4 IKEA

IKEA consistently strives to reduce the negative impact of its activities on the natural environment. The company has introduced a wide range of furniture made of renewable materials, such as wood from certified forest sources, which are managed while respecting the principles of sustainable forest management. In addition, the company also uses other ecological materials and experiments with innovative production technologies, such as 3D printing or the use of materials derived from biomass, to create more sustainable products. IKEA has committed to the development of renewable energy by installing solar panels on the roofs of its stores and warehouses. Thanks to this, the company producing electricity from renewable sources reduces the demand for energy from the power grid based on fossil fuels, which contributes to the reduction of greenhouse gas emissions. The company is committed to the development of energy-saving lighting technologies. LED lighting uses much less energy than traditional light bulbs, which allows for lower electricity bills and reduces CO_2 emissions related to the production of electricity. IKEA is also developing sustainable packaging technologies, such as packaging made from recycled materials or biodegradable packaging that has a lower impact on the environment (IKEA, 2023).

6.6.5 Siemens

Siemens is one of the largest technology companies in the world, known for its innovative solutions in the field of sustainable development. The company operates on many fronts, offering sustainable technologies for various economic sectors. The company is involved in the development of renewable energy technologies, such as solar, wind, water, and biomass. Siemens designs and implements solutions aimed at increasing the efficiency of energy production and reducing greenhouse gas emissions. The company designs and implements electromobility systems, including electric rail vehicles, electric buses, and infrastructure for charging electric vehicles. Thanks to this, Siemens contributes to reducing exhaust emissions in the transport sector and promotes sustainable public transport. Siemens offers innovative technological solutions for the healthcare sector that contribute to improving the quality of patient care and the efficiency of medical facilities. The company provides advanced diagnostic systems, medical devices, and IT solutions that help diagnose and treat diseases, while minimizing resource consumption and waste generation. Siemens offers intelligent solutions for industry that aim to increase production efficiency, minimize waste, and optimize the use of energy and raw materials. The company applies advanced technologies such as the Internet of Things (IoT), artificial intelligence, and digital management systems to help enterprises achieve sustainable and competitive production strategies (Siemens, 2024).

6.6.6 Walmart

Walmart, one of the world's largest retailers, has a significant commitment to the development and implementation of sustainable technologies. The company is working on multiple fronts to reduce its carbon footprint, minimize resource use, and promote more sustainable business practices. One of the key areas in which Walmart is taking action is the use of renewable energy. The company invests in photovoltaic installations on the roofs of its stores and distribution centers, allowing it to reduce the consumption of electricity from fossil fuels. Additionally, Walmart actively supports the development of energy storage technologies to store and use electricity from renewable sources more efficiently. Walmart is also focusing on reducing water use in its operations. The company introduces innovative technologies for watering plants in its gardens, uses water consumption monitoring systems in its stores, and promotes economical washing and cleaning solutions. Additionally, Walmart strives to minimize the amount of waste generated by its operations. The company uses advanced recycling technologies in its distribution centers and stores, promotes packaging reuse, and works with suppliers to reduce waste packaging. In transportation and logistics, Walmart uses advanced technologies to monitor and optimize delivery routes to reduce greenhouse gas emissions associated with the transportation of goods. The company is also developing programs to support the transportation of goods by rail and water, which helps reduce the use of fossil fuels (Corporate Walmart, 2021).

The companies presented above, through innovative approaches to energy, transport, production, and management contribute to reducing the negative impacts on the environment. Their commitment to sustainable business practices is key to building a greener and more efficient economy.

6.7 CONCLUSIONS

Introducing and developing sustainable technologies requires cooperation between the public, private, and civil society sectors. Investments in research and development, education, and innovation-friendly regulation are key to the effective implementation of sustainable technologies and achieving global sustainable development goals.

Sustainable technologies not only contribute to environmental protection and economic growth but also have a significant impact on improving the quality of life and strengthening communities around the world. Their development and implementation should take into account the broad social context to ensure that the benefits of these technologies are equally distributed and available to all.

Furthermore, it is crucial to recognize the importance of supporting decision-making processes in the development and implementation of sustainable technologies. Engaging diverse stakeholders, including marginalized communities and local residents, ensures that their perspectives and needs are taken into account. This approach not only increases social acceptance and adoption of sustainable technologies but also promotes equality and social justice. By prioritizing inclusion and participation, solutions to address the unique challenges faced by diverse communities, which contribute to building a more just and resilient society, can be created.

Fully integrating the economic, social, and environmental dimensions requires a holistic approach that takes into account economic, social, and ecological aims. The implementation of sustainable technologies requires cooperation between the public, private, and civil society sectors, and continuous monitoring and assessment of their impact in all these areas. This commitment can help build a more sustainable future where technologies not only accelerate progress but also improve the well-being of society and protect the natural environment.

REFERENCES

Al-Emran M., & Griffy-Brown C. (2023). The role of technology adoption in sustainable development: Overview, opportunities, challenges, and future research agendas. *Technology in Society*, 73, 102240

Anand A., & Bhambri, P. (2018). Orientation, scale and location invariant character recognition system using neural networks. *International Journal of Theoretical & Applied Sciences*, 10(1), 106–109.

Argyroudis S.A., Mitoulis S.A., Chatzi E., Baker J.W., Brilakis I., Gkoumas K., Vousdoukas M., Hynes W., Carluccio S., Keou O., Frangopol D.M., & Linkov I. (2022). Digital technologies can enhance climate resilience of critical infrastructure. *Climate Risk Management*, 35, 100387. https://doi.org/10.1016/j.crm.2021.100387

Avelar S., Borges-Tiago T., Almeida A., & Tiago F. (2024). Confluence of sustainable entrepreneurship, innovation, and digitalization in SMEs. *Journal of Business Research*, 170, 114346.

Bhambri P., Sinha V.K., & Jaiswal M. (2019). Change in iris dimensions as a potential human consciousness level indicator. *International Journal of Innovative Technology and Exploring Engineering*, 8(9S), 517–525.

Cherlinka V. (2024). Agricultural water management with sustainable methods, available on the website: https://eos.com/blog/agricultural-water-management/, (access: 07.01.2024).

Chovancová J., Majerník M., Drábik P., & Štofková Z. (2023). Environmental technological innovations and the sustainability of their development. *Ecological Engineering & Environmental Technology*, 24(4), 245–252.

Corbera E., Roth D., Work C. (2019). Climate change policies, natural resources and conflict: Implications for development. *Climate Policy*, 19(S1), S1–S7.

Davos Agenda. (2022). Digital solutions can reduce global emissions by up to 20%, available on the website: https://www.weforum.org/agenda/2022/05/how-digital-solutions-can-reduce-global-emissions/, (access: 17.01.2024).

de Mello Santos V.H., Campos T.L.R., Espuny M. et al. (2022). Towards a green industry through cleaner production development. *Environmental Science and Pollution Research*, 29, 349–370. https://doi.org/10.1007/s11356-021-16615-2

Prosumentklasteroze. (2024). Ekologiczna energia, available on the website: https://prosumentklasteroze.pl/energetyka-przyszlosci-innowacyjne-technologie-i-zrownowazone-rozwiazania-dla-srodowiska/, (access: 04.01.2024).

The House Project. (2023). Empowering local communities: Building stronger and sustainable futures, available on the website: https://thehouseproject.foundation/blogs/blog/empowering-local-communities, (access: 07.01.2024).

Gielen D., Boshell F., Saygin D., Bazilian M.D., Wagner N., & Gorini R. (2019). The role of renewable energy in the global energy transformation. *Energy Strategy Reviews*, 24, 38–50.

Google Environmental Report. (2023). available on the website: https://www.gstatic.com/gumdrop/sustainability/google-2023-environmental-report.pdf, (access: 12.01.2024).

Greene-Dewasmes G., & Wilson K. (2023). How developing countries can empower themselves to navigate the challenges of global cooperation, available on the website: https://www.weforum.org/agenda/2023/09/how-developing-countries-empower-themselves-to-navigate-the-challenges-of-global-cooperation-sdim23/, (access: 07.01.2024).

IKEA. (2023). IKEA sustainability report FY23, available on the website: https://www.ikea.com/global/en/images/IKEA_SUSTAINABILITY_Report_FY_23_20240125_1b190c008f.pdf.

Hrstandard. (2021). Inwestycje w technologie cyfrowe i zrównoważony rozwój – co dzięki nim zyskamy?, available on the website: https://hrstandard.pl/2021/07/26/inwestycje-w-technologie-cyfrowe-i-zrownowazony-rozwoj-co-dzieki-nim-zyskamy/, (access: 12.01.2024).

Netmetis. (2023). Jakie są zagrożenia wynikające z rozwoju technologii?, available on the website: https://www.netmetis.pl/jakie-sa-zagrozenia-wynikajace-z-rozwoju-technologii/ (access: 12.01.2024).

Jaumotte J., Oikonomou M., Pizzinelli C., & Tavares M.M. (2023). How pandemic accelerated digital transformation in advanced economies, available on the website: https://www.imf.org/en/Blogs/Articles/2023/03/21/how-pandemic-accelerated-digital-transformation-in-advanced-economies, (access: 04.01.2024).

Mouratidis L. (2021), Urban planning and quality of life: A review of pathways linking the built environment to subjective well-being. *Cities*, 115, 103229. https://doi.org/10.1016/j.cities.2021.103229.

Nogalski B., & Karpacz J. (2007). Przedsiębiorczość jako czynnik stymulacji aktywności innowacyjnej małych przedsiębiorstw – ujęcie regionalne. [w:] Bogdanienko J, Kuzel M., & Sobczak I. (red.), *Działalnoś ć innowacyjna przedsiębiorstw w warunkach globalnych*. A. Marszałek, Toruń, 117–128.

Park S., & Humphry J. (2019). Exclusion by design: intersections of social, digital and data exclusion. *Information, Communication & Society*, 22(7), 934–953.

Siemens. (2024). Scaling sustainability impact, available on the website: https://www.siemens.com/global/en/company/sustainability.html, (access: 07.01.2024).

Shahabuddin M., Uddin M.N., Chowdhury J.I. et al. (2023). A review of the recent development, challenges, and opportunities of electronic waste (e-waste). *International Journal of Environmental Science and Technology*, 20, 4513–4520. https://doi.org/10.1007/s13762-022-04274-w.

Sher F., Chen S., Raza A., Rasheed T., Razmkhah O., Rashid T., Rafi-ul-Shan P.M., Erten B. (2021). Novel strategies to reduce engine emissions and improve energy efficiency in hybrid vehicles. *Cleaner Engineering and Technology*, 2, 100074. https://doi.org/10.1016/j.clet.2021.100074.

Smith M. (2021). The 10 most in-demand green jobs right now—some pay over $100,000 a year, available on the website: https://www.cnbc.com/2021/10/19/the-10-most-in-demand-green-jobs-right-now.html, (access: 12.01.2024).

Nestle. (2023). Sustainable technologies, available on the website: https://www.nestle.com/about/research-development/sustainable-technologies, (access: 28.12.2023).

Tarapata J. (2015). Konsumpcja zrównoważona a proekologiczne zachowania konsumentów. *Nowoczesne Systemy Zarządzania*, 10(1), 51–59.

Forbes. (2022). Tesla: A history of innovation (and headaches), available on the website: https://www.forbes.com/sites/qai/2022/09/29/tesla-a-history-of-innovation-and-headaches/?sh=a0b923418735, (access: 12.01.2024).

Toniolo S., Pieretto C., & Camana D. (2023). Improving sustainability in communities: Linking the local scale to the concept of sustainable development. *Environmental Impact Assessment Review*, 101, 107126. https://doi.org/10.1016/j.eiar.2023.107126.

Ullo S.L., & Sinha G.R. (2020). Advances in smart environment monitoring systems using IoT and sensors. *Sensors*, 20(11), 3113. https://doi.org/10.3390/s20113113.

Corporate Walmart. (2021). Walmart to offer technologies and capabilities to help other businesses navigate their own digital transformation, available on the website: https://corporate.walmart.com/news/2021/07/28/walmart-to-offer-technologies-and-capabilities-to-help-other-businesses-navigate-their-own-digital-transformation, (access: 15.01.2024).

Williams A. (2024). Sustainable technology: Examples, benefits and challenges. https://www.forbes.com/sites/technology/article/sustainable-technology/?sh=1a4bb1ca9f15, (access: 21.01.2024).

World Economic Forum. (2023). Future of jobs report 2023, available on the website: https://www.weforum.org/publications/the-future-of-jobs-report-2023/, (access: 12.01.2024).

Zaman Q.U. (2023). *Precision agriculture technology: A pathway toward sustainable agriculture precision agriculture*. Academic Press, pp. 1–17.

7 Technological Sustainability Unveiled

A Comprehensive Examination of Economic, Social, and Environmental Dimensions

Pankaj Bhambri and Paula Bajdor

7.1 INTRODUCTION

The chapter elaborates on an in-depth exploration of the intricate relationship between technology and sustainability across three critical dimensions: economic, social, and environmental. In essence, this title encapsulates a holistic investigation into how technological advancements contribute to sustainable development by addressing the interconnected challenges facing our world.

Firstly, the emphasis on "Technological Sustainability" underscores the need to assess the long-term viability of technological solutions. It implies an exploration beyond mere technological innovation, focusing on how these innovations can be designed, implemented, and managed to endure over time. This includes considerations of resource efficiency, renewable energy integration, and minimizing the ecological footprint of technologies.

The addition of "A Comprehensive Examination" signals a commitment to a thorough and exhaustive analysis. This likely involves delving into various facets of sustainability within the technological realm. The chapter examines multiple case studies, theories, and frameworks to provide a well-rounded understanding of the topic. This thorough method ensures that the investigation is not confined to a single viewpoint but incorporates a wide range of factors.

The incorporation of "Economic, Social, and Environmental Dimensions" indicates the acknowledgment of the three fundamental aspects of sustainability, sometimes known as the triple bottom line. This means that the chapter analyzes how technical improvements might contribute to growth in the economy, social fairness, and preservation of the environment simultaneously. It acknowledges the

DOI: 10.1201/9781003475989-8

interconnectedness of these dimensions and the need for a balanced approach that does not sacrifice one for the sake of the others.

7.2 FOUNDATIONS OF AI AND IOT IN HEALTHCARE

Technological sustainability is built upon a foundation of environmental responsibility and resource efficiency. To ensure a positive environmental impact, it is crucial to conduct thorough environmental impact assessments (EIAs) before adopting new technologies (Heinberg, 2003). These assessments evaluate variables such as the utilization of resources, production of waste, and release of pollutants across every stage of the life of a technology. The adoption of the circular economy model enhances sustainability by creating goods that are specifically designed for recycling, reuse, and remanufacturing. This approach minimizes waste and prolongs the lifespan of resources (Chertow, 2000). Moreover, the use of clean and renewable energy sources is crucial in minimizing the carbon emissions caused by technology and addressing the environmental impacts associated with the limited use of fossil fuels. Resource efficiency entails the optimization of raw material, energy, and water utilization in the design and manufacture of technology, promoting a more environmentally friendly approach to resource usage.

Technological sustainability necessitates the incorporation of social fairness and inclusion, highlighting the significance of assessing societal consequences. This entails tackling concerns such as proficiency in using digital tools, availability of technology, and ensuring fair allocation of advantages across different communities. Collaboration and partnerships among governments, corporations, academia, and civil society are crucial for the exchange of knowledge, resources, and optimal methodologies (Tondon and Bhambri, 2017). Regulatory frameworks and standards set by governments and international organizations guide the development and use of technologies, establishing benchmarks for environmental performance (Porter and Van der, 1995). Ultimately, ethical considerations, education, and awareness play vital roles in shaping a sustainable technological landscape, ensuring that innovations are aligned with human values, respect for human rights, and a commitment to responsible and equitable technology use.

7.2.1 Conceptual Framework

The conceptual structure of technological sustainability is based on a comprehensive approach that takes into account the interaction between the economy, society, and the environment. At its core, this framework emphasizes the need for technologies to be designed, implemented, and managed in ways that minimize negative environmental impacts. This involves conducting thorough life cycle assessments to understand and address the entire spectrum of a technology's influence from raw material extraction and production to end-of-life disposal. Concepts such as resource efficiency, renewable energy utilization, and the integration of circular economy principles form the environmental pillar of the framework, guiding the development of technologies that contribute positively to ecological well-being.

In tandem with environmental considerations, technological sustainability encompasses a strong social and economic dimension. Social equity and inclusion are critical elements, ensuring that technological advancements benefit all segments of society. This involves addressing issues like digital access, promoting digital literacy, and minimizing the potential negative impacts on employment and communities (S., 2015). The economic dimension emphasizes the importance of innovation, research, and the creation of regulatory frameworks that incentivize sustainable practices. Additionally, ethical considerations play a pivotal role in this conceptual framework, emphasizing responsible technology use, privacy, and security (Panda et al., 2019). In summary, the conceptual structure of technological sustainability combines environmental stewardship, social fairness, and economic feasibility to establish a well-rounded and long-lasting approach to the advancement and implementation of technology.

7.3 ECONOMIC DIMENSIONS OF TECHNOLOGICAL SUSTAINABILITY

The economic aspects of technological sustainability involve a comprehensive strategy to promote economic prosperity while limiting negative effects on the environment and society. At its core, sustainable technologies are designed to optimize resource utilization and enhance efficiency, leading to cost savings over the long term (Zehner, 2012). This involves the incorporation of eco-friendly practices in manufacturing, energy production, and overall resource management. Companies adopting sustainable technologies often benefit from reduced operational costs, improved resource efficiency, and enhanced resilience to external economic shocks (Schumpeter, 1942). Moreover, the economic aspect of technological sustainability extends beyond individual organizations to broader economic systems, influencing job creation, market competitiveness, and the emergence of new industries (McDonough and Braungart, 2002). Investments in the advancement of sustainable technologies bolster economic growth by fostering innovation and placing economies as leaders in growing global marketplaces.

However, the economic dimensions of technological sustainability also involve addressing challenges such as the initial costs of adopting green technologies, potential disruptions to existing industries, and the need for supportive policies and incentives. It is crucial to achieve a harmonious equilibrium between immediate economic factors and long-term environmental objectives (Bansal and Song, 2017). Governments, businesses, and financial institutions play pivotal roles in shaping economic dimensions through the formulation of policies, incentives, and investment strategies that encourage the development and adoption of sustainable technologies. In essence, the economic dimensions of technological sustainability underscore the importance of aligning economic growth with ecological and social responsibility, emphasizing the creation of a resilient and inclusive economy that thrives in harmony with the planet's finite resources.

7.3.1 Sustainable Economic Development

Sustainable development in the economy is a growth strategy that aims to achieve a harmonious combination of economic prosperity, environmental conservation, and social fairness in order to fulfill the current generation's requirements without jeopardizing the ability of subsequent generations to fulfill their own requirements (Reddy and Williams, 2019). It involves integrating sustainable practices into various facets of economic activities, with the aim of creating a resilient and inclusive economy. Several key principles and aspects characterize sustainable economic development:

- Environmental Stewardship: Sustainable economic development prioritizes environmental conservation and minimizes negative impacts on ecosystems. This involves adopting practices that reduce pollution, promote energy efficiency, and conserve natural resources. Green technologies, renewable energy sources, and eco-friendly production methods are essential components of this approach.
- Social Equity and Inclusion: Social considerations are integral to sustainable economic development. It emphasizes the fair distribution of benefits, resources, and opportunities across society. Inclusivity involves ensuring that marginalized and vulnerable populations have access to economic opportunities, education, healthcare, and social services.
- Economic Resilience: Sustainable economic development aims to build resilient economies capable of withstanding external shocks (Potluri et al., 2019). Diversification of industries, investment in innovation, and the development of adaptive strategies contribute to economic stability in the face of challenges such as climate change, economic downturns, and technological disruptions.
- Long-Term Planning: Sustainable economic growth entails the implementation of strategic plans that take into account the repercussions of present activities on future generations. This necessitates a transition from decision-making focused on immediate profitability to tactics that prioritize fairness across generations and the conservation of natural resources.
- Community Engagement: Involving local people in decision-making processes is essential for achieving sustainable economic development. Communities should be given the opportunity to actively participate in the design and implementation of development programs that have a direct influence on their means of living, surroundings, and overall welfare.
- Innovation and Technology: Adopting cutting-edge technologies is crucial for achieving long-term economic growth. This encompasses the advancement and implementation of technologies that encourage the efficient use of resources, minimize harm to the environment, and improve production in a manner that can be sustained over time (UNDP, 2015).
- Green Jobs and Inclusive Growth: Sustainable economic development fosters the creation of green jobs that contribute to environmental sustainability. It also promotes inclusive growth by ensuring that economic benefits

are shared across different segments of the population, reducing income inequality.

- Corporate Social Responsibility (CSR): Businesses play a crucial role in sustainable economic development through responsible practices (Rana et al., 2019). Corporate Social Responsibility initiatives, ethical business conduct, and sustainable supply chain management contribute to positive social and environmental impacts.
- Policy and Governance: Effective policies and governance structures are essential for sustainable economic development. Governments need to create a regulatory framework that incentivizes sustainability, enforces environmental protection measures, and supports socially responsible business practices.

7.3.2 Economic Impact of Sustainable Technologies

The economic impact of sustainable technologies is substantial, influencing various aspects of business, industry, and the overall economy. Sustainable technologies, which are designed to minimize environmental impact while enhancing efficiency, have far-reaching effects on economic systems. Several key dimensions of the economic impact of sustainable technologies include:

- Cost Savings and Efficiency: One of the immediate economic benefits of sustainable technologies lies in cost savings for businesses. Technologies that optimize resource use, reduce energy consumption, and minimize waste contribute to improved operational efficiency. Businesses adopting sustainable practices often experience reduced utility costs, enhanced productivity, and lower operational expenses.
- Innovation and Job Creation: The development and adoption of sustainable technologies drive innovation, leading to the creation of new industries and job opportunities (Rifkin, 2014). As companies invest in research and development of green technologies, they stimulate economic growth and contribute to the emergence of a skilled workforce in sustainable industries such as renewable energy, clean technology, and eco-friendly manufacturing.
- Market Competitiveness: Sustainable technologies enhance the competitiveness of businesses in the global market. Consumers increasingly prefer products and services with lower environmental footprints, prompting companies to integrate sustainable practices into their operations to meet market demands. Being environmentally responsible can be a key differentiator, attracting environmentally conscious consumers and investors.
- Resilience to Regulatory Changes: Sustainable technologies position businesses to navigate and adapt to evolving regulatory landscapes. As governments implement environmental regulations and standards, companies that have already adopted sustainable technologies are better prepared to comply, avoiding potential legal and financial consequences.

- Resource Efficiency and Conservation: Sustainable technologies contribute to resource efficiency, reducing the strain on finite resources. This not only benefits the environment but also helps stabilize prices for raw materials, fostering economic stability. Technologies that promote circular economy principles, such as recycling and remanufacturing, further contribute to resource conservation.
- Investment Opportunities: The increasing focus on sustainability has attracted significant investment in green technologies. Sustainable startups and established companies in the renewable energy, energy efficiency, and environmental conservation sectors have access to funding, fostering economic growth and development.
- Brand Value and Reputation: Adopting sustainable technologies enhances a company's brand value and reputation (Arthur, 2009). Businesses with strong environmental and social responsibility credentials often enjoy increased customer loyalty, positive public perception, and a competitive edge in the market.
- Economic Diversity and Resilience: The integration of sustainable technologies supports economic diversity by fostering the growth of industries beyond traditional sectors. This diversification contributes to economic resilience, reducing dependence on a single industry and mitigating the impact of economic downturns in specific sectors.
- Global Supply Chain Resilience: Sustainable technologies contribute to the resilience of global supply chains by reducing vulnerabilities associated with resource scarcity, climate-related disruptions, and geopolitical risks. Companies that prioritize sustainability in their supply chains are better equipped to navigate global challenges.

7.3.3 Business Models for Sustainable Innovation

Business models for sustainable innovation play a crucial role in aligning economic success with positive environmental and social impacts (Rockström et al., 2009). These models are designed to drive innovation while addressing environmental challenges and promoting social responsibility. Several business models have emerged to integrate sustainability into various industries:

- Circular Economy Models: Emphasizing resource efficiency and waste reduction, circular economy models involve designing products for reuse, recycling, and remanufacturing. Companies adopting this model aim to create closed-loop systems where materials are continuously cycled, minimizing environmental impact and reducing reliance on finite resources.
- Product-as-a-Service (PaaS): PaaS models shift the focus from selling products to offering them as services. This encourages product durability, repairability, and efficient use of resources, as companies are motivated to design products that last longer and can be easily maintained, upgraded, or recycled.

- Sharing Economy Platforms: Sharing economy business models promote the shared use of goods and services, reducing overall consumption. Companies like Airbnb and Zipcar exemplify this approach, allowing individuals to access resources without the need for ownership, thus lowering the environmental footprint associated with excessive production (Bocken et al., 2014).
- Sustainable Supply Chain Models: Businesses are increasingly adopting sustainable supply chain models to ensure that their products are sourced, produced, and distributed in an environmentally and socially responsible manner (Bhambri et al., 2019). This involves collaborating with suppliers who adhere to ethical and sustainable practices.
- Impact Investing and Triple Bottom Line Models: Impact investing integrates financial returns with social and environmental outcomes. Triple bottom line models measure success based on economic, social, and environmental performance. Companies adopting these models prioritize people, planet, and profit, considering the broader impact of their activities.
- Open Innovation Platforms: Open innovation models encourage collaboration and knowledge sharing among businesses, researchers, and the public. By opening up the innovation process, companies can tap into a diverse range of ideas and expertise, accelerating the development of sustainable solutions.
- Biomimicry and Nature-Inspired Design: Biomimicry business models draw inspiration from nature to develop sustainable products and processes. By emulating natural systems and processes, companies aim to create innovative, eco-friendly solutions that align with the principles observed in the natural world (Stern, 2007).
- Energy-as-a-Service (EaaS): EaaS models provide energy services instead of selling energy products. This encourages the adoption of renewable energy sources and energy-efficient technologies, with companies offering solutions such as solar power generation, energy storage, and demand-side management.
- Regenerative Business Models: Regenerative models go beyond sustainability by aiming to restore and enhance ecosystems. Companies adopting regenerative practices focus on activities that contribute positively to the environment, aiming to leave a net-positive impact on natural systems.
- Blockchain for Transparency and Traceability: Blockchain technology is increasingly being used to enhance transparency and traceability in supply chains. This ensures that consumers can make informed choices about products with clear information on the sustainability and ethical practices involved in their production.

7.4 SOCIAL DIMENSIONS OF TECHNOLOGICAL SUSTAINABILITY

The social dimensions of technological sustainability pertain to the societal effects of technology, with a focus on fairness, availability, and the general welfare of people as well as communities (World Commission on Environment and Development, 1987). A crucial element is digital inclusion, which aims to prevent technological improvements from worsening pre-existing social disparities. Attempts are being undertaken to narrow the gap between those who have access to digital technology and those who do not by ensuring equal opportunities and promoting digital literacy among various groups of people. Social sustainability also involves addressing the potential displacement of jobs due to automation and technological advancements. Strategies such as reskilling and upskilling programs are implemented to equip individuals with the skills needed in the evolving job market, ensuring that the benefits of technology are shared inclusively.

Moreover, ethical considerations are paramount in the social dimensions of technological sustainability. This involves examining the impact of technology on privacy, security, and human rights. Technologies must be developed and implemented in ways that respect individuals' autonomy and protect their sensitive information. Additionally, the ethical use of emerging technologies, such as artificial intelligence, is a critical focus (Grin et al., 2010). Establishing guidelines and regulations to govern the responsible use of technology helps prevent the misuse of powerful tools and ensures that technological advancements align with societal values and ethical standards. In essence, the social dimensions of technological sustainability aim to create a technologically advanced society that prioritizes inclusivity, fairness, and ethical considerations, promoting the well-being of individuals and communities.

7.4.1 Social Impacts of Sustainable Technologies

Sustainable technologies have a significant influence on society by promoting inclusivity, fairness, and resilience. One key facet is the democratization of access to essential services, as sustainable technologies often bridge gaps in education, healthcare, and communication, particularly in underserved communities. The adoption of sustainable energy solutions not only mitigates the environmental impact but also addresses energy poverty, enhancing the quality of life for marginalized populations. Additionally, sustainable technologies contribute to job creation, fostering economic opportunities in emerging green sectors. The empowerment of local communities through the deployment of decentralized, eco-friendly solutions further strengthens social resilience. Nevertheless, it is imperative to confront possible obstacles, such as the requirement for digital literacy and guaranteeing universal access to the advantages of sustainable technologies, in order to optimize their favorable societal influence (Weizsäcker et al., 1997). Overall, the widespread integration of sustainable technologies contributes to building a socially just and environmentally conscious future.

7.4.2 Equity and Social Justice Considerations

Equity and social justice considerations are fundamental principles in ensuring that technological advancements and societal progress benefit all members of a community, regardless of background or socioeconomic status. In the context of technological sustainability, these considerations involve addressing and mitigating disparities in access to and the benefits derived from technology. This includes efforts to bridge the digital divide, ensuring that marginalized and underserved populations have equal opportunities to engage with and benefit from technological innovations. Social justice considerations also extend to the ethical development and deployment of technologies, aiming to prevent discriminatory practices and biases in algorithms or systems (GeSI, 2019). A commitment to equity and social justice in the technological landscape involves actively seeking to understand and rectify historical and systemic inequalities, fostering inclusive innovation that positively impacts diverse communities and promotes a fair and just society.

7.4.3 Community Engagement and Empowerment

Community engagement and empowerment are integral components of fostering sustainable development and inclusive decision-making processes. Efficient community involvement means actively including local citizens, stakeholders, and groups in the process of designing, executing, and assessing initiatives and policies that have a direct impact on their lives. Empowering communities entails equipping them with the necessary tools, resources, and information to actively engage in decision-making processes, thereby enabling them to determine their own futures. By valuing local knowledge and perspectives, community engagement not only enhances the effectiveness and sustainability of projects but also strengthens the social fabric and sense of ownership within communities. Empowering communities fosters a sense of responsibility and self-determination, creating a collaborative and resilient foundation for sustainable development.

7.5 ENVIRONMENTAL DIMENSIONS OF TECHNOLOGICAL SUSTAINABILITY

The environmental dimensions of technological sustainability center on the development and deployment of technologies that minimize negative impacts on the natural world. This involves adopting practices that promote resource efficiency, reduce pollution, and mitigate climate change (Young, 2002). Sustainable technologies aim to maximize the efficiency of material, energy, and water usage during their entire lifespan. This includes taking into account aspects such as environmentally friendly design, manufacturing methods, and proper disposal at the end of their life cycle. The focus is on utilizing renewable energy sources, decreasing emissions of greenhouse gases, and adopting principles from the circular economy to decrease waste and prolong the lifespan of items. In addition, sustainable developments in technology include the preservation of biodiversity, the maintenance of ecosystems, and the facilitation of a harmonious cohabitation between human endeavors and the intricate equilibrium of the Earth's systems of nature. In general, the environmental aspects of technological sustainability strive to develop technologies that have a positive impact on the planet's well-being and ability to recover.

7.5.1 Environmental Impact Assessment

Environmental Impact Assessment (EIA) is a methodical procedure employed to assess the probable environmental effects of proposed projects, programs, or policies (Elkington, 1997). It is the process of recognizing, forecasting, and evaluating the probable effects on the natural surroundings, encompassing aspects such as the quality of air and water, diversity of species, functioning of ecosystems, and well-being of humans. The main objective of EIA is to furnish decision-makers with extensive information in order to make well-informed decisions that strike a balance between development objectives and environmental preservation. EIA commonly includes the involvement of the public, ensuring that individuals and groups with a vested interest, such as local communities, are given the chance to provide their viewpoints. The early integration of environmental factors in the decision-making process is essential for encouraging sustainable development and reducing negative environmental impacts caused by human activities. EIA plays a vital role in achieving these objectives.

7.5.2 Eco-innovation and Green Technologies

Eco-innovation and green technologies represent pivotal drivers in the pursuit of sustainable development, offering innovative solutions to address environmental challenges while promoting economic growth. Eco-innovation encompasses the development and application of novel technologies, products, and processes that result in positive ecological outcomes. Green technologies, a key component of eco-innovation, focus on minimizing environmental impact through resource efficiency, reduced emissions, and the utilization of renewable energy sources. These technologies span various sectors, including energy, transportation, construction, and manufacturing, with the common goal of fostering environmental sustainability. By integrating eco-innovation and green technologies, societies can transition toward more sustainable practices, mitigating the ecological footprint of human activities and contributing to a resilient and harmonious coexistence with the natural environment.

7.5.3 Life Cycle Analysis of Sustainable Technologies

Life cycle analysis (LCA) is a thorough methodology that evaluates the ecological consequences of sustainable technologies across all stages of their existence, encompassing the extraction of raw materials, manufacture, utilization, and discard (United Nations, 2015). This analytical tool evaluates resource consumption, energy use, emissions, and waste generation, providing a holistic view of a technology's environmental footprint. LCA enables the identification of opportunities for improvement in design, manufacturing, and end-of-life processes to minimize adverse effects on the environment. By considering the complete life cycle, sustainable technologies can be optimized for resource efficiency, reduced environmental impact, and enhanced overall sustainability, ensuring that the benefits of technological innovation extend beyond performance metrics to include environmental responsibility and long-term viability.

7.6 INTEGRATION OF THE TRIPLE BOTTOM LINE

The incorporation of the triple bottom line (TBL) is a holistic approach to business that surpasses conventional profit-oriented strategies. The TBL framework encompasses three fundamental dimensions: economic, social, and environmental. It assesses an organization's performance by considering its influence on individuals, the environment, and financial gain. When businesses adopt this approach, they take into account not just their financial performance but also their social duty and environmental stewardship. Companies are encouraged to evaluate their impact on the environment, support fairness in society, and practice ethical behavior, all while remaining financially sustainable. Businesses aim to create a harmonious equilibrium between financial stability, societal well-being, and environmental sustainability by incorporating the TBL framework. This involves aligning their strategy with sustainable long-term goals and exhibiting a commitment to ethical business conduct.

7.6.1 The Triple Bottom Line Concept

The TBL, or triple bottom line, is a concept that highlights a more expansive and all-encompassing method of assessing corporate success. It takes into account three interrelated dimensions: the economy, society, and the environment. The TBL paradigm goes beyond the conventional emphasis on financial success and assesses the influence of an organization's operations on individuals, the environment, and economic gains. This comprehensive strategy necessitates that firms consider their social obligations, including community involvement, employee welfare, and ethical conduct, in addition to environmental conservation and economic sustainability (Jørgensen et al., 2019). The TBL idea emphasizes the significance of sustainability and social responsibility as a whole in promoting enduring prosperity via the harmonization of economic objectives with environmental and social variables. This ultimately leads to the development of a more comprehensive and accountable company model.

7.6.2 Balancing Economic, Social, and Environmental Objectives

Achieving sustainable development requires the intricate and important task of harmonizing social, environmental, and economic objectives. To attain this balance, a comprehensive strategy is necessary that acknowledges the interdependence of financial well-being, fairness in society, and responsible management of the environment. Businesses and policymakers should adopt measures that foster economic growth while mitigating adverse environmental effects and safeguarding the welfare of various people. This involves adopting sustainable business models, investing in clean technologies, and formulating policies that prioritize social inclusivity.

Striking this balance involves constant adaptation and innovation as organizations strive to enhance resource efficiency, create fair employment opportunities, and mitigate ecological footprints. Ultimately, a harmonious integration of economic, social, and environmental goals is crucial for building resilient, equitable, and environmentally sustainable societies for current and future generations.

7.6.3 Challenges and Opportunities in Achieving Integration

Achieving integration, whether in the context of technological systems, organizational structures, or societal frameworks, poses both challenges and opportunities. One significant challenge lies in overcoming existing silos and fragmentation within systems. In many cases, different departments, technologies, or sectors operate independently, hindering seamless collaboration and communication. Overcoming this challenge requires a concerted effort to break down barriers, foster interdisciplinary collaboration, and implement interoperable technologies. Integration efforts often face resistance from established structures and processes, necessitating organizational change management strategies to ensure that stakeholders are aligned with the integration goals. Moreover, ensuring data compatibility and standardization is crucial, as diverse systems may use different formats and protocols, making seamless integration complex and time-consuming.

Despite these challenges, achieving integration presents significant opportunities for efficiency, innovation, and holistic problem-solving. Integrated systems can streamline processes, reduce redundancy, and enhance overall operational efficiency (Tukker and Tischner, 2006). For example, in the context of technological sustainability, integrated solutions can optimize resource use, minimize waste, and improve the environmental performance of systems. Integration also facilitates innovation by allowing diverse perspectives and expertise to come together. Cross-disciplinary collaboration can lead to the development of novel solutions and breakthrough technologies that address complex challenges. Furthermore, integrated approaches are essential for addressing pressing global issues such as climate change, as they enable a coordinated and comprehensive response by bringing together diverse stakeholders, technologies, and strategies. In essence, while challenges exist, the opportunities presented by achieving integration are vast and can pave the way for more resilient, adaptable, and sustainable systems.

7.7 CASE STUDIES AND EXEMPLARY PRACTICES

The following are some examples of successful technological sustainability initiatives that were exemplary practices in terms of economic, social, and environmental dimensions.

7.7.1 Tesla's Electric Vehicles (EVs)

Tesla revolutionized the automotive business by effectively incorporating electric vehicles into the larger market. Despite initial doubts, Tesla's market valuation exceeded that of conventional automakers, demonstrating the economic feasibility of sustainable technologies. EVs play a significant role in mitigating air pollution and lowering reliance on fossil fuels, thereby fostering a better environment for communities. Tesla's impact has stimulated worldwide interest in sustainable transportation. EVs possess reduced carbon footprints in comparison to conventional vehicles, particularly when powered by renewable energy sources.

7.7.2 Circular Economy at Interface

Interface, a global carpet tile manufacturer, embraced a circular economy model, emphasizing the recycling and reuse of materials. This initiative has led to cost savings through reduced resource consumption and waste. Interface's commitment to sustainability resonates with environmentally conscious consumers, enhancing brand reputation. The company also engages in social responsibility initiatives, contributing to community well-being. By adopting recycled materials and reducing waste, Interface minimizes environmental impact. This approach aligns with global efforts to combat climate change and resource depletion.

7.7.3 Google's Data Center Sustainability

Google has invested heavily in renewable energy to power its data centers. This long-term commitment to clean energy not only aligns with sustainability goals but also stabilizes energy costs over time (Kates et al., 2001). Google's transparency about its sustainability efforts builds trust with users and stakeholders. The company also supports local communities through job creation and educational initiatives. By offsetting energy consumption with renewable sources, Google reduces its carbon footprint. The company has set a goal to operate on 24/7 carbon-free energy by 2030.

7.7.4 Sustainable Agriculture with Precision Farming

Precision farming technology, such as IoT sensors and data analytics, enhances the efficiency of resource utilization, resulting in increased crop yields and decreased expenses for farmers. These technologies enhance the capabilities of farmers, particularly in underdeveloped areas, by offering data-based analysis, boosting efficiency, and fostering food safety. Precision farming employs strategies to reduce the ecological consequences of agriculture through the reduction of water and chemical consumption, the prevention of soil deterioration, and the encouragement of sustainable land stewardship.

7.7.5 Circular Design in Fashion – Patagonia

Patagonia promotes a circular design approach by encouraging product repair, reuse, and recycling (Christensen, 2013). This strategy builds customer loyalty and extends the lifecycle of products. The company's dedication to ethical and environmental processes appeals to conscientious consumers, hence fostering a favorable societal influence within the fashion sector. By minimizing waste and emphasizing recycled materials, Patagonia reduces the environmental footprint of its products, addressing concerns related to fast fashion.

7.8 CHALLENGES AND FUTURE DIRECTIONS

In the pursuit of technological sustainability through a comprehensive examination of economic, social, and environmental dimensions, several challenges and future directions emerge. One major challenge lies in balancing economic viability with sustainable practices, as upfront costs for green technologies can be substantial. Additionally, ensuring social inclusivity and addressing potential job displacement during technological transitions remains a crucial concern. The integration of sustainable technologies also necessitates overcoming infrastructure limitations, especially in developing regions. On the environmental front, managing electronic waste and mitigating the environmental impact of manufacturing processes are ongoing challenges. Future directions should emphasize collaborative efforts among governments, industries, and communities to establish supportive policies, foster innovation, and promote sustainable consumption patterns. Strengthening international cooperation and leveraging emerging technologies, such as artificial intelligence and blockchain, could further enhance the monitoring and implementation of sustainable practices across global supply chains. Education and awareness initiatives will be pivotal in fostering a culture of sustainability, encouraging responsible consumption, and ensuring the long-term success of technological sustainability endeavors.

7.8.1 Current Challenges in Technological Sustainability

Several challenges persist in the realm of technological sustainability as shared below:

- High Initial Costs: Many sustainable technologies involve substantial upfront costs, posing a challenge for widespread adoption. Businesses and individuals may be deterred by the initial investment required for renewable energy systems, energy-efficient infrastructure, or sustainable manufacturing processes.
- Interconnectedness of Economic, Social, and Environmental Dimensions: Achieving a balance between economic growth, social equity, and environmental preservation remains a complex task (Meadows et al., 1972). Implementing sustainable technologies necessitates considering these dimensions comprehensively, and finding solutions that address all aspects simultaneously can be challenging.
- Global Supply Chain Complexities: The complexity of global supply chains introduces challenges in ensuring that raw materials are sourced sustainably, and products are manufactured, transported, and disposed of in environmentally friendly ways. Monitoring and regulating these processes on

a global scale require international cooperation and standardized sustainability practices.

- E-Waste Management: The rapid pace of technological advancement contributes to a growing electronic waste (e-waste) problem. Proper disposal and recycling of electronic devices pose significant challenges, as inadequate handling can lead to environmental pollution and health risks.
- Technological Transitions and Job Displacement: The adoption of sustainable technologies, like automation and artificial intelligence, has the potential to result in employment displacement in specific industries (Sachs, 2015). Ensuring a just transition for affected workers and communities, and providing retraining and reskilling opportunities, is a critical challenge for achieving social sustainability.
- Limited Infrastructure in Developing Regions: The implementation of sustainable technologies can be hindered by inadequate infrastructure in developing regions. Access to clean energy, efficient transportation, and advanced waste management systems may be limited, impeding the widespread adoption of sustainable practices.
- Policy and Regulatory Frameworks: Inconsistent or inadequate regulatory frameworks pose challenges to technological sustainability. Clear and supportive policies are essential to incentivize businesses and individuals to adopt sustainable practices and technologies.
- Technological Obsolescence: The rapid pace of technological innovation can lead to the quick obsolescence of products, contributing to a throwaway culture. Encouraging durable design, repairability, and circular economy practices is essential to minimize the environmental impact of technological advancements.

7.8.2 Anticipated Future Developments

Anticipated future developments in technological sustainability encompass a broad spectrum of innovations that aim to address pressing environmental, economic, and social challenges (Lovins et al., 1999). One prominent trend is the increasing integration of advanced technologies to enhance sustainability efforts. Artificial intelligence (AI) and machine learning, for instance, are anticipated to play a pivotal role in optimizing resource management, predicting environmental trends, and improving the efficiency of sustainable technologies. The rise of the Internet of Things (IoT) will likely lead to more interconnected and smart systems, facilitating data-driven decision-making in areas such as energy management, agriculture, and urban planning. The development and adoption of breakthrough materials, such as those derived from sustainable sources or designed for circular economies, will revolutionize product manufacturing and reduce reliance on finite resources. Innovations in renewable energy storage technologies and grid management are poised to make renewable energy sources more reliable and scalable, accelerating the transition to a low-carbon energy landscape. Additionally, advancements in sustainable transportation, including electric and autonomous vehicles, are expected to reshape the mobility sector, reducing emissions and promoting more efficient transportation systems. The increasing focus on environmentally friendly finance and the incorporation of criteria related to the environment, society, and governance (ESG) into investment choices indicate a transition toward more conscientious and sustainable business strategies. Overall, the anticipated future developments in technological sustainability converge on the idea of creating a more interconnected, intelligent, and environmentally conscious world. These developments have the potential to foster a more resilient and sustainable global ecosystem, addressing the complex challenges facing society. However, the successful implementation of these innovations will require collaborative efforts among governments, industries, and communities to ensure responsible and equitable adoption, considering the diverse needs and circumstances across different regions and sectors.

7.8.3 Recommendations for Advancing Technological Sustainability

Advancing technological sustainability requires a multifaceted approach, encompassing economic, social, and environmental considerations. Key recommendations to propel progress in this domain include:

- Policy Frameworks and Incentives: Authorities ought to formulate and implement stringent policies that provide incentives for enterprises to embrace sustainable technologies. This encompasses financial incentives, subsidies, and regulatory frameworks that incentivize ecologically sustainable practices. Definitive rules and standards can serve as a strategic plan for industries to harmonize their operations alongside sustainability objectives.
- Research and Development Investment: Increased investment in research and development is vital for the creation of innovative and sustainable technologies. Governments, private sectors, and academic institutions should collaborate to fund projects focused on renewable energy, circular economy solutions, and eco-friendly materials, fostering a culture of continuous improvement.
- Collaboration and Knowledge Sharing: Encouraging collaboration among industries, governments, and research institutions is essential. Platforms for sharing best practices, success stories, and failures can accelerate the adoption of sustainable technologies. Cross-sector partnerships can leverage expertise and resources to address complex sustainability challenges collectively.
- Education and Awareness Campaigns: Public awareness and understanding of technological sustainability are crucial. Educational programs should be implemented to inform consumers, businesses, and policymakers about the environmental, social, and economic implications of their choices. This can foster a demand for sustainable products and services and drive behavioral change.

- Circular Economy Practices: Promoting circular economy principles, such as product design for longevity, reuse, and recycling, is essential. Businesses should adopt practices that minimize waste, and governments can support initiatives that encourage the circular flow of materials, reducing the overall environmental impact of manufacturing and consumption.
- Technological Integration for Monitoring and Reporting: Leveraging technologies like artificial intelligence, blockchain, and IoT for monitoring and reporting can enhance transparency and accountability in supply chains. This technology-driven approach can help track the environmental and social impact of products and services, enabling informed decision-making and fostering a culture of responsibility.
- Capacity Building in Developing Regions: Efforts should be directed toward building technological and infrastructural capacity, particularly in developing regions. This includes providing access to sustainable technologies, training local communities, and ensuring that the benefits of sustainable practices are equitably distributed.
- Green Financing: Financial institutions should incorporate sustainability criteria into their lending and investment decisions. This can drive capital toward projects and businesses that prioritize technological sustainability. Green bonds, sustainable investment funds, and other financial mechanisms can support initiatives that align with environmental, social, and governance (ESG) criteria.
- Continuous Evaluation and Adaptation: Regular assessments and evaluations of sustainability initiatives are crucial. Flexibility and a willingness to adapt strategies based on evolving technological, economic, and environmental landscapes are essential to ensuring the continued effectiveness of sustainability efforts.
- Global Collaboration and Standards: International cooperation is paramount in addressing global sustainability challenges. Establishing and adhering to global standards for sustainable practices can create a level playing field and facilitate the exchange of knowledge and resources across borders.

REFERENCES

Arthur, W. B. (2009). *The nature of technology: What it is and how it evolves.* Free Press.

Bansal, P., & Song, H. C. (2017). Similar but not the same: Differentiating corporate sustainability from corporate responsibility. *Academy of Management Annals*, 11(1), 105–149.

Bhambri, P., Sinha, V. K., & Jaiswal, M. (2019). Change in iris dimensions as a potential human consciousness level indicator. In International Conference on Innovations in Communication, Computing and Sciences.

Bocken, N. M. P., Short, S. W., Rana, P., & Evans, S. (2014). A literature and practice review to develop sustainable business model archetypes. *Journal of Cleaner Production*, 65, 42–56.

Chertow, M. R. (2000). The IPAT equation and its variants. *Journal of Industrial Ecology*, 4(4), 13–29.

Christensen, C. M. (2013). *The innovator's dilemma: When new technologies cause great firms to fail.* Harvard Business Review Press.

Elkington, J. (1997). *Cannibals with forks: The triple bottom line of 21st-century business.* Capstone.

GeSI (Global e-Sustainability Initiative). (2019). #SMARTer2030: ICT Solutions for 21st Century Challenges.

Grin, J., Rotmans, J., & Schot, J. (2010). *Transitions to sustainable development: New directions in the study of long-term transformative change.* Routledge.

Heinberg, R. (2003). *The party's over: Oil, war, and the fate of industrial societies.* New Society Publishers.

Jørgensen, U., Gjølberg, M., & Pedersen, A. B. (2019). The interplay between sustainability innovation, corporate sustainability performance, and sustainability culture. *Journal of Business Ethics*, 157(2), 455–470.

Kates, R. W., Clark, W. C., Corell, R., Hall, J. M., Jaeger, C. C., Lowe, I., … Svedin, U. (2001). Sustainability science. *Science*, 292(5517), 641–642.

Lovins, A. B., Lovins, L. H., & Hawken, P. (1999). A road map for natural capitalism. *Harvard Business Review*, 77(3), 145–158.

McDonough, W., & Braungart, M. (2002). *Cradle to cradle: Remaking the way we make things.* North Point Press.

Meadows, D. H., Meadows, D. L., Randers, J., & Behrens III, W. W. (1972). *The limits to growth.* Universe Books.

Panda, S. K., Reddy, G. S. M., Goyal, S. B., Thirunavukkarasu, K., Bhambri, P., Rao, M. V., Singh, A. S., Fakih, A. H., Shukla, P. K., Shukla, P. K., & others. (2019). *Method for management of scholarship of large number of students based on blockchain.* IN Patent App. 201,911,034,937 A.

Porter, M. E., & Van der Linde, C. (1995). Toward a new conception of the environment-competitiveness relationship. *Journal of Economic Perspectives*, 9(4), 97–118.

Potluri, S., Tiwari, P. K., Bhambri, P., Obulesu, O., Naidu, P. A., Lakshmi, L., Kallam, S., Gupta, S., & Gupta, B. (2019). *Method of load distribution balancing for fog cloud computing in IoT environment.* IN Patent App. 201,941,044,511.

Rana, R., Chhabta, Y., & Bhambri, P. (2019). A review on development and challenges in wireless sensor network. In International Multidisciplinary Academic Research Conference (pp. 184–188).

Reddy, A. K. N., & Williams, R. H. (2019). *Energy and society: An introduction.* Routledge.

Rifkin, J. (2014). *The zero marginal cost society: The internet of things, the collaborative commons, and the eclipse of capitalism.* Palgrave Macmillan.

Rockström, J., Steffen, W., Noone, K., Persson, Å., Chapin III, F. S., Lambin, E. F., … Nykvist, B. (2009). Planetary boundaries: Exploring the safe operating space for humanity. *Ecology and Society*, 14(2), 32.

Sachs, J. D. (2015). *The age of sustainable development.* Columbia University Press.

Schumpeter, J. A. (1942). *Capitalism, socialism and democracy.* Harper & Brothers.

Steffen, W., Richardson, K., Rockström, J., Cornell, S. E., Fetzer, I., Bennett, E. M., … Sörlin, S. (2015). Planetary boundaries: Guiding human development on a changing planet. *Science*, 347(6223), 1259855.

Stern, N. (2007). *The economics of climate change: The Stern review.* Cambridge University Press.

Tondon, N., & Bhambri, P. (2017). Novel approach for drug discovery. *International Journal of Research in Engineering and Applied Sciences*, 7(6), 28–46.

Tukker, A., & Tischner, U. (2006). Product-services as a research field: Past, present and future. Reflections from a decade of research. *Journal of Cleaner Production*, 14(17), 1552–1556.

UNDP (United Nations Development Programme). (2015). *Sustainable development goals.*

United Nations. (2015). *Paris agreement.*

Weizsäcker, E. U., Lovins, A. B., & Lovins, L. H. (1997). *Factor four: Doubling wealth, halving resource use.* Routledge.

World Commission on Environment and Development. (1987). *Our common future.* Oxford University Press.

Young, O. R. (2002). *The institutional dimensions of environmental change: Fit, interplay, and scale.* MIT Press.

Zehner, O. (2012). *Green illusions: The dirty secrets of clean energy and the future of environmentalism.* University of Nebraska Press.

Part II

Technological Solutions for Sustainable Living

8 Renewable Energy Technologies

Izabela Rącka and Izabela Małecka

8.1 INTRODUCTION

Sustainable development has its roots in the concept of "sustainability", which was formulated at the turn of the 1980s and 1990s. The United Nations World Commission on the Environment and Development presented the very general sustainable development concept. Sustainable development includes activities that "meet the needs of the present without compromising the ability of future generations to meet their own needs" (WCED, 1987).

Sustainable development is complex and diverse and is related to many fields, such as science, engineering, ecology and environment, economics and business, philosophy, sociology, and others (Rosen, 2017). Sustainable development is particularly interesting from the perspective of urban development (Turcu, 2012; Strezov et al., 2017; Rodrigues & Franco, 2020).

Economic development is a long-term process of quantitative and qualitative changes taking place in the economy. Already in the 1970s, economists defined economic growth as a long-term increase in the ability to provide people with increasingly diverse goods. Back then, this growth was based on technological progress with necessary adjustments (Kuznets, 1973).

Social progress is related to both increasing prosperity and eliminating poverty, which can occur by reducing inequalities, enabling the achievement of a basic standard of living and building an inclusive society (Greve, 2017).

Equally important in sustainable development is the integrated promotion of environmental and ecosystems management. The world leader in setting environmental rules at the end of the 20th century was the European Union (Ballor, 2023). This does not change the fact that the environmental aspect is as important as other sustainable development factors and is inextricably linked to technological development. The development of technology causes constant changes in the use of energy sources use. They lead to the abandonment of sources with lower energy content and the use of sources with higher energy content.

DOI: 10.1201/9781003475989-10

8.1.1 Systematics of Energy Sources

Energy is a branch of industry dealing with electricity and heat production and distribution. It is divided into (Chauhan, 2006):

- conventional energy – production of electricity and heat by burning fuels, such as hard coal, brown coal, oil, gas, biogas, plant and animal biomass, peat, and oil shale;
- unconventional energy – obtaining electricity and heat from alternative sources, e.g., hydropower, tidal, geothermal, wind, sun, nuclear reactions, and ambient heat.

Conventional energy may prove insufficient to meet human needs in the future due to the depletion of non-renewable fuel resources. The development of unconventional energy is important for the natural environment because the extraction and combustion of fuels cause damage to nature – it destroys the natural environment and pollutes the air, water, and soil. However, unconventional energy is generally environmentally friendly.

In turn, energy sources are divided into:

- renewable – having the ability to self-regenerate (biofuels, solid biomass, liquid biomass, biogas, fuels from municipal waste, hydropower, geothermal, wind, sun, ambient heat); renewable sources are used primarily in unconventional energy; in conventional energy, renewable fuels (biofuels) are still rarely used;
- non-renewable – mineral raw materials (e.g., hard coal, brown coal, oil, gas), the resources of which will eventually run out, constitute the basis of conventional energy.

8.1.2 Evolution of Energy Sources Use

Until the Industrial Revolution, which took place in the 18th century, energy production was based solely on the work of human and animal muscles, as well as the use of biomass.

The industry development resulted in new needs. In the mid-19th century, there was a significant change in energy sources – coal use increased, and it was used mainly for power steam engines and power plants. Thermal energy was used to generate mechanical energy, especially in areas near coal deposits.

Although at the beginning of the 20th century, the major source of energy was still coal, crude oil, which is more energetic than coal, started to gain popularity. Later, the dominance of oil products became noticeable. They established a massive oil distribution system, which included pipelines, storage tanks, and bulk carriers for transporting liquids. This was problematic because oil prices affected the economies of individual countries. However, technological development enabled the use of natural gas, which turned out to be more efficient as a fossil fuel. A completely new form of energy began to be used – nuclear fission. The discussion on the need to change the structure of different energy sources was initiated at the end of the 20th century (Ragwitz et al., 2005).

Energy sources have undergone significant changes in the 21st century. As fossil fuels like coal and oil are gradually phased out, there i's a transition towards more efficient alternatives like natural gas. Subsequently, progress has been made in biotechnology, driven by the growing potential of fuels obtained from biomass. Wind and solar energy have certainly become some of the most popular modern energy sources. It is expected that nuclear energy may play a significant role in the use of energy sources, especially if nuclear fusion becomes commercially viable. Resources of conventional energy are decreasing, which is why hydrogen is one of the most promising solutions. Its main advantage is that when burned or used in fuel cells, its only by-product is water, making it an attractive zero-carbon energy source. Scientists around the world conduct intensive research on the use of hydrogen. Politicians are also interested in it because of minimizing the impact on the natural environment and promoting environmental protection (Johnston et al., 2005).

8.1.3 Disadvantages of Conventional Energy

Conventional energy involves the use of nonrenewable energy sources – thermal power plants are fired primarily with hard or brown coal, natural gas, or crude oil. The production of electricity in thermal power plants is associated with energy transformations in which heat production plays a leading role. Thermal power plants burning fossil fuels include (Qizi & O'G'Li, 2019):

- coal power plants (Central Europe, USA, China, South Africa);
- gas power plants (faster and cheaper to build and emit less air pollution than coal-fired power plants, but often more expensive to operate due to the high price of gas);
- oil-fired power plants (mainly in oil-rich countries on the Persian Gulf, e.g., Saudi Arabia);
- other power plants, e.g., those using oil shale (Estonia) or peat (Russia, Finland).

The main disadvantage of conventional energy is its low efficiency. The energy efficiency of coal-fired power plants is approximately 30-45%. Natural gas-fired power plants achieve an efficiency of 60%. Thermal power plants have an energy efficiency of approximately 85%.

Among various fuel materials, the most popular are wood, coal, liquid gas, and kerosene. Each of these raw materials has slightly different properties and requires an appropriate heating boiler for proper combustion. Coal stands out among them primarily because of its not too high price. Hard coal is mainly used for heating purposes, as it has higher combustion efficiency. Due to its low calorific value, brown coal is not transported over longer distances, and it has the best calorific value immediately after extraction. Power plants that consume almost all of the mined coal are located near mines.

Hard and brown coal resources are found all over the world, but they are very concentrated. It is estimated that ten countries in the world have as much as 88% of coal resources. The United States has the largest resources – approximately 23% of world coal reserves.

The transformation process in the coal-based energy sector, including mining and coal-based electricity, is constantly accelerating. In 2022, 55 million tonnes of hard coal were produced in the EU, i.e., 80% less than in 1990 (Eurostat, 2024). Unfortunately, in 2022 global coal mining increased by 8.2%. Coal producers in Asia are responsible for more than two-thirds of global coal mining, and in 2022, they increased their production by 11%. The largest producer of hard and brown coal is China, accounting for over half of the world's supply (Enerdata, 2024).

8.2 GREEN DEAL

The main reasons for the energy sources change are environmental protection and fossil energy resource depletion. The European Union noticed the need for such a change. The European Commission has approved legislative proposals to help adapt EU climate, energy, transport, and tax policies to lower net greenhouse gas emissions by at least 55% by 2030 compared to the quantity recorded 40 years earlier. The European Commission presented the "European Green Deal", which defined a new growth policy for Europe (Eur-Lex, 2019). Its main goal is to achieve net zero greenhouse gas emissions by 2050 and decouple economic growth from the use of natural resources. The European Green Deal aims to make Europe climate neutral before 2050.

Over 75% of EU greenhouse gas emissions come from energy production and use. Decarbonizing the EU's energy system is therefore crucial to achieving the 2030 climate goals and the EU's long-term strategy to become carbon neutral by 2050.

The EU currently has legally binding climate targets covering all key economic sectors. The whole package includes:

- emission reduction targets across multiple sectors;
- a goal to increase the carbon sequestration capacity of natural sinks, an emissions trading system that has been updated to reduce emissions and make polluters pay, as well as to generate investment in the green transition;
- social support for citizens and small businesses.

8.2.1 Sustainable Transport

The shift to greener mobility means green, accessible, and affordable transport and logistics solutions for European Union citizens, connecting rural and remote regions. Thanks to new carbon emission standards, all new passenger cars registered in Europe will be emission-free before 2035. An intermediate step toward zero emissions is to reduce the average emissions from new passenger cars by 55% and from new commercial vehicles by 50% by 2030. This will increase the chances of achieving emission-free road transport by 2050. The EU is working to create the infrastructure that will be needed to charge zero-emission vehicles on short and long routes. Mandatory targets will be set for the deployment of electric charging and hydrogen refueling infrastructure along European roads. The capacity of public charging points will be sufficient to meet the demand of the larger fleet of zero-emission cars that will enter the market. The number of private charging stations at home or at work will also increase (Tie & Tan, 2013). Greenhouse gas emissions charges are also imposed in the aviation sector. To promote sustainable aviation fuels, the minimum content of these fuels in kerosene blends offered by aviation fuel suppliers and delivered to EU airports has been increased.

Greenhouse gas emission charges also cover the maritime sector. To promote the take-up of renewable and low-carbon fuels, a target has been set to gradually reduce the average annual greenhouse gas emissions intensity from energy consumption on board ships.

8.2.2 Green Industrial Revolution

The green transformation represents a huge opportunity for European industry as it enables the creation of new markets for new green technologies and products. The Green Deal industrial plan aims to help Europe lead in industrial innovation and clean technologies. To achieve this, the plan is based on four main pillars:

- a predictable and simplified regulatory environment;
- easier financing;
- improving skills;
- facilitating open and fair trade while supporting resilient supply chains.

8.2.3 Ecological Energy System

Reducing greenhouse gas emissions by at least 55% by 2030 requires an increase in the share of energy from renewable sources, as well as higher energy efficiency. Assumptions regarding the change of the energy system are presented in the RE Power EU plan (Eur-Lex, 2022). It focused on three pillars:

- using more energy from renewable sources;
- energy saving;
- diversification of energy supplies.

The new renewable energy target is to increase renewable energy generation capacity by raising the binding 2030 target to at least 42.5% (with the ambition to reach 45%). Additionally, it is assumed that energy efficiency will improve by 11.7% by 2030, which means the need to save energy.

The use of renewable energy is essential to reduce greenhouse gas emissions in the EU and to reduce the EU's dependence on fossil fuels and imported energy, i.e., to contribute to the security of energy supply.

To avoid interruptions and energy shortages, EU policy aims to store gas and secure energy supplies from alternative sources. It is important to diversify suppliers, look for them outside the European Union, and become independent from gas supplies from Russia (Umar et al., 2022).

8.2.4 Renovation of Buildings

Renovation of buildings will contribute to saving energy – it will provide protection against heat or frost and help eliminate energy poverty. The Green Deal assumes increasing building renovation rates, which will improve energy efficiency and natural resource management. The quality of life of people who live in the buildings or stay there for other purposes on a daily basis will certainly increase (Bhambri & Gupta, 2018). Savings in heating houses in winter and cooling them in summer will reduce greenhouse gas emissions in Europe. Building renovation is therefore intended to improve the energy performance of buildings. This is possible by installing insulation, replacing old windows and doors, modernizing heating systems, and installing solar panels (Nowogońska & Mielczarek, 2021).

8.2.5 Nature Protection

Nature is essential to human existence and plays an important role in the fight against climate change. Restoring natural resources allows for carbon dioxide reduction. The EU's 2030 Biodiversity Strategy is a comprehensive long-term plan to protect nature and restore Europe's biodiversity, for the benefit of people, the climate, and the planet. The strategy provides for the following actions:

- increasing the area of Natura 2000 areas – a network of protected land and sea areas;
- reconstruction of natural resources;
- launching a financing system for activities related to biodiversity.

The reconstruction of natural resources will help limit the progression of global warming and mitigate the effects of violent natural disasters (floods, droughts, and heat waves).

The Green Deal also assumes activities focused on achieving the goals of the so-called forest strategy. It is intended to ensure the multifunctionality of European forests, as they are crucial in the fight against climate change. Forests absorb carbon dioxide, cool cities, protect against floods, and reduce drought effects. For this reason, it is important to strengthen their protection, recovery, and resilience. It is therefore important to keep some forests intact. This is especially important from the perspective of construction market entities, where wood is a raw material used for the production of buildings and their finishing.

8.2.6 Climate Protection

Climate change can only be stopped if countries around the world cooperate on this issue. The actions of the European Union alone are not sufficient in this respect. The European Green Deal is a perfect example of what should be done to achieve climate neutrality. Investments in renewable energy technologies and green transport, including maritime and air transport, will help reduce climate change on a global scale.

8.3 RENEWABLE ENERGY TECHNOLOGIES

At the same time, in line with the Green Deal, it is necessary to prioritize the energy efficiency of buildings and improve their energy performance. For this purpose, it is worth using renewable energy technologies. "Renewable energy technologies" is a general term for energy production using a renewable energy source such as solar, wind, water (hydro and tidal), biomass (biofuels and waste), and geothermal heat (Zepf, 2020).

Nowadays, the importance of renewable energy systems and their global contribution to electricity supply is increasing. This is due to concerns about the sufficiency of raw materials – both in quantity and quality – needed to produce the required clean energy systems.

The share of renewable energy sources in global energy production increased from about 20% to nearly 30% in the period 2010–2022. The greatest contribution to the production of energy from renewable sources is made by countries with large hydropower resources (Brazil, Colombia, Canada, New Zealand, Sweden, and Norway). In other countries, an important factor supporting the use of renewable energy sources is individual policies, including subsidies and investment incentives, as well as the falling costs of electricity production in photovoltaic and wind technologies (Maglio, 2022). In Europe, renewable energy sources account for as much as 43% of production, with the greatest impact on this situation coming from Great Britain, the Netherlands, Germany, and Turkey (Enerdata, 2024).

The use of systems utilizing renewable energy sources, however, involves certain material requirements. Biomass and water-based systems require durable materials such as concrete and rebar. Tidal systems should be corrosion-resistant, which is possible by using appropriate steel. Only solar and wind energy typically do not require the use of critical materials. Natural resources are used in only a few green energy systems, but for most of them, alternative system solutions are available (Zepf, 2020). Later in the chapter, the technologies used to supply new buildings with renewable energy sources are described.

8.3.1 Biomass

Biomass is all plant and animal products that can be used to create energy. These include both post-production plant waste and plants grown specifically for this purpose. These also include animal products and farm-related waste. Obtaining energy involves recovering valuable components from biomass produced in the processes of photosynthesis, primarily carbon. There are several ways to use biomass. Energy can be obtained using natural processes that occur in plant and animal products. Among other things, gases and biofuels produced during controlled putrefaction and fermentation processes are recovered. However, most energy is produced through the combustion of carbon contained in plants (Singh et al., 2016).

8.3.2 Geothermal

Geothermal energy involves using the earth's thermal energy to produce heat and electricity. It is obtained by drilling into naturally hot groundwater. Geothermal heating uses energy stored in the ground, which allows you to save up to 70% of heating costs. Geothermal energy is ecological heating using a free energy source, independent of raw material prices, political events, and crisis situations.

8.3.3 Hydroenergy Systems

Hydropower is a renewable energy source. The kinetic energy of flowing water is harnessed by hydroelectric power plants to generate electricity. Water flowing in the river becomes a source of energy and is used to produce electricity through dams, turbines, and tidal streams. When the water is obstructed at a high level, it plunges onto the turbine blades and sets them in motion. Through the turbines, the kinetic energy of the falling water is transformed into electricity by a generator. The height of the waterfall is a criterion that allows hydropower plants to be divided into slow, speed, and high speed. Transformer stations are typically found in hydroelectric power plants to distribute electricity from turbines to the grid.

8.3.4 Solar Technologies

A photovoltaic installation is a set of devices that obtain energy from the sun and convert it into alternating current. The key devices are solar panels, which are usually placed on the roofs of buildings, but there are some new solutions that allow the panels to be mounted on the walls of buildings, on balconies, etc. Solar roof tiles are also produced. Direct current from photovoltaic panels is transferred to the solar inverter, where it is converted into alternating current. Thanks to the higher voltage, free electricity obtained from the sun in a PV installation "displaces" grid electricity from the house, and its surplus goes to the grid.

8.3.5 Tidal Systems

By utilizing the regular ebb and flow of ocean water, a tidal power plant generates hydroelectric power. Electricity in this type of power plant is generated by the inflow and outflow of sea and ocean waters, which are created thanks to the tidal phenomenon. Tides are the result of the gravitational forces of the Moon and the Sun and are also caused by the centrifugal force due to the Earth's rotation around its center of gravity.

To obtain electricity from such tides, a dam equipped with turbines through which water flows is placed across the reservoir where they occur (bay, fjord). Such a power plant usually operates only for a few hours, twice a day – during high tide, when water flows into the reservoir, and during low tide, when it flows out of it.

8.3.6 Wind Energy Systems

Wind power plants produce electricity using generators, i.e., wind turbines powered by wind energy (Anand & Bhambri, 2018). This energy is considered ecologically clean and does not involve the burning of any fuel. Power plants may be based on single devices or take the form of large wind farms. Home wind farms are becoming more and more popular, meeting some of the needs of users of residential buildings (Abd Ali et al., 2020).

8.4 IMPACT ON CONSTRUCTION AND REAL ESTATE

The European Union guidelines, which are a model for countries from other areas of the world, accelerate the fight against the economy's negative impact on the natural environment. The Construction and Real Estate sectors must introduce profound changes to reduce carbon emissions and respond to society's current needs (Zimmermann & Gengnagel, 2023). Private and commercial entities operating in both sectors are unable to meet emission reduction goals on their own. Support from public funds is certainly necessary, and perhaps also activities in the form of public-private initiatives. However, it is probably investors who primarily influence the pace of change in both sectors. In preparation for the requirements of the Green Deal, they introduce factors related to the greenness of projects in the processes of assessing the profitability of investments. This is undoubtedly important for the possibility of obtaining financing for new investments, as the neutralization of environmental impact will increase the investor's bargaining power. In contrast, developers are facing growing demands for sustainable construction solutions from commercial and residential real estate clients.

The real estate market also includes older facilities that require modernization. Therefore, there is a need to use technologies and materials that will improve the technical and operational condition of buildings and reduce their emissions.

8.4.1 Sustainable Construction

Sustainable construction means activities aimed at reducing the negative impact of construction on the environment throughout the entire life cycle of buildings, starting from project preparation through the construction process to operation. This means designing, building, and using buildings with the future in mind. The buildings created in this way do not pollute the environment and are user-friendly. The goal of sustainable construction is to improve existing and shape new living and working conditions.

The greatest advantages of sustainable construction are its susceptibility to innovation and great development potential – working on new solutions can have a real impact on improving people's quality of life. It has a major impact on three priority areas of sustainable development (environment, economy, and society):

- Environment: Through the use of specific materials and appropriate operations, modern construction can contribute to environmental protection. The primary objective is to decrease energy and resource usage and minimize waste generation. Minimizing the consumption of electricity, water, and heating, as well as reducing the amount of sewage and exhaust emissions, significantly improves the climate. Therefore, sustainable construction focuses heavily on renewable energy sources, energy and water conservation, and the reuse of water (such as rainwater or greywater for watering plants).
- Economics: The use of emission-reducing materials translates into building maintenance costs. Lower expenses for utilities and the possibility of reusing materials and resources are important in reducing expenses related to the operation of the building. Great emphasis is also placed on reducing the costs of building facilities through economical management of raw materials and the use of those that are considered "green";
- Society: Sustainable construction is guided by concern for the comfortable use of objects and spaces. Citizens of developed countries spend most of their lives inside buildings, so comfort, health, and a friendly environment are key factors in the process of designing and constructing structures.

A sustainable building is a building that combines the three pillars of sustainable development and ensures a healthy microclimate inside the building, positively affecting the productivity of users in a cost-effective and environmentally friendly way. Green buildings are characterized by care for the natural environment and the economical use of raw materials throughout the entire construction cycle. The priority is to reduce water and energy consumption and the impact of building materials on the environment, all while maintaining high user comfort (Munaro et al., 2020).

The building should be warm in winter, have plenty of natural light, and protect against outside noise. It is necessary to ensure access for people of all ages and people with disabilities. Appropriate insulation and ventilation provide protection against the formation of mold and fungus, and materials that do not absorb dust (e.g., wood) support the use of the facility by allergy sufferers. Green construction, being user-friendly, should meet standards of quality and aesthetics, which affects the comfort of use.

A number of materials based on renewable plant materials are used in sustainable construction, such as (Sahlol et al., 2021)

- bamboo, lumber, straw;
- stone and recycled metals;

- industrial waste materials that are environmentally neutral, including brick and concrete waste;
- other renewable materials and plastics that are non-toxic and recyclable.

Features of sustainable buildings include (Mercader-Moyano & Esquivias, 2020):

- effective use of renewable energy sources (solar panels, use of non-toxic and ecological building materials, inclusion of green forms in the construction of buildings – green walls);
- use of technology that allows for appropriate recycling;
- taking into account technologies based on the reduction of air pollution;
- installing mechanisms to reduce the building's water footprint;
- the use of systems that allow effective penetration of sunlight and retain heat in rooms, which reduces the need for heating;
- consistency of the construction process with natural conditions and with full respect for the natural environment (ecologically friendly location);
- ensuring a healthy and comfortable environment for building users;
- responsibility toward the building's surroundings and location;
- flexibility and possibility of re-adaptation of the building and installations and devices in the building (resource protection and savings);
- the use of building management systems that monitor and control devices and installations in order to minimize the consumption of energy and other resources.

To determine whether a building is "green", existing green building rating systems can be used.

They provide building stakeholders with tools that enable the implementation of solutions in the field of pro-ecological design, construction, use, and maintenance of buildings. The advantage of certification systems is their universality – buildings are assessed according to the same principles in all countries. The most common rating systems for sustainable buildings include BREEAM, LEED, DGNB, HQE, and WELL.

Sustainable construction also includes the principles of designing new housing estates and development areas. In each case, it is important to design green areas intended for rest and recreation and to create better conditions for moving around by ecological means of transport (e.g., bicycle). The designed sustainable area is more willingly chosen for living and investing because it creates better living conditions. This, in turn, increases the value of the investment. Importantly, the main assumption of sustainable construction is care for future generations. Modern solutions should increase their development potential and leave them with a legacy of a non-devastated environment (Hossain et al., 2020).

8.4.2 Real Estate Market

Ensuring sustainable development and implementing the assumptions of the European Green Deal is successfully carried out in the real estate market. This will be a long-term process and will require the implementation of many new factors that will allow, among others, the reduction of carbon emissions and energy emissions. Energy inefficiency and lack of renovation are the main elements requiring improvement, especially when the goal is to reduce negative environmental impacts and overall sustainability.

Buildings account for 36% of carbon emissions and 40% of energy consumption in Europe. About three out of four buildings are energy inefficient. At the same time, only 1% of resources are subject to renovation annually. The assumptions of the European Green Deal, which come down to ESG (Environmental, Social responsibility, Corporate governance), are becoming a lasting trend influencing changes in the real estate market on a local and global scale. The strong trend toward sustainable construction is evidenced by the rapidly growing number of green building certificates in various market segments. The growing awareness of investors is also beginning to be reflected in the condition of the housing sector, which is recording the fastest development in the group of sustainable real estate.

Buildings with an environmental certificate are more attractive to investors because the certificate affects the transparency and standardization of transactions (Rącka, 2021). Thanks to the certificates, investors gain the opportunity to immediately learn many features of the building that would have to be verified at a later stage of real estate analysis. This facilitates the selection of preferred investments and translates into savings in time and costs related to the due diligence of purchased properties. When conducting investment activities, investors' awareness of the pursuit of climate neutrality is important, influencing the decisions they make. However, the level of risk remains crucial for investment decisions and is lower in the case of so-called "green real estate". The level of risk is directly related to the interest of tenants. Lease agreements for the best commercial properties are usually signed for 5–10 years. Therefore, what is important for investors is the ability of real estate to attract "attractive" and reliable tenants both in the near future and in the long term. Investors, therefore, take into account, among others: local environmental factors, development of the adjacent area, green areas, or access to public transport. Broadly speaking, the more modern, functional, and compliant with modern international standards the development, the higher the value of the property.

European regulations impose many obligations on commercial real estate entities, and failure to fulfill them results in difficulties in selling or renting buildings that do not meet standards. When selling or renting a building or premises, an energy performance certificate is required. An energy performance certificate is a document that specifies the amount of energy demand necessary to meet the energy needs related to the use of a building or part of it. This concerns energy used for heating and ventilation, preparation of domestic hot water and cooling, and in the case of non-residential buildings, also lighting.

The energy class of a building is a factor influencing real estate prices. The market notices the difference between class A and class G – it is expressed in the costs of maintaining the property. Offers for the sale or rental of real estate should include the indicator of the annual demand for various kinds of energy and energy sources, as well as carbon emissions amount.

The type of materials and technological solutions used to build the facility also affect the price. Lower operating costs of "green" buildings may result in their higher prices on the market. Already encountered solutions include:

- energy-saving modular buildings that are made of ready-made spatial prefabricates prepared in advance in the production plant;
- advanced ventilation systems, intelligent lighting, intelligent energy management;
- modern building materials, renewable raw materials, and solutions that minimize the negative environmental impact, such as ecological concrete, wood in construction, and thermal insulation based on renewable raw materials;
- green roofs and walls (development of free space for use as a green area, e.g., creating gardens).

Green investments are more expensive to implement, but those that are implemented contrary to the Green Deal guidelines – although they are cheaper – will be practically impossible to finance after some time. In a relatively short time, they may become unsellable on the secondary market, or their value will be much lower. Some banks already offer customers lower mortgage loan costs, rewarding buyers who decide to live in a building with above-standard energy efficiency parameters (Bertoldi et al., 2021).

Issues related to the European Green Deal are now included in real estate valuations. The leading, internationally recognized real estate valuation standards already include issues related to ESG, which correlate with the assumptions of the Green Deal and represent a set of criteria and standards that assess the impact of activities on three key areas: environment, society, and governance. These criteria are taken into account in real estate valuations based on the European Valuation Standards (TEGOVA, 2020), prepared by the European Group of Associations of Valuers, the so-called *Blue Book*, or the Global Standards (RICS, 2022) of the Royal Institute of Chartered Surveyors, the so-called *Red Book*.

The above factors affect the construction and real estate market, which can be reflected in the prices of building materials, the construction technologies used, and the differentiation of property prices. Implementing activities to ensure sustainable development in the construction and real estate markets is an action of the future. This is part of the fight for the climate, an efficient economy, a society without exclusion, and ethical business practices. The use of renewable energy technologies is also an opportunity for enterprises – taking efforts for sustainable development allows them to build a positive image of the company, increase competitive advantage, improve results, and make the company be perceived as serious and prospective business partners. Activities of this type also enjoy a positive reaction from the conscious society.

REFERENCES

Abd Ali, L.M., Al-Rufaee, F.M., Kuvshinov, V.V., et al. (2020). Study of hybrid wind–solar systems for the iraq energy complex. *Applied Solar Energy,* 56, 284–290. https://doi.org/10.3103/S0003701X20040027.

Anand, A., & Bhambri, P. (2018). Rotation, scale and translation invariant character recognition system using neural network. Punjab Technical University, Jalandhar (M.Tech. thesis).

Ballor, G. (2023), Liberal environmentalism: The public-private production of european emissions standards. *Business History Review*, 97(3), 575–601. https://doi.org/10.1017/S0007680523000272.

Bertoldi, P., Economidou, M., Palermo, V., Boza-Kiss, B., & Todeschi, V. (2021). How to finance energy renovation of residential buildings: Review of current and emerging financing instruments in the EU. *WIREs Energy & Environment*. 10, 384. https://doi.org/10.1002/wene.384.

Bhambri, P., & Gupta, O. P. (2018). Implementing machine learning algorithms for distance based phylogenetic trees. I.K.Gujral Punjab Technical University, Jalandhar (Ph.D. thesis).

Chauhan, D.S. (2006). *Non-conventional energy resources*. New Age International.

Enerdata (2024). Coal and lignite production. online: https://yearbook.enerdata.net/coal-lignite/coal-production-data.html [access: 2.01.2024]

Eur-Lex (2019). The European green deal. online: https://eur-lex.europa.eu/legal-content/EN/TXT/HTML/?uri=CELEX:52019DC0640 [access: 2.01.2024]

Eur-Lex (2022). REPowerEU Plan. online: https://eur-lex.europa.eu/legal-content/EN/TXT/?uri=COM%3A2022%3A230%3AFIN&qid=1653033742483 [access: 2.01.2024]

Eurostat (2024). Coal production and consumption statistics, online: https://ec.europa.eu/eurostat/statistics-explained/index.php?title=Coal_production_and_consumption_statistics [access: 2.01.2024]

Greve, B. (2017) How to measure social progress?. *Social Policy & Administration*, 51, 1002–1022. doi: 10.1111/spol.12219.

Hossain, M.S., Ng, T., Antwi-Afari, P., & Amor, B. (2020). Circular economy and the construction industry: Existing trends, challenges and prospective framework for sustainable construction. *Renewable and Sustainable Energy Reviews*, 130, 109948. https://doi.org/10.1016/j.rser.2020.109948.

Kuznets, S. (1973). Modern economic growth: Findings and reflections. *The American Economic Review*, 63(3), 247–258. http://www.jstor.org/stable/1914358.

Maglio, M. (2022, Winter). Visions of cities beyond the green deal: From imagination to reality. *Journal of Urban Regeneration & Renewal*, 15(2), 176–192(17).

Mercader-Moyano, P., & Esquivias, P.M. (2020). Decarbonization and circular economy in the sustainable development and renovation of buildings and neighbourhoods. *Sustainability*, 12(19), 7914. https://doi.org/10.3390/su12197914.

Munaro, M.R., Tavares, S.F., & Bragança, L. (2020). Towards circular and more sustainable buildings: A systematic literature review on the circular economy in the built environment. *Journal of Cleaner Production*, 260, 121134. https://doi.org/10.1016/j.jclepro.2020.121134.

Nowogońska, B., & Mielczarek, M. (2021). Renovation management method in neglected buildings. *Sustainability*, 13(2), 929. https://doi.org/10.3390/su13020929.

Qizi, J.N.D., & O'G'Li, S.R.A. (2019). Renewable sources of energy: advantages and disadvantages. *Д о с т и ж е н и я н а у к и и о б р а з о в а н и я* , 8–3(49), 230–243.

Rącka, I., (2021). Drivers of Subregional Housing Markets in Poland. *European Research Studies Journal*, XXIV(2B), 810–834. https://doi.org/10.35808/ersj/2266

Ragwitz, M., Schleich, J., Huber, C., Resch, G., Faber, T., Voogt, M., Coenraads, R., Cleijne, H., & Bodo, P. (2005). *Analyses of the EU renewable energy sources' evolution up to 2020 (FORRES 2020)*. Fraunhofer IRB Verlag.

RICS (2022). *Global standards*. RICS.

Rodrigues, M., & Franco, M. (2020). Measuring the urban sustainable development in cities through a Composite Index: The case of Portugal. *Sustainable Development*, 28, 507–520. https://doi.org/10.1002/sd.2005.

Rosen, M.A. (2017). Sustainable development: A vital quest. *European Journal of Sustainable Development Research*, 1, 2.

Sahlol, D.S., Elbeltagi, E., Elzoughiby, M., & Elrahman, M.A. (2021). Sustainable building materials assessment and selection using system dynamics. *Journal of Building Engineering*, 35, 101978, https://doi.org/10.1016/j.jobe.2020.101978.

Singh, R., Krishna, B.B., Kumar, J., & Bhaskar, T. (2016). Opportunities for utilization of non-conventional energy sources for biomass pretreatment. *Bioresource Technology*, 199, 398–407. https://doi.org/10.1016/j.biortech.2015.08.117.

Strezov, V., Evans, A., & Evans, T.J. (2017). Assessment of the economic, social and environmental dimensions of the indicators for sustainable development. *Sustainable Development*, 25, 242–253. doi: 10.1002/sd.1649.

TEGOVA (2020). *European valuation standards.*

Tie, S.F., & Tan, C.W. (2013). A review of energy sources and energy management system in electric vehicles. *Renewable and Sustainable Energy Reviews*, 20, 82–102. https://doi.org/10.1016/j.rser.2012.11.077.

Turcu, C. (2012). Local experiences of urban sustainability: Researching housing market renewal interventions in three English neighbourhoods. *Progress in Planning*, 78, 101–150.

Umar, M., Riaz, Y., & Yousaf, I. (2022). Impact of Russian-Ukraine war on clean energy, conventional energy, and metal markets: Evidence from event study approach. *Resources Policy*, 79, 102966. https://doi.org/10.1016/j.resourpol.2022.102966.

WCED World Commissionon Environment and Development. (1987). *Our common future.* Oxford University Press.

Zepf, V. (2020). The dependency of renewable energy technologies on critical resources. In: A. Bleicher & A. Pehlken (eds.), *The material basis of energy transitions* (pp. 49–70). Academic Press. https://doi.org/10.1016/B978-0-12-819534–5.00004–0.

Zimmermann, K., & Gengnagel, V. (2023). Mapping the social dimension of the European Green Deal. *European Journal of Social Security*, 25(4), 523–544. https://doi.org/10.1177/13882627231208698.

9 Eco-Tech Chronicles
Sustainable Living and Renewable Energy Solutions

Paweł Patyk

9.1 INTRODUCTION

The term "Eco-Tech" is an abbreviation of Ecological Technology and refers to a paradigm combining ecology with modern technology. It aims to create, develop, and implement technological solutions that support sustainable development and minimize the negative impact of human and industrial activities on the natural environment (Wu et al., 2021). Within this approach, technological innovations aim to increase the efficiency of using raw materials, reduce emissions of harmful substances, protect ecosystems, and promote sustainable use and management of natural resources.

In the Eco-Tech concept, the most important technologies are technologies that serve the implementation of ecological goals and sustainable development priorities. Therefore, energy efficiency and technologies that reduce energy consumption in various sectors, including construction, transport, and industry, are essential here. Equally important are renewable energy sources and the technologies that use them, such as solar, wind, water, geothermal, and biomass, to replace fossil fuels. Waste management and technological innovations in the recycling and circular economy are also critical, enabling the effective reuse of materials and minimization of waste. Sustainable production and consumption is another area in the Eco-Tech concept, with technologies supporting the design and production of sustainable products that are durable, easy to repair, and recyclable, as well as promoting consumption patterns that minimize negative environmental impacts. It is also necessary to protect and regenerate ecosystems using technological solutions to preserve biodiversity, restore damaged ecosystems, and manage natural resources effectively (Ahlborg et al., 2019; Weaver et al., 2017).

Therefore, the Eco-Tech concept combines advanced technologies with ecological principles, striving to balance technological development and environmental protection. This paradigm emphasizes that technological progress and innovation can and should support sustainable development efforts by helping to solve global environmental challenges such as climate change, pollution, and the degradation of natural resources.

 DOI: 10.1201/9781003475989-11

Often, solutions from the Eco-Tech concept promote and disseminate sustainable living, i.e., making conscious decisions that aim to reduce the negative impact of our activities on the natural environment while ensuring well-being and meeting the needs of current and future generations (Holden et al., 2014). This characterization indicates that sustainable living comes from sustainable development, or more precisely, from applying its concept in the everyday lives of individuals.

In sustainable living, renewable energy is one of the main components, playing a pivotal role in counteracting climate change and promoting an ecological approach to the use of resources. This is related to the increasing public awareness of the negative impact of conventional energy sources on the environment, such as CO_2 emissions and other pollutants causing the greenhouse effect and degradation of ecosystems (Twidell, 2021). Renewable energy, derived from natural and inexhaustible sources such as the sun, wind, water, and biomass, offers an alternative that minimizes harmful environmental effects and supports long-term stability and energy security (Nelson, 2011). The adoption of renewable energy in everyday life, through photovoltaic installations, heating systems based on heat pumps, or the use of green energy tariffs, is, therefore, an integral element of striving for a more sustainable and responsible living that harmonizes the needs of people with the protection of our planet.

9.2 THE MAIN ASPECTS OF SUSTAINABLE LIVING

Living is a culturally conditioned way of existence: patterns of life, motives of behavior, hierarchies of values, directives, and means of their implementation, i.e., elements of consciousness and the premises of existence inherent in everyday life. Living expresses the balance between aspirations, awareness of needs, preferences, and the ability to satisfy them, between the recognized, desired, and achievable values. Living can refer to both social communities and individuals, and it means a way of lifestyle. It means a set of everyday behaviors specific to a given community or individual. A characteristic way of distinguishing a given community or individual from others (Chaney, 2012). The whole that constitutes living consists of people's behaviors varied in scope and form, sequences of these behaviors aimed at specific goals, their motivations, assigned meanings, and values, as well as certain functions of things that are either results, goals, or instruments of these behaviors. The primary defining elements of living (Sobel, 2013) are the relationships between economic position, the ability to choose a specific living, and the importance of normative systems recognized by individuals or groups that influence these choices.

9.2.1 Characteristics of Sustainable Living

Sustainable living will, in turn, be understood as one in which an individual, taking into account social, economic, and environmental aspects, tries to balance them in daily existence with the process of meeting their own needs, taking into account the well-being of future generations (Black and Cherrier, 2010). The concept of sustainable living is sometimes used interchangeably with the category of ecological living, considering that this dimension, alongside the social and economic, is one of the constitutive elements of sustainable development. By ecological, we mean living that, more or less directly, concerns the relationship between humans and the surrounding natural environment. Sustainable living also appears in literature, i.e., one that provides harmony and stability, allowing a person to experience higher levels of health and happiness (Barr et al., 2010).

In sustainable living, the most important aspects include (Caradonna, 2022):

- Saving natural resources by minimizing the use of water, energy, and other natural resources through the effective use and selection of resource-saving technologies, i.e., water conservation practices, supporting sustainable forest management, and participating in tree planting initiatives.
- Reducing carbon dioxide emissions through sustainable mobility, i.e., reducing energy consumption from fossil fuels and preferring environmentally friendly means of transport, such as bicycles, public transport, or electric cars, and promoting and participating in car-sharing and bike-sharing systems.
- Recycling and reuse by separating waste, recycling raw materials such as paper, glass, metal, and plastic, and promoting the reuse of items to reduce the amount of waste going to landfills. It is simply about minimalism and waste reduction by limiting the consumption of unnecessary items, promoting "less is more", and implementing waste reduction, reuse, and recycling practices.
- Sustainable consumption is choosing sustainably produced products, supporting the local economy, avoiding single-use products, minimizing food waste, and preferring organic and certified products. In other words, it is also conscious consumption, i.e., choosing products and services that are produced sustainably, have a low carbon footprint, and are ethically sourced. It also includes avoiding wasting resources such as food and energy and preferring recycled products.
- Protecting nature by actively participating in environmental protection activities such as planting trees, protecting biodiversity, and supporting sustainable ecological projects. Additionally, implementing solutions at home and in the workplace that increase energy efficiency, such as thermal insulation, energy-saving devices, or systems based on renewable energy sources.
- Environmental education and awareness, i.e., increasing knowledge about human impact on the environment and ways to minimize it and promoting sustainable living among family, friends, and communities. Equally important is social involvement in the form of active participation in local communities and ecological education, aimed at raising awareness of the impact of our activities on the environment and ways to minimize it.
- Healthy eating results from preferring sustainable food sources, such as local, seasonal, and organic products, and reducing meat consumption in favor of a plant-based diet, thereby reducing the carbon footprint associated with food production.

Creating sustainable living means reevaluating how we live and rethinking how we shop, consume, and organize our daily lives. It is a process that requires commitment at the individual and community levels and covers various aspects of everyday life. Sustainable living requires a conscious approach to daily choices and actions, with an eye to the long-term impact on the planet and future generations. It is about protecting the environment and striving for a better quality of life, a healthier environment, and a fairer society. This type of living is about transforming societies and living harmoniously with the natural environment (Huckle and Wals, 2015). All private and professional human choices and activities – energy consumption, transport, food, waste, and communication – matter for the forms they take.

For sustainable living to become part of societies and everyday life, it must also become the subject of action at all levels in social and technological systems and institutions. These systems include efficient infrastructure, services, goods such as communication systems, and individual choices and actions that help reduce the use of natural resources, emissions, waste, and pollution while supporting equitable socio-economic development and conserving natural resources without compromising their performance (Tomislav, 2018). The role of governments is to provide appropriate conditions and infrastructure that will enable citizens to adopt sustainable living to socio-economic reality. The business sector can support the development of innovative solutions that bring people closer to sustainable living, as can the education and information systems, which are crucial to moderating the involvement of all residents, including youth, in sustainable choices and more conscious consumption.

9.3 SECTORAL INTEGRATION OF SUSTAINABLE LIVING

Integrating sustainable living into the everyday lives of societies at all levels requires coordinated action by various sectors, including governments, businesses, educational institutions, and citizens themselves.

9.3.1 Government and Public Policies

Governments and public policies play a fundamental role in promoting sustainable living by creating legal, economic, and social frameworks that enable and encourage sustainable practices (Doppelt, 2017; Hassan and Lee, 2015; Hess, 2014):

1. Creation and implementation of regulations: In this area, actions are taken to create and implement rules promoting sustainable development, such as emission standards (setting limits on emissions of harmful substances for industry, transport, and other economic sectors to reduce air pollution and the impact on climate change), energy efficiency requirements (implementation of standards for buildings, home, and industrial devices aimed at reducing energy consumption and promoting energy-saving technologies), and waste management (implementing regulations on recycling, composting, and limiting waste production, including bans or restrictions on single-use products).

2. Investments in sustainable infrastructure: Such investments include public transport and mobility (developing efficient and accessible public transport systems, cycling infrastructure, and pedestrian areas to reduce dependence on passenger cars), renewable energy (financing solar, wind, hydropower, and other renewable energy projects and initiatives, supporting their integration with the national energy system), and recycling, waste management, and water purification systems (construction and modernization of infrastructure for segregation, recycling, and safe waste disposal and water purification, promoting the circular economy).
3. Financial and tax support for sustainable practices and technologies, including subsidies and tax breaks for companies and consumers choosing green bonds and financing. As part of subsidies and tax breaks, companies and consumers receive financial incentives for investing in green technologies, energy-saving devices, solar panels, building insulation, etc. However, green bonds are issued as part of green options, or funds dedicated to sustainable development projects are created, supporting financing pro-ecological initiatives and sustainable infrastructure.
4. Education and awareness campaigns to raise awareness of sustainable development and promote sustainable choices among citizens. Here, educational programs introduce topics related to sustainable development and environmental protection in school and academic curricula to develop ecological awareness from an early age. Public information and educational campaigns are also organized on sustainable living, climate change, the consequences of wasting resources, and methods of reducing them, aimed at increasing awareness and social involvement.
5. International cooperation, where there is significant participation of countries in international agreements (active participation in global initiatives and agreements on climate change, environmental protection, and sustainable development) and exchange of knowledge and experience (cooperation with other countries and international organizations in the exchange of best practices, technologies, and solutions supporting sustainable development).

Government action to promote sustainable living is vital to achieving broad-based ecological, social, and economic goals. By creating favorable regulatory conditions, infrastructure investments, and financial and educational support, governments can significantly contribute to transforming societies toward sustainable development.

9.3.2 Business Sector

The business sector promotes sustainable living by developing, implementing, and distributing sustainable technologies, products, and services. Its commitment is essential to accelerate the transition to more sustainable practices in both production and consumption (Doppelt, 2017; Gil-Garcia et al., 2016; Lu et al., 2015):

1. Develop and implement sustainable technologies and services that minimize negative environmental impacts and are available to consumers. This area is implemented through product innovations. They involve creating products designed to have minimal environmental impact, are energy efficient, made from renewable or recycled materials, and are easy to recycle at the end of their life. Sustainable services supporting sustainable living are also offered here, e.g., car-sharing systems, solutions for sustainable mobility, household renewable energy, and energy efficiency consulting.
2. Corporate practices consistent with the principles of sustainable development, such as supply chain management (implementation of sustainable development principles throughout the supply chain, from raw materials to the final product, including cooperation with suppliers on a fair trade basis and the selection of raw materials from responsible sources), reliable and sustainable production (minimizing the consumption of energy, water, and raw materials in production processes, reducing waste through recycling and reusing materials, and using cleaner technologies), and sustainable consumption along with corporate social responsibility (CSR that contributes to the development of local communities, environmental protection, and promotion of ecological education).
3. Investments in research and development focused on innovations supporting sustainable development, including eco-technologies and solutions improving resource efficiency (investing in research and development of products and processes that offer sustainable solutions, including clean energy technologies, sustainable materials, and production methods that minimize environmental impact). Cooperation with research institutes is also significant here (partnership with universities, research centers, and startups to support innovation and technology transfer for sustainable development).
4. Standardization and reporting: In this aspect, it is essential to obtain international ecological certificates, such as ISO 14001, Fair Trade, or LEED, which confirm the company's commitment to sustainable development, as well as the regular publication of reports on the impact of the company's activities on the environment, society, and economy, including achievements in the fields of emission reduction, waste management, and resource efficiency.
5. Engaging and educating consumers through sustainable marketing (promoting sustainable products and services through transparent communication about ecological and social benefits, encouraging consumers to make informed choices) and loyalty programs and initiatives (creating loyalty programs and campaigns that reward sustainable consumer choices and promote active participation in environmental activities).

Through its activities, the business sector can significantly contribute to promoting sustainable living. Companies that integrate sustainable development into their strategies and operations not only contribute to environmental protection but also build a stronger market position, increase their competitiveness, and respond to the growing expectations of consumers and business partners regarding sustainable operations.

9.3.3 Education and Information Exchange Sector

Education and information exchange play a key role in promoting sustainable living, enabling people to understand the impact of their actions on the environment and how they can contribute to sustainable development. Education and communication support sustainable living through the following (Griggs et al., 2013; Servaes, 2022; Wu et al., 2018):

1. Integrate education by incorporating sustainability and living topics into curricula at all levels of education, from primary to higher education. This applies to theoretical knowledge and practical skills regarding environmental protection, energy efficiency, sustainable agriculture, and recycling principles. Additional support includes workshops, courses, and training addressed to various social groups to raise awareness and skills in sustainable living.
2. Information and awareness campaigns that increase knowledge about sustainable living and the consequences of human activities on the environment. Here, support is provided by implementing media and social campaigns promoting sustainable practices, such as saving water, waste segregation, using public transport, or reducing meat consumption. These campaigns can use a variety of communication channels, including television, radio, internet, and social media. Influencers and opinion leaders are also used to promote sustainable living and engage more people in environmental protection activities.
3. Providing platforms and tools for easy access to information about sustainable practices and choices. The development of online platforms and mobile applications offering tips on sustainable living, carbon footprint monitoring tools, energy efficiency at home, and recipes for balanced meals is helpful in this area. Open databases and repositories containing educational materials, research, and case studies on sustainable development are also created and are available to teachers, students, researchers, and the general public.
4. Interinstitutional and international cooperation involves building partnerships between educational institutions, non-governmental organizations, the business sector, and public administration to exchange knowledge and experiences in sustainable developmen. Participating in international educational programs and projects that promote global cooperation and exchange of best practices related to sustainable development is also essential.
5. Development of future competencies, i.e., shaping the skills of critical thinking, problem-solving, and innovation in the context of sustainable development, preparing young generations to actively participate in creating

a sustainable future, and also educating in the area of circular economy and green technologies, educating future leaders and innovators capable of creating solutions that support sustainable development.

Education and information exchange are essential to building awareness and engaging society in activities for sustainable living. Providing knowledge, tools, and inspiration enables individual and collective action to protect the environment, promote social justice, and support sustainable economic development.

9.3.4 Individual Choices and Actions

Individual choices and actions are also crucial in promoting sustainable living, as everyday human decisions directly impact the environment, communities, and the economy. These choices and actions contribute to shaping sustainable living (Abson et al., 2017; Griggs et al., 2013; Hannigan, 2022):

1. Conscious consumer decisions, such as choosing organic and certified products (preferring products with ecological certificates such as Fair Trade, Rainforest Alliance, or organic products that are produced sustainably), reducing food waste (planning purchases, using leftover food, and storing food in a way that minimizes food waste), reducing meat consumption (reducing meat consumption in favor of a plant-based diet, which can significantly reduce the carbon footprint of food production), and preferring sustainable energy sources.
2. Adopting sustainable practices at home and work, e.g., recycling, saving water and energy, and using public transport. Energy-saving activities are essential here – using energy-saving devices, switching devices from standby mode, using LED lighting, and insulating the house to reduce energy demand; water management – installing water-saving devices such as aerators, rainwater collection systems, and applying practices that reduce water consumption; recycling and composting – waste segregation, recycling, and composting of organic waste to reduce the amount of waste going to landfills.
3. Sustainable mobility, i.e., using public transport, bicycles, or walking, means choosing means of transport with a lower carbon footprint for daily commuting and choosing ecological vehicles – considering purchasing an electric or hybrid vehicle if using a car is necessary.
4. Engaging in social activities for the environment, such as participation in local ecological initiatives, i.e., active participation in local environmental protection projects, such as tree planting, ecological clean-up campaigns, or community initiatives promoting sustainable development. Support for a sustainable local economy is also essential here, which means buying products from local producers and supporting local businesses that apply the principles of sustainable development.

5. Educate and share knowledge by increasing knowledge of sustainability, using available educational resources to learn more about the impact of individual choices on the environment. Inspire and educating others, i.e., sharing knowledge and experiences about sustainable living with family, friends, and the community to inspire change.

Although each decision may seem insignificant when multiplied by millions of people, it significantly impacts the future. Everyone can contribute to building a more sustainable future through conscious choices and actions. This shows sustainable development involves significant initiatives, policies, and everyday individual decisions and actions.

9.3.5 Cross-sector Cooperation

Achieving sustainable living on a large scale requires cooperation between all the sectors mentioned, each playing its role and working together toward common goals. For example, governments can create favorable regulatory and financial conditions, the private sector can provide innovative solutions and products, educational institutions can shape awareness and knowledge, and individual citizens can make sustainable choices that support and promote these innovations.

Cross-sector collaboration is critical in promoting sustainable living, combining different actors' resources, knowledge, and skills – governments, the business sector, educational institutions, non-governmental organizations, and individual citizens. This approach makes it possible to address challenges related to sustainable development more effectively by using synergies between different sectors. Cross-sector collaboration can support sustainable development in the following forms (Gil-Garcia et al., 2016; Hess, 2014; Servaes, 2022):

1. Public–private partnerships: Governments can work with businesses to finance and implement infrastructure projects (development of sustainable infrastructure), such as renewable energy sources, efficient public transport, and waste management systems. The private sector also often has the resources and know-how necessary to develop technological innovations, which public policies and financing can support.
2. Cooperation between non-governmental organizations and businesses: NGOs can work with companies to create and implement sustainability programs supporting sustainable practices in supply chains and internal operations. NGOs can also help companies obtain eco-certifications and develop transparent sustainability reports.
3. Cooperation between educational institutions and other sectors: Universities and research institutions can collaborate with businesses and governments on scientific research (R&D) focused on solutions for sustainable development. Educational institutions can also work with companies and non-governmental organizations to create education and training programs that prepare future generations to take on sustainable development challenges.

4. Joint initiatives and dialogue platforms: This area supports creating platforms for dialogue and knowledge exchange (multistakeholder platforms) that connect different sectors and enable cooperation on common sustainable development goals. Joint projects and awareness campaigns are also organized for the general public to promote sustainable practices and living.
5. Engaging the public and cooperating with local communities: Cross-sector collaboration can support the development of community projects such as local recycling systems, green infrastructure initiatives, and educational programs that directly engage residents.

Cross-sector collaboration is essential to achieving sustainable development because no single sector alone can address the challenges of environmental protection, social justice, and economic growth. By joining forces, different sectors can jointly create innovative Eco-Tech solutions that contribute to building a more sustainable future.

9.4 RENEWABLE ENERGY FOR A SUSTAINABLE LIVING

Combining sustainable living with renewable energy is crucial in reducing human activity's negative impact of human activity on the environment. By combining sustainable living with renewable energy, we contribute to protecting the environment and reducing greenhouse gas emissions, support the development of a green economy, and promote technological innovations that can benefit future generations.

9.4.1 Installation of Renewable Energy Systems in Individual Households

Installing renewable energy systems in households is a critical element of sustainable living and energy transformation, enabling not only the reduction of carbon footprint but also contributing to increased energy independence and lower energy bills.

One renewable energy system is solar panels (photovoltaics), which convert sunlight directly into electricity, i.e., produce green energy. They can be installed on the roofs of houses, residential buildings, and public buildings, in gardens, or as elements integrated with buildings (e.g., photovoltaic roof tiles). Their use brings benefits in the form of lower electricity bills, the ability to sell surplus energy to the grid, and an increase in real estate value (Sampaio and González, 2017). Additionally, many countries offer subsidies, tax breaks, or other financial support for households installing solar panels.

Another system is wind turbines. The installation of small wind turbines is possible on private or community land, especially in areas with high potential and suitable wind conditions, to produce electricity for domestic needs. Households benefit here by generating their electricity, reducing dependence on external suppliers, and the possibility of obtaining subsidies for the energy generated (Ellabban et al., 2014). However, in this case, appropriate permits are usually required, and the location must be appropriately selected in terms of the strength and direction of the winds.

Another system is heat pumps, which use energy from the ground, water, or air to heat and cool buildings energy-efficiently. They can also heat domestic water.

They are particularly effective when used in well-insulated buildings and can significantly reduce conventional energy consumption, reduce CO_2 emissions, and demonstrate long-term operational cost savings (Moriarty and Honnery, 2012). In this area, many government and local programs are available to support the installation of heat pumps in homes.

Renewable energy systems also include solar thermal collectors to heat utility water and support central heating systems. Like solar panels, they can be mounted on roofs or other sunny places. They contribute to reducing the consumption of fossil fuels for heating, lowering energy bills, and increasing the energy efficiency of buildings (Ellabban et al., 2014). In their case, similar to photovoltaic panels, there are various forms of obtaining financial support for this investment.

Biogas, as a local source of energy production, is also a system for obtaining renewable energy. However, in this case, its installation requires appropriate knowledge and compliance with safety standards. Small biogas installations can convert organic waste into biogas, which can then be used for cooking, heating, or electricity production (Scarlat et al., 2018). This reduces organic waste and greenhouse gas emissions.

Installing renewable energy systems in households requires an initial investment but offers long-term economic and environmental benefits. By choosing renewable energy, households reduce their energy bills, contribute to the fight against climate change, and promote sustainable living by using eco-tech solutions.

9.4.2 Using Renewable Energy for a Sustainable Living

Using renewable energy in everyday life is vital to transitioning to a more sustainable way of living. Integrating renewable energy into everyday life not only helps reduce the negative impact of human activity on the environment but can also lead to financial savings in the long run. By making conscious consumer choices and actively promoting green energy, everyone can contribute to accelerating the energy transition and increasing sustainability at local and global levels.

Households can benefit from green energy tariffs in this respect. By choosing to use the services of renewable energy suppliers, energy production from sources such as wind, sun, or biomass is supported. This is one of the easiest ways to increase the share of renewable energy in everyday energy consumption without requiring direct investment in infrastructure (Ellabban et al., 2014). Additionally, by choosing green tariffs, consumers can put pressure on the energy market, promoting the development of renewable sources and increasing their share in the energy mix.

Electric mobility is also essential for individual sustainability. Using electric cars, bicycles, or scooters charged with energy from renewable sources reduces dependence on fossil fuels and greenhouse gas emissions. Using charging stations powered by renewable energy or installing one's charging station at home with a solar panel ensures an ecological use cycle of the electric vehicle (Sampaio and González, 2017).

In everyday life, using solar energy through portable solar chargers to power mobile phones, laptops, and other small devices is also possible. Installing solar garden lamps or lighting systems that use solar energy to illuminate outdoor spaces is also beneficial. Using solar collectors to heat domestic water can significantly reduce electricity or gas consumption (Sampaio and González, 2017).

Social support for local renewable energy projects is essential. It is possible to join local energy cooperatives that invest in renewable energy projects, enabling co-ownership and use of green energy. Membership in such cooperatives allows the use of green energy and the sharing of profits from sales. It also contributes to increasing social acceptance of renewable energy projects and increases community members' ecological and energy awareness.

Participating in crowdfunding of renewable energy projects, including local wind farms, solar farms, or small hydropower plants, is also possible. Investing in such projects directly contributes to increasing the share of green energy and promotes innovative solutions in renewable energy sources. By participating in crowdfunding, individual investors can feel greater involvement and influence in developing sustainable energy in their environment(Moriarty and Honnery, 2012).

It is also necessary to actively support or initiate community projects focused on renewable energy, such as photovoltaic installations in schools, health centers, or community areas. It is also possible to purchase a green certificate or shares in renewable energy. This allows the offset of one's own energy consumption by buying guarantees of origin, which confirm that the equivalent of the energy consumed was produced from renewable sources (Sampaio and González, 2017). This is a way to support sustainable energy production, even if using renewable energy directly is impossible. Some projects offer the opportunity to purchase shares in renewable energy installations, ensuring long-term benefits from producing and selling green energy.

Sustainable living can also be promoted through appropriate energy management. For this purpose, investing in energy-saving devices with a high energy class that minimizes energy consumption is beneficial – they use less energy to perform the same job compared to less efficient models. Similarly, replacing traditional light bulbs with energy-saving light sources such as LEDs offers longer life and significantly lower energy consumption (Al-Ali et al., 2011). Intelligent energy management systems are also used, i.e., the introduction of building automation systems and smart thermostats to optimize energy consumption, reduce waste, and increase energy efficiency.

Intelligent home management systems (Smart Home) function similarly. Intelligent thermostats automatically adjust heating and cooling to actual needs, reducing energy consumption. Additionally, these systems enable remote control of lighting, heating, and household appliances, optimizing energy consumption even during the absence of household members (Al-Ali et al., 2011).

Equally important activities are educating and promoting knowledge about renewable energy. In this regard, conducting or participating in educational activities about the advantages and benefits of using renewable energy, including workshops, online courses, and information campaigns, is possible. It is also important to publish and share personal experiences related to using renewable energy at home and in everyday life, which may inspire others to take similar actions and use Eco-Tech solutions in their sustainable living.

9.5 CHALLENGES AND BARRIERS TO ECO-TECH SOLUTIONS

Although eco-tech solutions offer significant benefits for the environment and society, they also encounter challenges and barriers that may inhibit their widespread adoption and implementation.

The first thing to consider here is costs. Many eco-tech technologies, such as solar panels, wind turbines, and intelligent energy management systems, require significant upfront investments, which can be a barrier for households and small businesses (Undheim, 2023). In addition, lack of access to appropriate sources of financing or subsidies may hinder investments in eco-tech solutions, especially in lower-income regions.

Lack of awareness and lack of knowledge are also barriers. Low public awareness of the benefits of eco-tech and sustainable living may limit interest and adoption of these solutions. In turn, the lack of access to reliable information and guides on selecting and implementing eco-tech solutions may be a barrier for many people (Abdollahi, 2016).

Technological and infrastructural limitations are also significant. Some eco-tech solutions are still in development and are so immature that they may not be practical or reliable enough to meet user needs. Unfortunately, in some areas, there is a lack of appropriate infrastructure to support eco-tech solutions, e.g., charging stations for electric vehicles or distribution networks capable of managing renewable energy (Undheim, 2023), which prevents the use of opportunities.

Also, complicated or unclear legal regulations may hinder the development and implementation of eco-tech technologies, and insufficient political support and lack of incentives, such as tax breaks or subsidies, may even inhibit investments in eco-tech. Another problem is the natural social resistance to change and attachment to traditional lifestyles, making adopting new, sustainable practices challenging (Gil-Garcia et al., 2016). Differences in access to eco-tech technologies caused by social and economic inequalities may further deepen divisions and limit equitable, sustainable development.

Economic sustainability in the form of an expected return on investment is also needed, as in some cases, the long-term financial benefits of investing in eco-tech may not be immediately visible, discouraging potential investors (Undheim, 2023). Additionally, there is the investment risk associated with the dynamically changing eco-tech market and the potential risk associated with new, untested solutions.

Overcoming these challenges requires joint action by governments, the private sector, educational institutions, and local communities. Integrated strategies are therefore necessary, including financial support, education, infrastructure development, flexible regulation, and the promotion of public awareness and engagement to enable wider adoption and implementation of eco-tech solutions for sustainable living.

9.6 PROSPECTS FOR ECO-TECH SOLUTIONS

Despite existing challenges and barriers, the prospects for eco-tech solutions in sustainable living are promising. Technological progress, growing ecological awareness of society, and changes in global and local policies create an environment conducive to further developing and integrating ecological technologies. The key trends that may shape the future of eco-tech solutions are (Yordanova, 2022):

- Technological innovations and cost reduction: Advances in science and technology lead to the development of new, more effective, and cheaper eco-tech solutions. The falling costs of technologies such as photovoltaics and energy storage make them more accessible to consumers.
- Increased integration of energy systems: The development of smart energy grids and energy management systems will enable better use of renewable energy, optimization of consumption, and integration of various energy sources into more sustainable and flexible energy systems.
- Sustainable mobility: The rise of electric vehicles, the development of charging infrastructure, and innovations in public transport and urban mobility will continue the paradigm shift in sustainable mobility.
- Developing sustainable cities and communities: Concepts such as "green construction", "zero-emission cities", and "smart cities" are gaining importance, promoting the integration of eco-tech solutions in spatial planning, construction, and management of urban resources.
- Policies and regulations supporting sustainable development: Global and local climate agreements, environmental policies, and regulatory initiatives will continue to support the development and implementation of eco-tech solutions by stimulating investment, research and development, and setting environmental standards.
- Growing awareness and social involvement: Increasing ecological awareness among consumers and citizens, especially younger generations, drive demand for sustainable products and services, motivating companies and institutions to invest in eco-tech solutions.
- Increasing importance of the circular economy: Business models based on circular economy principles, promoting the reduction, reuse, recycling, and regeneration of materials, will increasingly influence production, consumption, and waste management.
- International and cross-sector cooperation: The increase in international cooperation and cross-sector partnerships to solve global environmental and climate challenges creates opportunities for exchanging knowledge, technologies, and best practices in eco-tech.

Therefore, the prospects for eco-tech solutions are optimistic, and their development and implementation will be vital to achieving sustainable development and building resilience to climate change. However, this requires joint action on many levels – from individual choices to global political strategies.

9.7 CONCLUSIONS

Eco-technologies and renewable energy are vital in global efforts to combat climate change and environmental degradation. By reducing dependence on fossil fuels, reducing greenhouse gas emissions, and promoting resource efficiency, these innovative solutions prioritize protecting our planet for future generations. Their development and implementation are crucial to achieving the sustainable development goals and ensuring long-term ecological, economic, and social stability.

Active public participation in promoting sustainable living is essential to achieving broad-based change. Conscious consumption decisions, support for sustainable local and global initiatives, and commitment to pro-ecological practices are the foundation for a sustainable future. Education and awareness-raising about the benefits of eco-technologies and renewable energy are crucial for mobilizing society to take action and support green innovation.

The future of green technologies looks promising, with the potential to significantly impact the world's energy future. Innovations in renewable energy, energy efficiency, electric mobility, and the circular economy are opening up new opportunities for sustainable development. As technology costs fall and public awareness and political support increase, eco-technologies are becoming more available and attractive, which could accelerate their wide-scale adoption.

Despite the positive prospects, challenges such as the need for further research, infrastructure development, and overcoming regulatory and economic barriers must be addressed to exploit the potential of eco-technologies fully. Collaboration between governments, the private sector, academic communities, and citizens is critical to overcoming these obstacles and achieving the common goal of a sustainable future.

Eco-technologies and renewable energy are a response to the climate crisis and a path to sustainable development that combines technological progress with environmental protection. Adopting and promoting sustainable living by societies worldwide is crucial to building a resilient and sustainable world for future generations.

REFERENCES

Abdollahi, M. (2016). The Impact of Sustainable Development on Eco-Tech Architecture. *Bulletin de la Société Royale des Sciences de Liège*, 85, 1371–1377.

Abson, D. J., Fischer, J., Leventon, J., Newig, J., Schomerus, T., Vilsmaier, U., ... Lang, D. J. (2017). Leverage points for sustainability transformation. *Ambio*, 46, 30–39.

Ahlborg, H., Ruiz-Mercado, I., Molander, S., & Masera, O. (2019). Bringing technology into social-ecological systems research—motivations for a socio-technical-ecological systems approach. *Sustainability*, 11(7), 2009.

Al-Ali, A. R., El-Hag, A., Bahadiri, M., Harbaji, M., & El Haj, Y. A. (2011). Smart home renewable energy management system. *Energy Procedia*, 12, 120–126.

Barr, S., Shaw, G., Coles, T., & Prillwitz, J. (2010). 'A holiday is a holiday': Practicing sustainability, home and away. *Journal of Transport Geography*, 18(3), 474–481.

Black, I. R., & Cherrier, H. (2010). Anti-consumption as part of living a sustainable lifestyle: daily practices, contextual motivations and subjective values. *Journal of Consumer Behaviour*, 9(6), 437–453.

Caradonna, J. L. (2022). *Sustainability: A history*. Oxford University Press.

Chaney, D. (2012). *Lifestyles*. Routledge.

Doppelt, B. (2017). *Leading change toward sustainability: A change-management guide for business, government and civil society*. Routledge.

Ellabban, O., Abu-Rub, H., & Blaabjerg, F. (2014). Renewable energy resources: Current status, future prospects and their enabling technology. *Renewable and Sustainable Energy Reviews*, 39, 748–764.

Gil-Garcia, J. R., Zhang, J., & Puron-Cid, G. (2016). Conceptualizing smartness in government: An integrative and multi-dimensional view. *Government Information Quarterly*, 33(3), 524–534.

Griggs, D., Stafford-Smith, M., Gaffney, O., Rockström, J., Öhman, M. C., Shyamsundar, P., ... Noble, I. (2013). Sustainable development goals for people and planet. *Nature*, 495(7441), 305–307.

Hannigan, J. (2022). *Environmental sociology*. Taylor & Francis Group.

Hassan, A. M., & Lee, H. (2015). Toward the sustainable development of urban areas: An overview of global trends in trials and policies. *Land Use Policy*, 48, 199–212.

Hess, D. J. (2014). Sustainability transitions: A political coalition perspective. *Research Policy*, 43(2), 278–283.

Holden, E., Linnerud, K., & Banister, D. (2014). Sustainable development: Our common future revisited. *Global Environmental Change*, 26, 130–139.

Huckle, J., & Wals, A. E. (2015). The UN decade of education for sustainable development: Business as usual in the end. *Environmental Education Research*, 21(3), 491–505.

Lu, Y., Nakicenovic, N., Visbeck, M., & Stevance, A. S. (2015). Policy: Five priorities for the UN sustainable development goals. *Nature*, 520(7548), 432–433.

Moriarty, P., & Honnery, D. (2012). What is the global potential for renewable energy?. *Renewable and Sustainable Energy Reviews*, 16(1), 244–252.

Nelson, V. C. (2011). *Introduction to renewable energy*. CRC Press.

Sampaio, P. G. V., & González, M. O. A. (2017). Photovoltaic solar energy: Conceptual framework. *Renewable and Sustainable Energy Reviews*, 74, 590–601.

Scarlat, N., Dallemand, J. F., & Fahl, F. (2018). Biogas: Developments and perspectives in Europe. *Renewable Energy*, 129, 457–472.

Servaes, J. (2022). Communication for development and social change. In G. Gonçalves, E. Oliveira (Eds.), *The Routledge handbook of nonprofit communication* (pp. 23–31). Routledge.

Sobel, M. E. (2013). *Lifestyle and social structure: Concepts, definitions, analyses*. Elsevier.

Tomislav, K. (2018). The concept of sustainable development: From its beginning to the contemporary issues. *Zagreb International Review of Economics & Business*, 21(1), 67–94.

Twidell, J. (2021). *Renewable energy resources*. Routledge.

Undheim, T. (2023). *Eco Tech: Investing in Regenerative Futures*. Taylor & Francis Group.

Weaver, P., Jansen, L., Van Grootveld, G., Van Spiegel, E., & Vergragt, P. (2017). *Sustainable technology development*. Routledge.

Wu, C. H., Tsai, S. B., Liu, W., Shao, X. F., Sun, R., & Wacławek, M. (2021). Eco-technology and eco-innovation for green sustainable growth. *Ecological Chemistry and Engineering S*, 28(1), 7–10.

Wu, J., Guo, S., Huang, H., Liu, W., & Xiang, Y. (2018). Information and communications technologies for sustainable development goals: state-of-the-art, needs and perspectives. *IEEE Communications Surveys & Tutorials*, 20(3), 2389–2406.

Yordanova, Z. (2022). Friends with benefits or enemies–eco-innovation and emerging technology. *SpecialusisUgdymas*, 2(43), 3516–3528.

10 Sustainable Agriculture and Food Technologies

Leszek Szczupak

10.1 INTRODUCTION

Sustainability in agriculture refers to practices that make it possible to meet today's needs for quality food production, environmental protection, and public health, while ensuring the preservation of the ability of current and future generations to meet their needs. It is characterized by a balance between the economic, environmental, and social aspects of food production. Sustainable agriculture practices include agroecology, organic farming, permaculture, biodiversity conservation, and the development of precision agriculture, among others. Sustainability in agriculture is based on a number of principles that are outlined below.

10.2 ENVIRONMENTAL PROTECTION

Agriculture in accordance with the principles of sustainable development should minimize the negative impact on the environment through, among other things, sustainable water management, which can be carried out in several ways (Kolodziejczak Anna, 2012),

- The land should be properly irrigated to avoid excessive leaching of nutrients into the groundwater. The use of drip or canal irrigation systems can help reduce water loss and limit water erosion. Using these practices helps maintain its proper quality.
- Effective irrigation management is of particular importance. Farmers should monitor the water needs of plants and supply the right amount of water. The implementation of soil moisture monitoring systems and the use of automatic irrigation systems are expected to help reduce excessive water use.
- Agricultural producers should strive to avoid excessive use of fertilizers and pesticides (Mateo-Sagasta et al., 2017). The use of excessive chemicals can contribute to groundwater and surface water pollution. Farmers should follow recommendations on the use of chemical fertilizers and crop protection products.
- Farmers should also promote sustainable practices in crop protection to minimize the use of chemicals. An important element is the protection and

DOI: 10.1201/9781003475989-12

restoration of coastal areas. Agriculture and cultivation along streams, rivers, and other coastal areas can significantly affect the pollution of these waters.
- Agricultural producers can help protect the environment by reforesting uncultivated farmland, creating conservation areas, and organizing biodiversity activities. These activities help filter water and maintain ecological balance.
- Protecting these areas through permanent vegetation plantings can significantly reduce erosion and pollution of these waters.

Introducing these practices in agriculture can help minimize negative environmental impacts, maintain sustainable water management, and protect soil from erosion and degradation.

Of particular importance in applying the principles of sustainable agriculture is the monitoring and testing of soil quality by farmers. Systematic monitoring of soil pH, chemical composition, and other parameters can help identify problems before they become more serious. Based on this information, farmers can make informed decisions about the use of fertilizers and other substances to maintain optimal soil conditions. Keep in mind that the above practices should be tailored to the local environment and the soil conditions present. Implementing restrictions on the use of pesticides and fertilizers can be done in accordance with the principles of sustainable agriculture, which include crop diversification. One of the most important principles of sustainable agriculture is to avoid monocultures, i.e., growing one crop species over a large area. Instead, it is advisable to introduce the distribution of different plant species in fields, which will allow pests and diseases to be controlled without the need for chemical protection. Another way to reduce the use of pesticides is to practice mixed cropping, that is, growing several plant species in the same field. Such crops are more resistant to infection and pests, resulting in less need for pesticides. Another principle of sustainable agriculture recommends keeping the soil healthy by increasing its organic structure. This can be achieved through the use of natural fertilizers such as compost, manure, and green manures. The use of such natural fertilizers helps to improve soil structure, improve water retention, and increase the availability of nutrients for plants. Balancing ecosystems is also a very important principle. Introducing elements such as habitats for friendly insects and birds or natural buffer strips into agriculture helps maintain the balance of ecosystems. This, in turn, minimizes the occurrence of pests that can threaten crops and require the use of pesticides. Also, by regularly monitoring agricultural fields, the presence of pests or diseases can be detected early and appropriate protective measures can be taken without the need for massive use of chemicals. Undoubtedly, a key action is proper training, counseling, and education of farmers in sustainable agriculture. Familiarizing agricultural producers with alternative methods of cultivation, plant protection, and fertilization will allow the introduction of restrictions on the use of pesticides and fertilizers while maintaining the productivity and health of agricultural ecosystems.

There are a number of ways to promote sustainable farming practices among agricultural producers. Education and raising awareness raising of farmers in this regard is particularly important (agriculture.ec.europa.eu/common-agricultural-policy_en). Environmental education can be increased by holding workshops, conferences, and seminars where experts and farmers share knowledge about sustainable farming practices. Educational campaigns, instructional videos, and informational materials can also be created to ensure that farmers have access to information about best practices in this area. Another element is to legislate policies and regulations that promote sustainable cultivation practices, for example, through the introduction of organic certification systems, financial incentives for farmers who use sustainable principles on their farms, or regulations that limit the use of chemicals. Another element that supports the development of sustainable farming principles is working with local communities, NGOs, companies, and other stakeholders to promote sustainable growing practices. Joint projects, information campaigns, or support for local farming initiatives can be organized. The creation of platforms and networks that allow farmers to share experiences, best practices, and innovations related to sustainable farming practices should also be supported. Study tours and meetings of farmers from different regions can also be organized to enable mutual learning. In addition to informing and educating farmers, efforts should also be made to inform and educate consumers about the benefits of buying and eating food from sustainable agricultural practices. Labels, certifications, and advertising campaigns that highlight sustainable aspects of production can be created. Another element is to financially support investment in farming activities based on sustainable agricultural practices by providing funds, grants, and low-interest loans (Kolodziejczak Anna, 2012), You can also create training and advisory programs for farmers to help them make changes in their practices. It is important to promote sustainable farming practices as a holistic and comprehensive process that encompasses all aspects of agricultural production, from renewable energy sources and biodiversity conservation to minimizing waste and optimizing the use of natural resources.

10.3 SOCIAL RESPONSIBILITY

Sustainable agriculture should take into account and respect workers' rights. Sustainable agriculture should be based on labor standards and principles that ensure safety and decent working conditions for workers. There should be minimum requirements for maximum working hours, health care, availability of drinking water and sanitation, and restrictions on the use of hazardous chemicals. It is important to provide workers with adequate training and education to enable them to perform their duties safely and responsibly. This can include training on safety, the use of agricultural tools and machinery, and knowledge of sustainable cultivation. Workers in the agricultural sector should have access to health care, including doctors and specialists. Employers should provide health insurance for their employees and arrange for regular check-ups among their workers. Agricultural workers should be paid fairly for their work. Employers should provide wages that comply with local standards and laws, and provide opportunities for promotion and professional development. It is important to monitor and inspect working conditions in the agriculture sector to make sure they meet established standards. Certification organizations can also award certificates to farmers who respect workers' rights, which helps promote fair practices and responsible labor attitudes.

Sustainable agriculture should also take into account the needs of local communities (Bakshi et al., 2021). It is important to involve these communities in agricultural decision-making, take into account their opinions, and promote a sustainable economy based on local resources. Implementing these methods can help ensure that the sustainable agriculture that is introduced will respect workers' rights and contribute to the development of rural communities. It will listen to and involve the local community, and contribute to improving living conditions in rural areas.

Sustainable agriculture plays a key role in improving living conditions in rural areas by allowing the production of healthy and nutritious food. This is extremely important for rural areas where there may be limited access to grocery stores or supermarkets. High-quality, locally produced food can contribute to better nutrition and health for local rural communities. Sustainable agriculture promotes a variety of crops and animal husbandry that not only increase productivity but also minimize production costs. As a result, farmers can increase income both through higher yields and with less spending on pesticides and other chemicals. This additional income can help raise the standard of living in rural areas through better education, health care, and other community needs. Sustainable agriculture aims to minimize the negative impact on the environment. By using techniques such as pre-seeding, organic fertilization, and reducing water, air, and soil consumption, agriculture can contribute to protecting biodiversity, keeping groundwater clean, and improving soil quality. This contributes to the long-term development of rural areas and improves the living conditions of local communities. This agriculture can contribute to the creation of new jobs in rural areas. Improved agricultural infrastructure, technology development in organic farming, and local farmers' markets can create employment opportunities for people living in these areas. This contributes to reducing poverty and migration in rural areas and ultimately improves living conditions for communities. All in all, the principles of sustainable agriculture can help improve living conditions in rural areas by providing healthy food, increasing farmers' incomes, protecting the environment, and creating new jobs.

10.4 ECONOMIC EFFICIENCY

Agriculture should be based on efficient and sustainable business practices that ensure stability and economic viability for farmers and food producers. Sustainable agriculture refers to farming practices that aim to meet current food needs while ensuring the ability of future generations to meet their own needs. This means that these practices should be ecologically, socially, and economically sustainable. In the context of efficiency and sustainable business practices, such agriculture provides an opportunity to achieve stability and economic viability for farmers and food producers. Sustainable farming practices can help reduce production costs by reducing the use of energy, water, and other resources. For example, the use of organic fertilizers can reduce the cost of purchasing chemical fertilizers, and the spread of precision agriculture can help avoid excessive use of pesticides, leading to lower costs and consequently an improved financial situation for agricultural producers. In addition, sustainable agriculture can significantly improve crop productivity and quality. The use of agricultural technologies that minimize the risk of pests and diseases consequently leads to healthier crops. This, in turn, offers the possibility of increased income for farmers and food producers. In the long term, sustainable agriculture can ensure the sustainability of the food production system, eliminating negative environmental impacts. Improving soil quality, protecting biodiversity, and reducing greenhouse gas emissions are just a few examples of how sustainable farming practices can help protect the planet. As a result, farmers and food producers who follow sustainable business practices have a better chance of long-term economic success. Using sustainable practices in agriculture and combining them with methods to improve efficiency provides the opportunity for stability and profit, while increasing environmental care and contributing to development and community (Klekotko Marta, 2015).

The introduction of the principles of sustainable development in agriculture is aimed at ensuring the long-term assurance of food production while minimizing the negative effects on the environment and society (Komorowska, D. 2014). An important element in the development of sustainable agriculture is local closed-loop food systems. These are food production models that aim to minimize environmental impact on the one hand and maximize support for the local community on the other. They involve the production, processing, and distribution of food within a given region, without heavy reliance on imports and exports of food products. Methods to support local closed-loop food systems can include promoting organic and/or traditional crops, such as organic farming, horticulture, or community gardening, and supporting local farmers through purchases directly from producers, including at local farmers' markets. Another element in spreading the principles of sustainable agriculture is to support and promote community initiatives such as local buying groups (CSA – Community Supported Agriculture). This involves a system in which a consumer, by paying a subscription or financial contribution, receives a regular supply of food straight from a local farmer. Another element is the promotion of local food processing, such as the making of cheeses, juices, jams, and other regional specialties. Related to this is the need for workshops, training, and educational campaigns that raise public awareness of the benefits of the local food system and the promotion of healthy eating principles. Public policies that support local food systems are also particularly important. This can include redirecting public funds to support local production, creating regulations for organic farming, facilitating the sale of local food products, or promoting them in public institutions such as schools or hospitals. Supporting local, closed-loop food systems is important because it helps reduce greenhouse gas emissions, reduce water consumption, minimize waste and improve food quality. In addition, such systems help increase the food sovereignty of the local community and support the local economy.

Another element of sustainability is promoting diversification of protein intake and changing eating habits. These are key elements that can contribute to reducing the environmental footprint of the food system. They include the principles of introducing dietary diversity and contributing to reducing the consumption of meat and other animal products (Rachna and Bhambri, 2021). Diversifying protein intake means swapping high-energy and environmentally intensive animal products for plant-based protein sources, such as legumes, nuts, seeds, and grain products. These foods are rich in protein and provide many other essential nutrients, such as fiber, vitamins, and minerals. Promoting a diverse protein intake can lead to a reduced ecological carbon footprint, reduced use of natural resources, and improved health. Meat production is intensive in terms of water consumption, greenhouse gas emissions, and environmental degradation. Reducing meat consumption and replacing it with plant-based alternatives, such as legumes, vegetables, and fruits, can effectively lower the environmental footprint. In addition, reducing meat consumption can bring health benefits, such as reducing the risk of heart disease, obesity, and diabetes. To effectively promote diversification of protein intake and change eating habits, action is needed at multiple levels. This can be achieved through education and information campaigns that encourage people to choose more balanced dietary options. Support from public institutions, such as schools, hospitals, and workplaces, can also play an important role in promoting healthy and balanced dietary choices. Moving toward greater consumption of plant-based protein and reducing meat consumption can help reduce natural resource use and greenhouse gas emissions. Education and support from public institutions are key to promoting these changes.

At the same time, with the development of sustainable agriculture principles, innovative technologies should be introduced and promoted in the food industry that are designed to lead to more sustainable food production and distribution. For example, the introduction of smart systems for tracking and monitoring the flow of food commodities can help identify and minimize food waste by better managing their delivery and expiration dates. Other innovative technological solutions can include new methods of storing, transporting, and packaging food that minimize waste and ensure a longer shelf life for food products. One example of such solutions is the use of packaging based on plant-based polymeric materials. Traditional food packaging often contains non-renewable plastic, which leads to environmental pollution. Replacing traditional plastic packaging with biodegradable packaging made from plant-based polymers such as polylactic acid (PLA) or cellulose significantly reduces environmental pollution because such packaging is biodegradable and less harmful (Poushpi Dwivedi et al., 2019). Plant polymers have the ability to decompose in the natural environment, which means that they can biodegrade under the influence of microorganisms and environmental factors. This is important for waste reduction and environmental protection. The production of plant-based polymers typically has a lower impact on CO_2 emissions than traditional plastics. The plants from which these polymers are made absorb carbon dioxide during growth, which contributes to neutralizing CO_2 emissions. The production of plant-based polymers can be more economical than the production of traditional plastics, due to the greater availability of plant-based raw materials such as sugar or starch. Plant-derived polymers have similar properties to traditional plastics, making it easier to process them into different forms and shapes, such as films, bottles, bags, and so on. These polymers are safer for health, as they do not contain toxic or harmful substances for organisms (Poushpi Dwivedi et al., 2019). Packaging made from plant-derived polymers is available in different variants that can be used for various applications, such as food packaging, cosmetic packaging, medical packaging, and so on. The production and processing of plant-based polymers require less energy consumption compared to the production of traditional plastics, which benefits the environment. Although plant-based polymers can be biodegradable, some of them can also be recycled. This means they can be reprocessed and used in the production of new products, reducing raw material consumption and waste generation.

10.5 SUMMARY

Sustainable agriculture is an approach to food production that takes into account the protection of the environment, human health, community development, and animal welfare. The article discusses the principles of sustainable agriculture and its importance in ensuring food security, reducing greenhouse gas emissions, and protecting biodiversity. It also mentions the need to change the model of agricultural production to a greener and more efficient one that will promote the sustainable development of the agricultural sector. Measures taken within the framework of sustainable agriculture include the use of agroecological techniques, the introduction of environmentally friendly vegetation, and the promotion of local products, among others. The need to combine the development of sustainable agriculture with the creation of consumer awareness in this regard is also presented. Local communities have the opportunity to support sustainable agriculture through the promotion of local agricultural products, support for farmers in the evaluation of farming practices, or public education on the benefits of sustainable agriculture.

The article's authors also emphasize the need for financial support for farmers practicing sustainable agriculture and the need for a legislative framework that encourages the development of this form of production. Cooperation between local communities with local governments, and NGOs can help promote sustainable agriculture and generate benefits for both the environment and local communities.

An example of the use of ecological packaging and the impact of its use on agricultural development and on improving environmental protection is also presented. The introduction of green packaging is crucial to the development of sustainable agriculture. Improving soil quality, protecting the environment, and increasing production efficiency are key goals of sustainable agriculture, and green packaging can make an important contribution to achieving these goals.

An example of the use of ecological packaging and the impact of its use on agricultural development and on improving environmental protection is also presented. The introduction of green packaging is crucial to the development of sustainable agriculture. Improving soil quality, protecting the environment, and increasing production efficiency are key goals of sustainable agriculture, and green packaging can make an important contribution to achieving them. Introducing green packaging into sustainable agriculture requires the cooperation of producers, suppliers, and consumers. Investing in new technologies and packaging solutions can bring multiple benefits to the economy, health, and the environment. Therefore, green packaging plays an important role in the development of sustainable agriculture.

The article's conclusions point to the need for further efforts in sustainable agriculture to ensure the sustainability of the agrarian sector and environmental protection.

REFERENCES

Anne-Kristin Løes. (2019). *The role of organic farming in sustainable agricultural production.* PJH 300: Sustainable Production Systems in Agriculture. www.agriculture.ec.europa.eu/common-agricultural-policy_en

Bakshi, P., Bhambri, P., & Thapar, V. (2021). A review paper on wireless sensor network techniques in Internet of Things (IoT). In International Conference on Contemporary Issues in Engineering & Technology.

Benkeblia Noureddine. (2022). *Climate change and agriculture: Perspectives, sustainability and resilience.* John Wiley & Sons Ltd.

Durham, T., & Mizik, T. (2021). Comparative economics of conventional, organic, and alternative agricultural production systems. *Economies*, 9(2), 64.

Dwivedi, P., Mishra, P. K., Mondal, M. K., & Srivastava, N. (2019). Non-biodegradable polymeric waste pyrolysis for energy recovery. *Heliyon,* 5,123–143.

Kaur, S., & Chauhan, B. S. (2023). *Chapter 2 – Challenges and opportunities to sustainable crop production.* Plant Small RNA in Food Crops.

Klekotko Marta (2015). Civil society, and sustainable rural development A cognitive approach. www.researchgate.net/publication/347380871

Kolodziejczak Anna. (2012). *Development of sustainable agriculture in Poland.* Regional Development and Regional Policy.

Komorowska, D. (2014). The regularities of agricultural development and the development of modern agriculture. *Zeszyty Naukowe SGGW W Warszawie - Problems of World Agriculture*, 14(3), 42–55.

Mateo-Sagasta, J., Zadeh, S. M., Turral, H., & Burke, J. (2017). *Agricultural water pollution: A global overview, executive summary.* FAO; Colombo, Sri Lanka: International Water Management Institute (IWMI). CGIAR Research Program on Water, Land and Ecosystems (WLE).

Rachna, C., & Bhambri, P. (2021). Various approaches and algorithms for monitoring energy efficiency of wireless sensor networks. In *Lecture notes in civil engineering* (Vol. 113, pp. 761–770). Springer.

Transforming Food and Agriculture to Achieve. (2018). https://www.fao.org/3/I9900EN/i9900en.pdf.

Transforming Food and Agriculture to Achieve the Sustainable Development Goals: 20 interrelated actions to guide policymakers, Technical Reference Paper. (2018). http://www.fao.org/3/I9900EN/i9900en.pdf.

11 Smart Solutions for Sustainable Farming and Food Systems

Anandhi Kandhaswamy, Umadevi Pongia, and P.F. Mishel

11.1 INTRODUCTION

Considering a staggering almost 800 million individuals being hungry now and an additional two billion predicted by 2050, it is imperative that the world's food and agriculture systems improve in order to enhance food production in a sustainable manner. Many obstacles, such as soil salinity in arid regions, hinder agricultural output, which lowers crop productivity (Wibowo et al., 2016). The amount and quality of crops are also impacted by the climate, and it may make the soil more susceptible to desertification (Marvin et al., 2022). Surveying land resources for use in agricultural development in dry locations is therefore the main focus. One of the key pillars of national income in emerging nations is the agriculture industry. Therefore, a key concern in bolstering the national economies of those nations is the adoption of new technology to enhance the agriculture sector.

Digital technologies, defined as having economic, social, and environmental elements, have been hailed as a potentially revolutionary approach to improve the performance and sustainability of agricultural production systems. The agrifood industry may help farmers, customers, and society at large by becoming more inclusive, efficient, and environmentally sustainable through the use of machine learning and artificial intelligence (AI) (Xiong et al., 2018). Digital technologies can help to raise on-farm productivity, improve resource use efficiency, and support climate resilience. Improvements in primary production, supply chain and logistics performance, and reductions in food losses and waste are especially notable, provided that digital technologies can be implemented. Moreover, the COVID-19 pandemic has increased attention to both the need for and utility of digital technologies including within the agrifood sector and has catalysed the introduction and adoption of digital technologies (Bhambhani et al., 2021). While developed countries have led much of the innovation and early adoption of digital technologies, the potential impact of these technologies on agriculture in developing countries is massive given the economic, social, and environmental roles played by their agricultural sectors (Putra et al., 2021).

 DOI: 10.1201/9781003475989-13

According to recent studies, innovations are necessary to accomplish sustainability goals and sustainable performance. Decision-making processes may be significantly impacted by the overlapping of data and information, the lack of sufficient understanding of human resources, and the unpredictable nature of climate events (Hassoun et al., 2022). This could result in less-than-ideal solutions and impede the sustainability agenda. Artificial intelligence (AI) technologies are able to solve complicated issues, process vast volumes of data, and uncover knowledge that would otherwise remain hidden. However, AI's usefulness goes beyond data processing. Nonetheless, it can help multi-stakeholder decision-making processes toward sustainability by pointing out evidence-based remedies for environmental and climatic degradation issues that are not skewed by the interests of any one person or group. AI can assist the various stakeholders in the food, energy, and water sectors in achieving the UN 2030 Agenda (Spanaki et al., 2022). Certain researchers have highlighted how artificial intelligence (AI) has the potential to transform corporate practices in a sustainable and socially conscious manner, in addition to how information is generated and used for decision-making. Furthermore, a few academics have emphasised how important digital technologies are to the food, energy, and water industries (Shoeb et al., 2024). By examining AI and machine learning as crucial instruments for developing sustainable business models in farming and smart food systems, incorporating the ideas of gene technology and drone applications, this study seeks to close this gap.

11.2 TECHNOLOGIES TO ASSIST SMART FARMING

11.2.1 Machine Learning

As technology has developed, machine learning has proven to be an invaluable tool in the agricultural industry. It is a well-known tool for smart agriculture decision assistance. Throughout the past 20 years, a number of machine learning models have been effectively applied to automate the process of smart farming (Feng et al., 2019). Depending on the nature of the research challenge, these models might be either descriptive or predictive. Depending on a number of variables, including data variety, soil quality, environmental conditions, and demography, machine learning seeks to construct high-performance models to handle numerous issues of smart farming (Sridhar et al., 2023). In the literature, several machine learning models have been presented. However, it is still difficult to choose the best machine learning model to manage the massive amounts of data on the underlying platform. Furthermore, the process of building an effective machine learning model is contingent upon several variables, including the distribution of the data, the climate, fertilisers, weather, and seed type (Kootstra et al., 2021). Four general categories can be used to categorise machine learning: supervised learning, unsupervised learning, semi-supervised learning, and reinforced learning.

1. Supervised learning: Models are trained with labelled data to map input and output variables in supervised learning. The issues of classification and regression are resolved through supervised learning. Numerous techniques

are available for classification in the current environment, including logistic regression, random forests, support vector machines, and decision trees (Spanaki et al., 2022). Regression analysis is done using methods including decision trees, random forests, gradient boosting, and linear regression.

2. Unsupervised learning: Data points in unsupervised learning are typically unlabelled and unclassified (Singh et al., 2021). Finding hidden patterns without the help of humans is the goal. Unsupervised learning algorithms, to put it simply, find the patterns on their own datasets (Hassoun et al., 2022). In the areas of association, dimensionality reduction, grouping, and anomaly detection, unsupervised methods are helpful. Popular unsupervised methods include hierarchical clustering, Gaussian mixture, k-means, and k-mode.
3. Semi-supervised learning: Learning that is semi-supervised. Both labelled and unlabelled data are used to train semi-supervised machine learning models (Marvin et al., 2022). These algorithms are typically applied to complex and challenging-to-understand data. Different relationships need to be built for data distribution in order to accommodate the unlabelled data. There are various assumptions that must be made in order to accomplish this, including manifold, continuity, and cluster.
4. Reinforcement learning: Learning through reinforcement. Predicting output is typically the goal of this kind of learning. Two models are commonly utilised in the application of these techniques: Q learning and the Markov decision process. Through the use of the hit-and-trial method in their learning strategy, reinforcement machine learning deals with intelligent agents in the learning environment (Olan et al., 2022). As of right now, it uses a range of gaming technology and adheres to video game conventions, like using user feedback to make the next game better.

11.2.1.1 Smart Farming with Machine Learning

As farming grows deeper, smart farming has become a useful tool for addressing contemporary issues. At the regional and national levels, decision-makers still have difficulties making quick decisions. Nonetheless, in order to ascertain the kind and timing of the crops, farmers want an accurate prediction model (Sridhar et al., 2023). This section summarises a number of publications that present significant discoveries and research findings related to machine learning and precision agriculture. The field of smart farming has greatly benefited from machine learning (Ricroch, 2019). It has been extensively used to automate and improve smart agricultural operations, including minimising food waste, adapting to climate change, organising missions, controlling water resources, detecting diseases automatically, and optimising the use of pesticides.

11.2.1.2 Farming Using Artificial Intelligence

Researchers are dedicated to integrating artificial intelligence technology throughout the entire agricultural production process, and at the moment, artificial intelligence agriculture is driving modern agriculture's development path. Preproduction,

TABLE 11.1
Illustrating the Application of AI in Agriculture

S. No	Approach	Strength	Limitations	Reference
1.	Artificial neural networks (ANN)	Excellent service and reduces trial and error	Massive information is needed	Ben Ayed, R., and Hanana, M. (2021)
2.	Sensor machine learning	Saves time and eradicates weeds with resistance	Costly. Regular usage of large machinery will lower soil production	Javaid et al. (2023)
3.	Forecasting and weed control	Excellent forecast accuracy and rate of adaptability	Requires knowledge of big data usage	Singh et al. (2022)
4.	Learning Vector Quantization (LVQ)	High rate of weed identification and quick processing	The AI's performance was impacted by the data entry technique	Jha et al. (2019)
5.	ANN	Estimates agricultural yield	Only considers weather while determining crop yield	Bhat, S.A., and Huang, N.F. (2021)
6.	Computer vision system (CVS)	Carries out tasks quickly and is able to multitask	Detection based on dimensions that could impact beneficial species	Vincent et al. (2019)
7.	ANN	Can forecast the activity of soil enzymes. Anticipates and categorises soil structure with accuracy	Limited to measuring a few soil enzymes. It gives classification a higher priority than enhancing soil performance	Dozono et al. 2021)

midproduction, and postproduction are the three phases that make up the agricultural production process (Bhambri et al., 2021). In order to serve as a guide for the quick and efficient development of intelligent agricultural technology and the advancement of agricultural industrialisation and modernisation, a brief overview of the application of artificial intelligence technology used in the preproduction and postproduction stages of agricultural production is given in Table 11.1.

11.2.2 The Agricultural Production Process before and after Reproduction

The goal of using artificial intelligence in preproduction is to make it easier for agricultural labourers to understand the requirements of the agricultural landscape.

Preproduction is the first stage of the agriculture supply chain and deals with crop output, soil properties, and irrigation requirements (Tsolakis et al., 2023). ML algorithms help stakeholders and farmers anticipate crop yields and improve smart farming practices by using input data, such as equipment needs, nutrients, and fertilisers. The following are the major AI experimentsdiscussed.

11.2.2.1 Testing the Quality of Soil Nutrients, Crop Yield, and Quality Forecast

Crop growth during the preproduction phase of agriculture is largely dependent on the quality of soil fertiliser. An important way to determine whether the soil is contaminated is to measure the amount of minerals, nutrients, and other elements present. The main focus of soil substances is on plant leftovers, which significantly shield the soil (Olan et al., 2022). They not only increase soil structure and lessen soil erosion, but they also bring nutrients and improve soil quality for replanting (Bose et al., 2021). However, there is no denying that the labor-intensive and time-consuming nature of manually classifying the different types of residues and their coverage rates over vast tracts of land is also subjective and uneven, and the accuracy of the results is uncertain due to the differing experiences of observers (Di Vaio et al., 2020). In this context, a number of dependable, automated techniques have been proposed by people using machine vision, image processing, and other technologies. Tao et al. (2021), for instance, created the MSCU-net + C deep learning approach, which was used to classify various coverage rates, measure the classification accuracy index, and draw the residual coverage area of maize on high spatial resolution satellite remote sensing images (Di Vaio et al., 2020).

Identification of the primary nutrients that are supplied by the soil for crop growth: carbon (C), nitrogen (N), phosphorus (P), and potassium (K). Soil salinisation is a phenomenon of soil deterioration that is directly associated with the sustainable development of agriculture. Thus, it is crucial to carefully track the salinisation of the soil. The "Internet of Things + machine learning" approach has become popular in recent years for forecasting soil salinity. Wang et al. (2021) and Wei et al. (2020) both employed the combination of "multispectral image acquisition + machine learning prediction model" to predict soil salinity. Chaganti et al. (2019) optimised fertiliser use decisions on farms using technologies including machine learning, image processing, and the Internet of Things. A rice management expert system based on the positive chain and deterministic factor technique was created by Adi and Isnanto (2020). It includes consultation for seed selection and pest detection.

11.2.2.2 Forecasting Crop Quality and Management

It is a crucial challenge for agricultural decision-makers and one of the most difficult problems in precision agriculture. Smart farms use automated irrigation systems that keep an eye on and manage water tanks, open irrigation systems, and chamber irrigation systems. Crop management with artificial intelligence (AI) starts at seeding and continues with crop development, harvesting, storage, and distribution monitoring (Kakani et al., 2020). Demand-driven supply chains, automation of farming processes, and crop health management are just a few of the uses of AI in agriculture (Sangwan Singh et al., 2021). AI assists in resolving crop selection concerns and increasing net yields throughout the season. A comprehensive review of the literature on crop yield forecasting was conducted by Van Klompenburg et al. (2020), who focused on precision agriculture and looked at related work using hyperspectral remote sensing for crop yield prediction and estimation. They summarised and analysed the literature from three perspectives: prediction feature selection, machine learning algorithm selection, and evaluation parameter selection. Deep learning techniques for fruit detection and yield calculation were also observed using AI. For identifying and managing crop diseases, deep learning and few-shot learning are used (Misra et al., 2020).

11.2.2.3 Reducing Time on Breeding Programme Duration

Depending on the plant reproductive systems, genome editing could dramatically accelerate breeding programmes from 7 to 25 years to as little as 2–3 years (Ricroch, 2019). Because of its target specificity, achieving a specific genetic combination doesn't require going through multiple plant generations. Farmers, customers, and breeders all profit from this advancement (Shoeb et al., 2024). Many plant species now have access to bioinformatics resources and advanced statistical analyses on genomic and phenotypic variation within and between varieties, strains, or breeds. These resources are a valuable link for metabolic pathway profiling, comparative genome analysis, and the identification of genes of interest and their expression patterns (Klerkx, L., & Rose, D. 2020). Recently, gene technology (genome editing and genetic engineering) has been utilised by plant and animal breeders, especially to target improvements in traits for which there is no within-species or within-breed genetic diversity (Bhambhani et al., 2021). Our daily lives will be altered by the emerging uses of genome editing technologies in agriculture for producing livestock and plants.

11.2.2.4 Harvest of Agricultural Products

Activities related to crop harvesting include picking, stacking, threshing, handling, cleaning, and transporting (Herrero et al., 2021). These tasks are typically labour-intensive, repetitive, and enormous in quantity. The harvesting of agricultural products must increase the degree of agricultural automation and roboticisation due to the impact of the labour shortage and the rising demand for agricultural products. To perform tasks using AI like camera calibration (Wang et al., 2019), target recognition and localization, target background recognition (Feng et al., 2019), 3D reconstruction (Kootstra, 2021), robot behaviour planning based on visual positioning (Wibowo et al., 2016), and avoiding complex factors interference localisation (Xiong et al., 2018) are all currently performed by fruit harvesting robots. Four robotic arms installed on the platform are used to harvest fruit since they can reach the fruit that is suspended above the vehicle. These harvesting robots use deep learning algorithms, multi-feature fusion techniques, and single-feature vision techniques for item detection.

11.3 SMART TECHNOLOGIES FOR FOOD PROCESSING

The processing cluster, which includes a variety of methods such as heating, chilling, milling, smoking, boiling, and drying for agricultural products, is the third stage in the food supply chain. It is feasible to generate food products of high quality and quantity while minimising resource usage by using a set of effective criteria throughout this stage (Vagsholm et al., 2020). SVM and ANN models support modern ML-based food processing technologies that detect nitrosamine in red meat food samples. Furthermore, AI-powered cucumber harvesting robots have been created (Sgroi, F., & Marino, G., 2022). These robots include hardware components such as an autonomous vehicle, a manipulator, and an end-effector, together with computer vision technologies that enable them to accurately detect and image the maturity of cucumbers (Kakani et al., 2020). Moreover, AI has demonstrated its efficacy in enhancing the physical fields of microwave, radio frequency, infrared radiation, and ultrasonic fields used in the drying process of fresh foods, fruits, and vegetables (Lezoche et al., 2020). For instance, utilising artificial intelligence (AI) to detect and control the drying process online helps cut down on energy use, minimise uneven drying, enhance sensory assessment, and minimise nutritional loss.

11.4 DISTRIBUTION AND CONSUMPTION OF FOOD

In the agriculture food chain's distribution phase, getting consumers safe, high-quality food is the main priority. To reduce damage and maintain food quality, machine learning algorithms help with processes including inventory management, transportation, storage, and consumer analytics (Lamine, C. 2020). AI improves supply chain management, facilitates food safety testing and monitoring at every level, and allows product tracking and safety assurance. It is impossible to overestimate the significance of food safety, and one way to do this is by classifying, identifying, and recognising the quality of food products using a variety of techniques, including image processing (IP) (Lajoie-O'Malley et al., 2020). IP systems assess the product's size, shape, and texture using document scanners, X-rays, near-infrared spectroscopy, and ultrasound (Hubeau et al., 2017). This method can also be used to grade quality and find flaws in product packaging. The application of AI has significantly advanced the field of sustainable food systems, especially in the areas of food grading and quality control in the food business (McGreevy et al., 2022). Food grading is an essential process that assesses the safety and quality of food products based on predetermined standards that include dimensions, colour, texture, flavour, freshness, and defect-free status (Miranda et al., 2021). This procedure has been improved and automated by artificial intelligence (AI) technologies, which include robotics, computer vision, and machine learning. Artificial intelligence (AI) minimises human error, lowers labour costs, decreases waste, and simultaneously increases production, profitability, and customer satisfaction by enabling real-time inspection, categorisation, and sorting of food products (Hassoun et al., 2023).

In keeping with the previous discussion about artificial intelligence's role in food composition analysis and its wider effects on sustainable food systems, it is important to note that notable initiatives such as the Food Label Information Programme (FLIP) 2020 demonstrate the usefulness of AI in creating extensive databases of food composition (El Bilali et al., 2021). The goal of FLIP 2020, which took place in three stages between May 2020 and February 2021, was to gather information from Canadian food and beverage container labels that were provided by significant online grocery retailers. Python-based webpage scraping and AI-enhanced optical character recognition (OCR) were used in this data collection (Gardeazabal et al., 2023; Spendrup, S.; &Fernqvist, F. 2019).

An instructive example is the Snap-n-Eat application, which uses a support vector machine (SVM) classifier to identify food items and estimate energy and nutritional consumption from mobile device images. When users take photos of their food, the technology removes the background and isolates the pertinent area (Spendrup, S., & Fernqvist, F. 2019). These parts are analysed by a linear SVM classifier, which uses features from different scales and locations to identify the meal (Gardeazabal et al., 2023). Determining the portion size and estimating the food's calorie and nutritional content are the results of this process.

11.5 RESEARCH CHALLENGES AND FUTURE DIRECTIONS

Implementation of smart monitoring systems: Smart farming's technological advancements have been severely hampered by the agriculture sector. One of the main problems in the agricultural area is the intelligent monitoring of crop health on huge farmlands (McGreevy et al., 2022). Swarm intelligence is needed to accomplish this and make agricultural systems intelligent (Mahdad et al., 2022). A number of agricultural applications, including water and soil, irrigation systems, asset tracking, greenhouse farming, remote control, and diagnosis, are monitored by swarm intelligence. Robots can be used as swarms in this system to keep an eye on crop diseases (Klerkx, L., & Rose, D. 2020). To alert farmers to the situation and allow for timely intervention, a warning messaging system might be created (El Bilali et al., 2021).

Agricultural autonomous systems: In the last ten years, a variety of autonomous systems have been created to boost output and dependability, somewhat replacing humans in the process. Agriculture-related operations are becoming more intelligent thanks to new robotics and autonomous systems. These technologies are used in many smart farming production patterns, including autonomous farming, automated husbandry, aerial spraying, biodiverse farming, 3D food printing, plant factories, and aerial spraying (Sgroi, F., & Marino, G. 2022). Nonetheless, there are still a lot of unanswered problems, like autonomous course navigation, precise identification, and operational effectiveness (Miranda et al., 2021). However, smart farming still requires some intelligent systems, such as precise work allocation, item localisation, and human–robot interaction.

Security challenge: Among all other technologies, swarm intelligence has some special security issues. Swarm systems have several security challenges, including the physical capture of robots in farms, communication, mobility, identification and authentication security, key management, and intrusion detection. Swarm intelligence is managed and analysed with an emphasis on solution-based techniques that yield suggestive answers, including organisational structure, situational prediction, process optimisation, and enhancement of various activities (Herrero et al., 2021). Moreover, to create a safe system that can oversee integrity and secrecy in an agricultural setting, it means that in order to ensure the authenticity of data gathered from dependable sources, an authentication system needs to be built.

11.6 CONCLUSION

Establishing sustainable development (SD) is necessary to balance environmental preservation, social protection, and profit. AI can assist companies, governments, and civil society in addressing this issue and the global obstacles to sustainable agricultural methods and differences in the resources available for agricultural production (Hassoun et al., 2023). The need for smart farming solutions is driven by comparable problems with natural supplies, the negative effects of climate change, an ageing labour force, and environmental concerns. In order to improve sustainable agriculture practices, all nations are seeking smart farming solutions. Their methods differ, though, and the key factors influencing these variances are government policies, farming practices, farm size, and technological readiness (Hubeau et al., 2017). The diverse approaches to smart farming solutions and their sharing of achievements and failures may offer fresh perspectives on how to help less fortunate countries where smart farming initiatives are now in progress advance both smart farming and the food system (Rachna et al., 2022).

This research highlighted the value of intelligent agriculture in enhancing and raising agricultural output to help close the gap between supply and demand for food. Since AI links every element of smart systems – both within and outside of the agricultural sector – it is regarded as the foundation of smart agriculture technology. AI applications in agriculture include a wide range of tasks like pest management, irrigation, harvesting, and farm monitoring (Lajoie-O'Malley et al., 2020). The use of intelligent agriculture technology benefits developing nations, including certain Egyptian strategies that can aid in the development of the agriculture sector, mark the start of the dissemination of such technology, and ensure sustainability in agriculture in such nations (Lamine, C. 2020; Wang et al., 2021). Lastly, since the goal of these clever technologies is to boost production, governments in third-world countries ought to encourage small farmers to use them and enhance the effective utilisation of water and land resources (Lezoche et al., 2020).

REFERENCES

Adi, K., & Isnanto, R. R. (2020, April). Rice crop management expert system with forwarding chaining method and certainty factor. In *Journal of physics: Conference series* (Vol. 1524, No. 1, p. 012037). IOP Publishing.

Ben Ayed, R., & Hanana, M. (2021). Artificial intelligence to improve the food and agriculture sector. *Journal of Food Quality, 2021*, 1–7.

Bhambhani, S., Kondhare, K. R., & Giri, A. P. (2021). Advanced genome editing strategies for manipulation of plant specialized metabolites pertaining to biofortification. *Phytochemistry Reviews, 13,* 1–19.

Bhambri, P., Singh, M., Jain, A., Dhanoa, I. S., Sinha, V. K., & Lal, S. (2021). Classification of the GENE expression data with the aid of optimized feature selection. *Turkish Journal of Physiotherapy and Rehabilitation, 32*(3), 1158–1167.

Bhat, S. A., & Huang, N. F. (2021). Big data and ai revolution in precision agriculture: Survey and challenges. *IEEE Access, 9,* 110209–110222.

Bose, M. M., Yadav, D., Bhambri, P., & Shankar, R. (2021). Electronic customer relationship management: Benefits and pre-implementation considerations. *Journal of Maharaja Sayajirao University of Baroda, 55*(01(VI)), 1343–1350.

Chaganti, S. Y., Ainapur, P., Singh, M., & Oktaviana, S. (2019, September). Prediction based smart farming. In *2019 2nd international conference of computer and informatics engineering (IC2IE)* (pp. 204–209). IEEE.

Di Vaio, A., Boccia, F., Landriani, L., & Palladino, R. (2020). Artificial intelligence in the agri-food system: Rethinking sustainable business models in the COVID-19 scenario. *Sustainability, 12*(12), 4851.

Dozono, K., Amalathas, S., & Saravanan, R. (2022). The impact of cloud computing and artificial intelligence in digital agriculture. In *Proceedings of* sixth international congress *on* information *and* communication technology*: ICICT 2021, London* (Vol. 1, pp. 557–569). Springer Singapore.

El Bilali, H., Strassner, C., & Ben Hassen, T. (2021). Sustainable agri-food systems: Environment, economy, society, and policy. *Sustainability, 13*(11), 6260.

Feng, J., Zeng, L., & He, L. (2019). Apple fruit recognition algorithm based on multi-spectral dynamic image analysis. *Sensors, 19*(4), 949.

Gardeazabal, A., Lunt, T., Jahn, M. M., Verhulst, N., Hellin, J., & Govaerts, B. (2023). Knowledge management for innovation in agri-food systems: A conceptual framework. *Knowledge Management Research & Practice, 21*(2), 303–315.

Hassoun, A., Aït-Kaddour, A., Abu-Mahfouz, A. M., Rathod, N. B., Bader, F., Barba, F. J., ... Regenstein, J. (2023). The fourth industrial revolution in the food industry—Part I: Industry 4.0 technologies. *Critical Reviews in Food Science and Nutrition, 63*(23), 6547–6563.

Hassoun, A., Prieto, M. A., Carpena, M., Bouzembrak, Y., Marvin, H. J., Pallares, N., ... Bono, G. (2022). Exploring the role of green and Industry 4.0 technologies in achieving sustainable development goals in food sectors. *Food Research International, 53,* 112068.

Herrero, M., Thornton, P. K., Mason-D'Croz, D., Palmer, J., Bodirsky, B. L., Pradhan, P., ... Rockström, J. (2021). Articulating the effect of food systems innovation on the sustainable development goals. *The Lancet Planetary Health, 5*(1), e50–e62.

Hubeau, M., Marchand, F., Coteur, I., Mondelaers, K., Debruyne, L., & Van Huylenbroeck, G. (2017). A new agri-food systems sustainability approach to identify shared transformation pathways towards sustainability. *Ecological Economics, 131,* 52–63.

Javaid, M., Haleem, A., Khan, I. H., & Suman, R. (2023). Understanding the potential applications of artificial intelligence in agriculture sector. *Advanced Agrochem, 2*(1), 15–30.

Jha, K., Doshi, A., Patel, P., & Shah, M. (2019). A comprehensive review on automation in agriculture using artificial intelligence. *Artificial Intelligence in Agriculture, 2,* 1–12.

Kakani, V., Nguyen, V. H., Kumar, B. P., Kim, H., & Pasupuleti, V. R. (2020). A critical review on computer vision and artificial intelligence in food industry. *Journal of Agriculture and Food Research, 2,* 100033.

Klerkx, L., & Rose, D. (2020). Dealing with the game-changing technologies of agriculture 4.0: How do we manage diversity and responsibility in food system transition pathways?. *Global Food Security, 24*, 100347.

Kootstra, G., Wang, X., Blok, P. M., Hemming, J., & Van Henten, E. (2021). Selective harvesting robotics: Current research, trends, and future directions. *Current Robotics Reports, 2*, 95–104.

Lajoie-O'Malley, A., Bronson, K., van der Burg, S., & Klerkx, L. (2020). The future (s) of digital agriculture and sustainable food systems: An analysis of high-level policy documents. *Ecosystem Services, 45*, 101183.

Lamine, C. (2020). *Sustainable agri-food systems: case studies in transitions towards sustainability from France and Brazil*. Bloomsbury Publishing.

Lezoche, M., Hernandez, J. E., Díaz, M. D. M. E. A., Panetto, H., & Kacprzyk, J. (2020). Agri-food 4.0: A survey of the supply chains and technologies for the future agriculture. *Computers in Industry, 117*, 103187.

Mahdad, M., Hasanov, M., Isakhanyan, G., & Dolfsma, W. (2022). A smart web of firms, farms and internet of things (IOT): Enabling collaboration-based business models in the agri-food industry. *British Food Journal, 124*(6), 1857–1874.

Marvin, H. J., Bouzembrak, Y., Van der Fels-Klerx, H. J., Kempenaar, C., Veerkamp, R., Chauhan, A., ... Tekinerdogan, B. (2022). Digitalisation and artificial intelligence for sustainable food systems. *Trends in Food Science & Technology, 120*, 344–348.

McGreevy, S. R., Rupprecht, C. D., Niles, D., Wiek, A., Carolan, M., Kallis, G., ... Tachikawa, M. (2022). Sustainable agrifood systems for a post-growth world. *Nature Sustainability, 5*(12), 1011–1017.

Miranda, B. V., Monteiro, G. F. A., & Rodrigues, V. P. (2021). Circular agri-food systems: A governance perspective for the analysis of sustainable agri-food value chains. *Technological Forecasting and Social Change, 170*, 120878.

Misra, N. N., Dixit, Y., Al-Mallahi, A., Bhullar, M. S., Upadhyay, R., & Martynenko, A. (2020). IoT, big data, and artificial intelligence in agriculture and food industry. *IEEE Internet of Things Journal, 9*(9), 6305–6324.

Olan, F., Liu, S., Suklan, J., Jayawickrama, U., & Arakpogun, E. O. (2022). The role of artificial intelligence networks in sustainable supply chain finance for food and drink industry. *International Journal of Production Research, 60*(14), 4418–4433.

Putra, A. B. N. R., Mizar, M. A., Masek, A. B., Nagari, P. M., Habibi, M. A., & Subandi, M. S. (2021, October). The hybrid technology of quadcopter spreader drone with crystal Fertilizer as an unmanned aerial vehicles innovation. In *2021 7th international conference on electrical, electronics and information engineering (ICEEIE)* (pp. 1–5). IEEE.

Rachna, R., Bhambri, P., & Chhabra, Y. (2022). Deployment of distributed clustering approach in WSNs and IoTs. In *Cloud and fog computing platforms for Internet of Things* (pp. 85–98). Chapman and Hall/CRC.

Ricroch, A. (2019, August). Global developments of genome editing in agriculture. In *Transgenic research* (Vol. 28, pp. 45–52). Springer International Publishing.

Sangwan Singh, Y., Lal, S., Bhambri, P., Kumar, A., & Dhanoa, I. S. (2021). Advancements in social data security and encryption: A review. *Natural Volatiles & Essential Oils, 8*(4), 15353–15362.

Sgroi, F., & Marino, G. (2022). Environmental and digital innovation in food: The role of digital food hubs in the creation of sustainable local agri-food systems. *Science of The Total Environment, 810*, 152257.

Shoeb, E., Venkataraman, S., Badar, U., & Hefferon, K. (2024). Genome editing for crop biofortification. *Applications of Genome Engineering in Plants, 22*, 91–121.

Singh, M., Bhambri, P., Dhanoa, I. S., Jain, A., & Kaur, K. (2021). Data mining model for predicting diabetes. *Annals of the Romanian Society for Cell Biology*, 25(4), 6702–6712.

Singh, T., Bhadwaj, H., Verma, L., Navadia, N. R., Singh, D., Sakalle, A., & Bhardwaj, A. (2022). Applications of AI in agriculture. *Challenges and Opportunities for Deep Learning Applications in Industry 4.0, 181.*

Spanaki, K., Karafili, E., Sivarajah, U., Despoudi, S., & Irani, Z. (2022). Artificial intelligence and food security: swarm intelligence of AgriTech drones for smart AgriFood operations. *Production Planning & Control, 33*(16), 1498–1516.

Spendrup, S., & Fernqvist, F. (2019). Innovation in agri-food systems–a systematic mapping of the literature. *International Journal on Food System Dynamics, 10*(5), 402–427.

Sridhar, A., Balakrishnan, A., Jacob, M. M., Sillanpää, M., & Dayanandan, N. (2023). Global impact of COVID-19 on agriculture: role of sustainable agriculture and digital farming. *Environmental Science and Pollution Research, 30*(15), 42509–42525.

Tao, W., Xie, Z., Zhang, Y., Li, J., Xuan, F., Huang, J., … Yin, D. (2021). Corn residue covered area mapping with a deep learning method using Chinese GF-1 B/D High resolution remote sensing images. *Remote Sensing, 13*(15), 2903.

Tsolakis, N., Schumacher, R., Dora, M., & Kumar, M. (2023). Artificial intelligence and blockchain implementation in supply chains: A pathway to sustainability and data monetisation? *Annals of Operations Research, 327*(1), 157–210.

Vågsholm, I., Arzoomand, N. S., & Boqvist, S. (2020). Food security, safety, and sustainability—getting the trade-offs right. *Frontiers in Sustainable Food Systems, 16,* 234–243.

Van Klompenburg, T., Kassahun, A., & Catal, C. (2020). Crop yield prediction using machine learning: A systematic literature review. *Computers and Electronics in Agriculture, 177,* 105709.

Vincent, D. R., Deepa, N., Elavarasan, D., Srinivasan, K., Chauhdary, S. H., & Iwendi, C. (2019). Sensors driven AI-based agriculture recommendation model for assessing land suitability. *Sensors, 19*(17), 3667.

Wang, J., Peng, J., Li, H., Yin, C., Liu, W., Wang, T., & Zhang, H. (2021). Soil salinity mapping using machine learning algorithms with the Sentinel-2 MSI in arid areas, China. *Remote Sensing, 13*(2), 305.

Wang, Y., Yang, Y., Yang, C., Zhao, H., Chen, G., Zhang, Z., ... Xu, H. (2019). End-effector with a bite mode for harvesting citrus fruit in random stalk orientation environment. *Computers and Electronics in Agriculture, 157,* 454–470.

Wei, G., Li, Y., Zhang, Z., Chen, Y., Chen, J., Yao, Z., ... Chen, H. (2020). Estimation of soil salt content by combining UAV-borne multispectral sensor and machine learning algorithms. *PeerJ, 8,* e9087.

Wibowo, T. S., Sulistijono, I. A., & Risnumawan, A. (2016). End-to-end coconut harvesting robot. In *2016 International Electronics Symposium (IES)* (pp. 444–449).

Xiong, J. T., Lin, R., Liu, Z., He, Z. L., Tang, L. Y., et al. (2018). The recognition of litchi clusters and the calculation of picking point in a nocturnal natural environment. *Biosystems Engineering, 166,* 44–57.

12 Sustainable Transportation and Mobility

Andrii Galkin

12.1 INTRODUCTION

"Sustainable transport and mobility" refers to a broad spectrum of transportation issues that are sustainable in terms of social, environmental, and climate impacts. Components of assessing the sustainability of transportation include specific modes of transport used for road, water, or air transport; energy sources; and infrastructure used for transportation placement (roads, railways, air routes, waterways, canals, and terminals). The assessment also involves transportation operations and logistics, as well as transit-oriented development. The sustainability of transport is largely measured by the efficiency of the transportation system's operation and its impact on the environment and climate.

Urban freight transportation or the movement of goods within cities is defined as having an unstable impact on:

People, such as the consequences of road traffic accidents, noise pollution, visual intrusion, odors, vibrations, and the effects of (local) emissions, such as NOx and PM_{10} on public health.

Profit, such as inefficiency (especially for carriers) due to regulations and restrictions, traffic jams, and reduced city accessibility.

The *planet*, such as the contribution of transport to global emissions of pollutants (CO_2) and the consequences of global warming.

The concept of sustainable mobility is derived from the term sustainability, which means the ability to meet current human needs without harming the environment and the interests of future generations. A well-known example of the application of this term is the "Sustainable Development Goals", which include (Bernstein, 2017):

- Promoting a healthy lifestyle and well-being for all ages.
- Creating sustainable infrastructure, fostering comprehensive and sustainable industrialization and innovation.
- Reducing inequality within and among countries.
- Ensuring inclusiveness, safety, resilience, and sustainability of cities and human settlements.

 DOI: 10.1201/9781003475989-14

- Promoting responsible consumption and production.
- Taking urgent action to combat climate change and its impacts.

Examples of countries where the mobility of the population has reached a high level of intensity and sustainability include Switzerland, Sweden, Germany, and other EU countries. According to a sociological study conducted in 2016 in 28 EU member countries, 82% of Europeans believe that their place of residence has good transportation connectivity, while only 5% stated that their locality lacks it altogether (SDG, 2015). The level of transportation accessibility is not related to living in rural or urban areas. For instance, Austria has one of the highest proportions of rural population (40%) and one of the highest levels of transportation coverage.

Therefore, sustainable mobility refers to the ability of people to engage in activities in different places with minimal negative impact on the environment, economy, and spatial planning of territories. Such mobility (Fiorello et al., 2016):

- Ensures accessibility and satisfaction of the developmental needs of individuals, companies, and society in a safe manner and in accordance with the needs of the community and the ecosystem.
- Is accessible, fair, and efficient.
- Provides a choice of transportation modes and supports competition and territorial development.
- Reduces the volume of harmful emissions and waste for easier absorption.
- Utilizes renewable energy sources to the extent possible, minimizes land impact, and reduces noise levels.

The digital age has brought about a new era of possibilities for transportation. The integration of advanced technologies such as the Internet of Things (IoT), artificial intelligence (AI), big data (BD), digital twin (DT), and synchromodality (S) analytics has revolutionized how we perceive and manage transportation systems (Sakti et al., 2023). A clustering diagram illustrating the various components involved in assessing the sustainability of transportation, including modes of transport, energy sources, infrastructure, operations, logistics, and their impacts on people, profit, and the planet, is presented in Fig. 12.1.

At the core of this exploration is the interplay among key transportation stakeholders – suppliers, transporters, retailers, and consumers. Their interactions are pivotal in shaping the urban transport ecosystem, emphasizing the significant potential of advanced technologies and data analytics (Przybylska et al., 2023). These tools are crucial for optimizing transport operations, minimizing environmental footprints, and responding adeptly to the changing needs of city dwellers.

The importance of robust policy frameworks and the transformative power of community engagement draws inspiration (Busch et al., 2021). These elements are essential in steering the societal transition toward more sustainable transportation modes. Illustrated through a series of global case studies, the text presents a variety of strategies deployed by cities to tackle transportation challenges (Gracias et al., 2023).

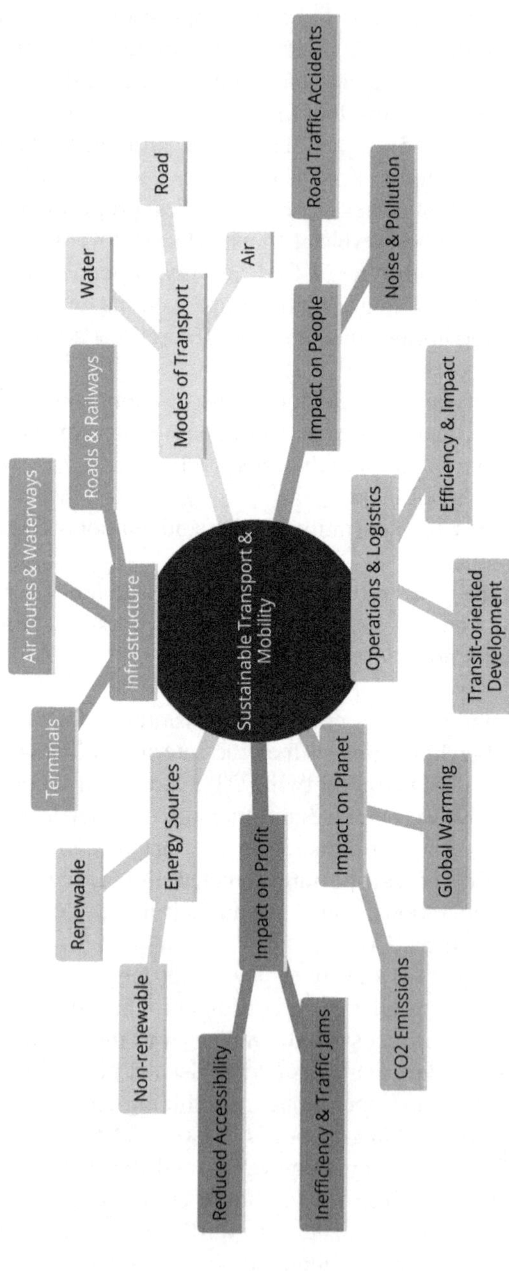

FIGURE 12.1 Assessment of sustainable transport and mobility.

12.2 THE DIGITAL TRANSFORMATION OF URBAN TRANSPORT

The digital transformation is just beginning to reshape urban transportation. The future urban mobility landscape will be significantly shaped by the integration of cutting-edge technologies. The deployment of 5G networks, for instance, will enable more efficient and reliable communication between vehicles and infrastructure, facilitating real-time data exchange and enhancing the capabilities of autonomous and connected vehicles (Bello et al., 2020).

As technologies continue to evolve, we can expect to see even more innovative and transformative solutions emerge. Some of the key trends that are likely to shape the future of urban transport include:

1. The Rise of Mobility-as-a-Service (MaaS) and Logistics-as-a-Service (LaaS) platforms is poised to revolutionize urban transport by integrating a variety of transportation options into a single service. Users will have access to a seamless blend of public transit, ride-sharing, and private vehicles, offering unparalleled flexibility and choice (Cohen et al., 2018). AI-powered platforms can match real-time passenger demand with available transportation options, such as taxis, buses, and shuttles. This on-demand approach can reduce the number of empty vehicles on the road and improve the overall efficiency of transportation systems (Jiang et al., 2017).
2. Smart city infrastructure will connect transportation systems with other city services, such as energy, lighting, and waste management. This integration will create more efficient and sustainable urban environments where transportation is seamlessly intertwined with other aspects of city life. IoT sensors and cameras can collect vast amounts of data on traffic conditions, vehicle movements, and passenger behavior. This data can be analyzed using AI and big data analytics to provide real-time insights into transportation patterns and identify potential problems (Gonzalez et al., 2008). Data analytics will be at the heart of urban mobility systems, enabling cities to harness vast amounts of data from various sources, including traffic sensors, GPS devices, and mobile applications (Ma et al., 2019). By analyzing this data, urban planners and transport authorities can gain valuable insights into travel patterns, congestion hotspots, and public transit usage, informing more effective and dynamic transport policies and infrastructure investments (Chien and Ding, 2018).
3. AI and ML will play a central role in optimizing transportation systems in real time. These technologies will provide personalized travel recommendations, predict traffic patterns, and manage the demand for transportation services efficiently (Zheng et al., 2019). AI algorithms can use historical data and real-time insights to predict future transportation demand and identify potential bottlenecks. This information can be used to optimize traffic flow, reduce congestion, and improve the efficiency of public transportation systems (Zheng et al., 2017). AI and ML algorithms will become increasingly sophisticated, offering predictive analytics for traffic

management, personalized travel recommendations, and automated diagnostics for vehicle maintenance (Li et al., 2019). These technologies will not only streamline transport operations but also contribute to the creation of safer and more reliable mobility services.

4. Connected and autonomous vehicles (CAVs) are set to become increasingly commonplace in urban areas. The development of connected and autonomous vehicles (CAVs) has the potential to revolutionize urban transportation. CAVs can communicate with each other and with infrastructure, enabling them to operate more safely and efficiently. They can also adapt their routes and schedules in real time to optimize traffic flow and reduce congestion. They offer a safer, more efficient, and more convenient form of transportation that has the potential to transform the way people move within cities (Fagnant and Kockelman, 2015).

Despite the many benefits, the digital transformation of urban transport also presents some challenges:

- Data privacy and security: The collection and analysis of vast amounts of personal data raise concerns about privacy and security. It is important to develop robust data governance frameworks to protect individual privacy and ensure the responsible use of data (Van den Berg et al., 2017).
- Infrastructure investment: The implementation of digital transportation solutions often requires significant investment in infrastructure, such as sensors, communication networks, and data centers. This can be a challenge for cities with limited resources (Huang et al., 2019).
- Public acceptance: New technologies, such as CAVs, may raise concerns about safety, job displacement, and social equity. It is important to engage the public in discussions about the potential benefits and risks of these technologies and address their concerns (Shladover, 2018).

The digital transformation of public and freight transport is a complex and multifaceted process. It will require collaboration between governments, businesses, and academia to address the challenges and harness the opportunities presented by these technological advancements.

12.3 KEY POLICY STRATEGIES FOR THE DIGITAL TRANSFORMATION

The foundation of sustainable urban mobility is laid by robust policy frameworks that not only encourage technological innovation and infrastructural development but also prioritize public safety, environmental sustainability, and social equity. These frameworks should be dynamic and capable of adapting to the rapid advancements in transportation technology and the changing needs of urban populations. Key policy strategies:

1. Support for clean energy vehicles: Policies that facilitate the adoption of electric and other clean energy vehicles are essential. This can include tax incentives for purchasers, subsidies for manufacturers, and grants for research and development in clean vehicle technologies (Liu et al., 2021).
2. Investment in sustainable infrastructure: Government investment in sustainable transportation infrastructure, such as cycling lanes, pedestrian zones, and electric vehicle charging stations, is crucial. Public funding can also support the expansion of public transit systems to underserved areas, enhancing accessibility and equity.
3. Integrated public transport networks**:** Policies should promote the integration of different modes of public transport to create seamless, efficient, and user-friendly networks. This includes coordinating timetables, integrating fare systems, and providing real-time travel information to users.

The regulatory environment must be agile to accommodate the introduction of autonomous vehicles, drone deliveries, and other innovations that are set to redefine urban mobility. This involves:

1. Safety standards for autonomous vehicles: Establishing comprehensive safety standards and testing protocols for autonomous vehicles to ensure they can safely coexist with traditional vehicles and pedestrians.
2. Regulations for drone deliveries: Developing clear guidelines for the commercial use of drones, including flight paths, altitude restrictions, and privacy protections, to ensure the safe and respectful deployment of these technologies.
3. Data privacy and cybersecurity measure**s:** Implementing strict regulations to protect the privacy and security of data generated by smart transportation systems. This includes ensuring transparency in data collection and use, as well as robust protection against cyber threats.

Policies must ensure that advancements in urban mobility benefit all segments of the population, particularly those who are most vulnerable or have been historically underserved by transportation systems. This includes:

1. Affordable access to transportation: Implementing pricing policies that make sustainable transportation options affordable for all income levels, such as subsidized fares for low-income residents.
2. Accessibility for persons with disabilities: Ensuring that all transportation infrastructure and services are fully accessible to persons with disabilities, in compliance with relevant legislation and best practices.

Public–private partnerships (PPPs) can play a pivotal role in accelerating the development and deployment of sustainable transportation solutions (Shrivastava et al., 2021). Governments can collaborate with technology companies, vehicle manufacturers, and service providers to pilot new technologies, finance infrastructure projects, and deliver innovative services to the public.

For policy frameworks and regulatory measures to be effective, they must have the trust and support of the public. This requires transparent decision-making processes, active engagement with community stakeholders, and clear communication about the benefits and implications of new transportation policies and technologies.

12.4 COMMUNITY ENGAGEMENT AND BEHAVIORAL CHANGE

Community engagement stands as a cornerstone in the edifice of sustainable urban mobility. It transcends mere consultation, involving residents, local businesses, and community groups in an active and meaningful dialogue about the design, development, and deployment of mobility solutions. This collaborative process ensures that transportation initiatives are not only reflective of the community's needs but also gain broader social acceptance and commitment.

The strategies outlined in Table 12.1 are designed not only to promote a shift in individual transportation choices but also to foster a broader cultural transformation toward sustainable mobility. By highlighting the interconnected roles of awareness, infrastructure, incentives, and community engagement, the table serves as a roadmap for policymakers, urban planners, and community leaders seeking to implement effective measures for encouraging sustainable transportation behaviors.

Community engagement in sustainable mobility initiatives not only addresses transportation challenges but also enhances social cohesion. By working together toward common goals, residents develop a stronger connection to their community and a shared sense of purpose. Improved mobility options also contribute to a higher quality of urban life, making cities more livable, accessible, and equitable.

Table 12.1 serves as a cornerstone for discussions on innovative and practical approaches to enhancing urban mobility, underpinning the critical analysis of sustainable transportation strategies within this chapter.

Engaging communities in the development and implementation of mobility initiatives provides a deeper understanding of local dynamics and potential obstacles. This insight is invaluable in creating flexible and adaptive solutions that can withstand the evolving challenges of urban environments. Moreover, when community members are actively involved in shaping their transportation landscape, they are more likely to support and sustain these initiatives over the long term.

12.5 DYNAMIC CONSUMER DEMANDS AND INTEGRATED DISTRIBUTION

In today's dynamic and interconnected world, consumer demands are constantly evolving. Consumers are increasingly demanding personalized, convenient, and sustainable products and services (Fornell et al., 2020). This presents a significant challenge for businesses, as it requires them to adapt their operations to meet the changing needs of their customers.

TABLE 12.1
Comprehensive Strategies for Fostering Sustainable Urban Mobility

Strategy Category	Description	References
Awareness campaigns	Utilizing both traditional and digital media to enlighten the public on the benefits of sustainable transport, employing narratives that connect with daily experiences.	Grilli and Curtis 2021)
Infrastructure improvements	Enhancing urban infrastructure to support sustainable mobility, such as bike lanes and pedestrian zones, emphasizing safety and connectivity.	Axon et al. (2018); Davidich et al. (2022)
Incentive programs	Implementing incentive schemes, including financial benefits for public transit users and perks for cyclists, to encourage a shift from private vehicle usage.	Schultz (2014)
Car-sharing and bike-sharing schemes	Promoting shared mobility services to reduce private vehicle dependency, alleviate congestion, and lower urban emissions.	Mouratidis (2022)
User-friendly design for transport	Prioritizing pedestrian-friendly urban design to foster vibrant, interconnected communities through safe and well-lit public spaces.	Oeschger et al. (2020)
Public participation in planning	Engaging the community in the transportation planning process to ensure that initiatives reflect the needs and preferences of local residents, fostering a sense of ownership and accountability toward the transportation system.	Maltese, et al. (2023)

In the context of distribution, understanding and responding to dynamic consumer demands is crucial for achieving sustainable transportation and mobility. Traditional distribution models, which often focus on mass production and one-size-fits-all approaches, are no longer sustainable in the face of today's evolving consumer landscape.

12.5.1 Data-Driven Approaches to Integrated Distribution Strategies

Data-driven approaches offer a powerful tool for businesses to adapt to dynamic consumer demands and develop integrated sustainable distribution strategies. By collecting, analyzing, and utilizing real-time data on consumer behavior, sales trends, and supply chain dynamics, businesses can gain insights into the evolving needs of their customers and optimize their distribution operations accordingly (Chen et al., 2012).

12.5.2 Key Components of Integrated Distribution Strategies

Integrated distribution strategies should prioritize the integrated efficiency of all actors in the supply chain, including end-consumer expenditures. This requires a holistic approach that encompasses:

- Demand forecasting: Accurately forecasting consumer demand is essential for optimizing inventory levels and reducing waste. Data-driven demand forecasting tools can help businesses anticipate future demand patterns and make informed decisions about production, transportation, and storage (Chopra and Meindl, 2015).
- Route optimization: Efficient route optimization can significantly reduce transportation costs and emissions. By analyzing real-time traffic data, traffic patterns, and delivery schedules, businesses can optimize delivery routes and reduce the number of empty miles driven (Braekers et al., 2016).
- Fleet management: Optimizing fleet management can improve vehicle utilization, reduce fuel consumption, and enhance overall fleet efficiency. This includes using telematics data to monitor vehicle performance, maintenance schedules, and driver behavior (Teixeira et al., 2017).
- Last-mile delivery: Last-mile delivery is often the most expensive and inefficient part of the distribution process. Data-driven approaches can help businesses optimize last-mile delivery routes, utilize alternative delivery methods, and reduce delivery times (Piplani et al., 2020).

12.5.3 Tangible Benefits of Aligning Distribution with Consumer Demands

Aligning distribution with dynamic consumer demands can lead to a number of tangible benefits for businesses, including:

- Improved customer satisfaction: By meeting customer demand with personalized and convenient delivery options, businesses can enhance customer satisfaction and loyalty (Rao and Vemuri, 2017).
- Reduced costs: Data-driven distribution strategies can help businesses reduce costs by optimizing inventory levels, transportation routes, and last-mile delivery (Rushton et al., 2017).
- Improved efficiency: Integrated distribution strategies can improve overall supply chain efficiency, leading to increased productivity and reduced waste (Christopher et al., 2016).
- Enhanced sustainability: By reducing transportation emissions and using more sustainable delivery methods, businesses can contribute to a more sustainable future (Seuring and Gold, 2013).

12.5.4 Case Studies

Several case studies have demonstrated the effectiveness of data-driven approaches to integrated distribution strategies.

- Amazon: Amazon's use of data analytics to optimize delivery routes has resulted in significant cost savings and improved customer satisfaction (Hosseini et al., 2019).
- IKEA: IKEA's use of predictive demand forecasting has helped the company reduce inventory levels and improve product availability (Sodhi and Tang, 2019).
- UPS: UPS's use of telematics data to monitor vehicle performance has led to increased fuel efficiency and reduced emissions (Simchi-Levi et al., 2019).
- DHL: DHL's use of last-mile delivery optimization tools has improved delivery times and reduced transportation costs (Dai et al., 2017).

Dynamic consumer demands present a significant challenge for businesses, but they also offer an opportunity to develop more efficient and sustainable distribution strategies (Bhambri et al., 2023). By harnessing the power of data analytics and adopting integrated distribution approaches, businesses can meet the evolving needs of their customers while reducing their environmental impact (Wu et al., 2019). The future of distribution lies in aligning supply chains with the dynamic demands of consumers and the need for a sustainable future.

12.6 GLOBAL PERSPECTIVES ON SUSTAINABLE URBAN MOBILITY

The quest for sustainable urban mobility has led various cities around the world to adopt innovative strategies that harmonize technological advancements, policy initiatives, and community engagement. These case studies showcase the diversity and effectiveness of approaches in different geographical and socio-economic contexts, offering valuable lessons for cities aiming to transition toward more sustainable transportation systems.

12.6.1 Electrification of Public Transport in Scandinavian Cities

Scandinavian cities, renowned for their commitment to sustainability and quality of life, have made significant strides in electrifying their public transport fleets (Laugesen et al., 2020). Cities like Oslo, Stockholm, and Copenhagen have invested heavily in electric buses, trams, and ferries, reducing greenhouse gas emissions and improving air quality (Vulkan et al., 2019). The success of these initiatives can be attributed to comprehensive policy frameworks that include subsidies for electric vehicles, investment in charging infrastructure, and public–private partnerships to foster innovation. The electrification of public transport in these cities demonstrates the feasibility and benefits of transitioning away from fossil fuels in urban mobility.

12.6.2 Extensive Cycling Networks in the Netherlands

The Netherlands stands as a paragon of cycling infrastructure, with cities like Amsterdam and Utrecht boasting extensive networks of bike lanes, secure parking facilities, and bike-sharing programs (Rietveld and Daniel, 2004). This cycling culture is underpinned by policies that prioritize bicycle travel over car usage in urban planning, stringent safety regulations, and public campaigns that encourage cycling as a healthy, environmentally friendly mode of transport (Fishman et al., 2014). The Dutch approach illustrates how dedicated infrastructure, combined with supportive policies and cultural acceptance, can make cycling a dominant form of urban mobility.

12.6.3 Integrated Public Transport in Singapore

Singapore's approach to sustainable urban mobility centers on an integrated public transport system that seamlessly connects buses, trains, and light rail (Shaheen and Cohen, 2020). The city-state's Land Transport Authority has implemented a holistic strategy that includes the expansion of public transport networks, real-time information systems for commuters, and incentives for using public transport (Hensher, 2010). Singapore's strict vehicle ownership policies, coupled with substantial investment in public transport, have curbed car usage and promoted a shift toward more sustainable modes of transportation. This case study exemplifies how regulatory measures and infrastructural development can work in tandem to enhance urban mobility.

12.6.4 Pedestrian-Friendly Urban Design in Barcelona

Barcelona's innovative "superblocks" concept reimagines urban space by restricting traffic in certain areas to create pedestrian-friendly zones (Bosch and Garriga, 2017). This initiative has not only improved walkability but also facilitated the development of green spaces, recreational areas, and community hubs within the city (Marti and March, 2019). By prioritizing pedestrians and cyclists over vehicles, Barcelona has fostered a more livable urban environment that encourages active transportation and reduces carbon emissions. The superblocks model showcases the potential of urban design in transforming cityscapes into more sustainable and human-centered spaces.

12.6.5 Smart Mobility Solutions in Tokyo

Tokyo, a megacity facing significant transportation challenges, has turned to smart mobility solutions to enhance efficiency and sustainability (Ishida et al., 2020). The city has implemented advanced traffic management systems, the widespread use of electric and hybrid vehicles, and innovative last-mile delivery solutions using autonomous drones and robots (Kamijo et al., 2018). Tokyo's commitment to technological innovation, combined with strict environmental regulations, has positioned it as a leader in smart urban mobility. This case study highlights the role of cutting-edge technology in addressing the complex demands of urban transport in densely populated cities.

12.7 RESILIENCE TRANSPORT AND MOBILITY

The COVID-19 pandemic brought about significant shifts in the mobility patterns of residents and their consumer behavior, as supported by recent research findings (Pivtorak et al., 2022). Transitioning to remote work, a surge in online goods delivery services, concerns regarding personal safety and health, and a growing environmental consciousness advocating resource conservation have collectively contributed to alterations in mobility (Nikiforiadis et al., 2022). These changes have impacted both residents' travel choices and supply chains (Beckers et al., 2021).

Disruption in the transport system refers to significant interruptions or disturbances in the normal operations and services of transportation networks. These disruptions can stem from various sources, including natural disasters, technological failures, accidents, and socio-political events, leading to widespread implications for urban mobility, economic activities, and societal well-being.

Causes of transport system disruption:

- Natural disasters: Events like hurricanes, earthquakes, and floods can severely damage infrastructure, as seen in the case of Hurricane Katrina, which disrupted transportation networks in New Orleans and the broader Gulf Coast region (Thompson et al., 2017).
- Technological failures: System malfunctions or cybersecurity breaches can lead to significant disruptions. An example includes the 2017 system outage at British Airways, which led to the cancellation of flights globally, affecting over 75,000 passengers (*The Guardian*, 2017).
- Accidents: Incidents such as vehicle collisions, train derailments, or bridge collapses can cause immediate and severe disruptions (Hezam et al., 2020).
- Socio-political events: Strikes, protests, terrorism, and war conflict can also lead to disruptions. The 2023 rail transport strikes in Germany and Belgium significantly impacted daily commutes, showcasing the vulnerability of transport systems to labor disputes (*Euronews*, 2024). In today's context of the war in Ukraine and other conflicts around the world, both passenger and freight transportation systems have experienced substantial transformations. Passenger capacity has significantly diminished, and demand characteristics have shifted considerably. The destruction of infrastructure, such as roads, bridges, and railway lines, can cripple transportation networks (Cairns et al., 2012). Moreover, the safety of passengers and transportation workers becomes a paramount concern. The need to evacuate civilians from conflict zones and ensure the continued operation of essential services poses significant challenges (Ishfaq et al., 2020). By addressing the unique challenges faced by both public and freight transportation systems in these contexts and implementing sustainable and resilient practices, societies can enhance their ability to adapt and thrive in the face of adversity. Resilience

measures, such as redundancy in transportation routes and disaster preparedness plans, can help transportation systems withstand and recover from disruptions (Zheng et al., 2021).

Disruptions in transport systems pose significant challenges that require a multifaceted approach to mitigation and adaptation. By understanding the causes, impacts, and effective strategies for managing disruptions, policymakers and transportation planners can enhance the resilience and sustainability of urban mobility networks.

Strategies for improving the resilience ability:

1. Enhancing infrastructure resilience: Investing in robust and adaptable infrastructure can minimize the impact of disruptions. The concept of "resilient design" emphasizes the importance of flexible systems that can withstand and quickly recover from disruptions.
2. Implementing advanced monitoring and management systems: Technologies like real-time monitoring and predictive analytics can help in the early detection and management of potential disruptions.
3. Developing comprehensive emergency preparedness plans: Effective planning and coordination among various stakeholders are essential for managing disruptions. The establishment of emergency operations centers and clear communication strategies can enhance response efforts.
4. Fostering community engagement, transport policy, and public awareness: Educating the public about alternative transportation options and emergency procedures can improve community resilience and reduce the societal impact of disruptions.

12.8 CONCLUSION

The transformative effects of the digital age on urban transportation emphasize the pressing need for sustainable and efficient mobility solutions. It underscores the challenges faced by traditional transportation systems, such as congestion, pollution, and inefficiency, which have been exacerbated by rapid urbanization and increasing demand for transportation services. However, the integration of advanced technologies like IoT, AI, BD, DT, and S offers a promising avenue for addressing these challenges.

The future of urban mobility is shaped by digital transformation, policy frameworks, community engagement, dynamic consumer demands, and global perspectives. The integration of advanced technologies offers solutions to pressing urban transportation challenges, but it also necessitates robust policy frameworks and regulatory considerations. Community engagement and behavioral change are pivotal for promoting sustainable mobility, while data-driven approaches in distribution align supply chains with evolving consumer demands. Global case studies demonstrate the effectiveness of diverse approaches in achieving sustainable urban mobility. As cities continue to evolve and adapt to the demands of the digital age, they must prioritize sustainability, equity, and resilience to ensure the well-being of urban populations and the environment.

ACKNOWLEDGEMENTS

This research was supported by the European Union's Marie Skłodowska-Curie Actions under the MSCA4Ukraine funding scheme (project number 1232812, "Crowd-shipping initiatives for sustainable development of urban logistics".

REFERENCES

Axon, S., Morrissey, J., Aiesha, R., Hillman, J., Revez, A., Lennon, B., ... Boo, E. (2018). The human factor: Classification of European community-based behaviour change initiatives. *Journal of Cleaner Production*, 182, 567–586.

Beckers, J., Weekx, S., Beutels, P., & Verhetsel, A. (2021). COVID-19 and retail: The catalyst for e-commerce in Belgium? *Journal of Retailing and Consumer Services*, 62, 102645.

Bello, O., Afolayan, J. A., & Hossain, M. A. (2020). 5G Wireless communication systems: Prospects and challenges. *Sustainability*, 12(21), 8797.

Bernstein, S. (2017). The United Nations and the governance of sustainable development goals. In *Governing through goals: Sustainable development goals as governance innovation* (pp. 213–240).

Bhambri, P., Rani, S., Balas, V. E., & Elngar, A. A. (2023). *Integration of ai-based manufacturing and industrial engineering systems with the Internet of Things*. CRC Press.

Bosch, J., & Garriga, N. (2017). The "superblocks" of Barcelona: A novel urban design in the streets of a traditional European city. *Cities*, 60, 274–284.

Braekers, K., Ramaekers, K., & Van Nieuwenhuyse, I. (2016). Traveling salesman problems with profits. *Transportation Science*, 50(4), 1332–1348.

Busch, H., Ruggiero, S., Isakovic, A., & Hansen, T. (2021). Policy challenges to community energy in the EU: A systematic review of the scientific literature. *Renewable and Sustainable Energy Reviews*, 151, 111535.

Chen, Z., Ma, L., & Wang, W. (2012). Big data analytics with applications. In *2012 12th International conference on data mining* (pp. 703–710). IEEE.

Chien, S., & Ding, Y. (2018). Big data in urban transportation: Perspectives, applications, and lessons learned from the BDUTMS initiative. *IEEE Transactions on Intelligent Transportation Systems*, 19(12), 3932–3943.

Chopra, S., & Meindl, P. (2015). *Supply chain management: Strategy, planning, and operation*. Pearson.

Christopher, M., Peck, H., & Towill, D. (2016). A taxonomy for selecting global supply chain strategies. *International Journal of Logistics Management*, 8(2), 1–17.

Cohen, B., Kietzmann, J., & Van Zyl, C. (2018). Ride On! Mobility business models for the sharing economy. *Organization & Environment*, 31(3), 262–288.

Dai, H., Zhang, M., & Xu, L. (2017). A tabu search algorithm for the multi-objective electric vehicle routing problem with time windows and recharging stations. *Transportation Research Part C: Emerging Technologies*, 78, 32–49.

Davidich, N., Galkin, A., Davidich, Y., Schlosser, T., Capayova, S., Nowakowska-Grunt, J., ... & Thompson, R. (2022). Intelligent decision support system for modeling transport and passenger flows in human-centric urban transport systems. *Energies*, 15(7), 2495.

Fagnant, D. J., & Kockelman, K. (2015). Preparing a Nation for Autonomous Vehicles: Opportunities, Barriers and Policy Recommendations. *Transportation Research Part A: Policy and Practice*, 77, 167–181.

Fiorello, D., Martino, A., Zani, L., Christidis, P., & Navajas-Cawood, E. (2016). Mobility data across the EU 28 member states: Results from an extensive CAWI survey. *Transportation Research Procedia*, 14, 1104–1113.

Fishman, E., Washington, S., & Haworth, N. (2014). Bike share: A synthesis of the literature. *Transport Reviews*, 34(4), 496–520.

Fornell, C., Rust, R. T., & Dekimpe, M. G. (2020). The effect of customer satisfaction on consumer spending growth. *Journal of Marketing Research*, 57(6), 1019–1034.

Gonzalez, M. C., Hidalgo, C. A., & Barabási, A. L. (2008). Understanding individual human mobility patterns. *Nature*, 453(7196), 779–782.

Gracias, J. S., Parnell, G. S., Specking, E., Pohl, E. A., & Buchanan, R. (2023). Smart cities—a structured literature review. *Smart Cities*, 6(4), 1719–1743..

Grilli, G., & Curtis, J. (2021). Encouraging pro-environmental behaviours: A review of methods and approaches. *Renewable and Sustainable Energy Reviews*, 135, 110039.

Hensher, D. A. (2010). The value of travel time savings in Singapore: Stated preference research. *Transport Policy*, 17(6), 417–423.

Hezam, I. M., Nayeem, M. K., & Lee, G. M. (2020). A systematic literature review on mathematical models of humanitarian logistics. *Symmetry*, 13(1), 11.

Hosseini, S., Ivanov, D., & Dolgui, A. (2019). Review of quantitative methods for supply chain resilience analysis. *Transportation Research Part E: Logistics and Transportation Review*, 125, 285–307. https://www.euronews.com/travel/2024/01/22/europes-summer-travel-strikes-when-where-and-what-disruption-you-can-expect-in-august

Huang, X., Ding, Y., & Lu, X. (2019). A smart city application: A fully distributed sensor network based on blockchain technology. *Future Generation Computer Systems*, 100, 44–53.

Ishfaq, R., Zheng, C., Haider, J., & Zhang, Y. (2020). Urban transport challenges in post-conflict scenarios: A case study of Peshawar, Pakistan. *Journal of Transport Geography*, 85, 102684.

Ishida, T., Tanaka, Y., & Ikeda, K. (2020). Toward smart and sustainable urban mobility: Lessons from Tokyo. *Transportation Research Interdisciplinary Perspectives*, 8, 100253.

Jiang, S., Zhao, S., & Yin, H. (2017). Investigating the predictability of transportation mode patterns using smartphone-based mobile sensing. *Pervasive and Mobile Computing*, 37, 26–40.

Kamijo, N., Cao, J., & Akiyama, Y. (2018). Autonomous vehicle applications for urban mobility. *Transportation Research Procedia*, 35, 40–47.

Laugesen, N., Ehrenberg, A., & Wolter, L. (2020). Public transport electrification in Scandinavian cities: A comparative review. *Transportation Research Procedia*, 45, 123–129.

Li, X., Zhang, X., Zhao, J., & Yuan, Y. (2019). A review of big data architecture and technology from the perspective of Internet of Things. *IEEE Access*, 7, 36584–36597.

Litman, T. (2017). *Autonomous vehicle implementation predictions: Implications for transport planning*. Victoria Transport Policy Institute.

Liu, X., Sun, X., Zheng, H., & Huang, D. (2021). Do policy incentives drive electric vehicle adoption? Evidence from China. *Transportation Research Part A: Policy and Practice*, 150, 49–62.

Ma, Z., Ma, W., & Xu, C. (2019). Big data analytics in intelligent transportation systems: A survey. *IEEE Transactions on Intelligent Transportation Systems*, 20(10), 3833–3850.

Maltese, I., Marcucci, E., Gatta, V., Sciullo, A., & Rye, T. (2023). Challenges for public participation in sustainable urban logistics planning: The experience of Rome. In *Public participation in transport in times of change* (pp. 77–95). Emerald Publishing Limited.

Marti, C., & March, H. (2019). Superblocks in Barcelona: A successful urban transformation to improve the living conditions of citizens. *Frontiers in Sustainable Cities*, 1, 2.

Mouratidis, K. (2022). Bike-sharing, car-sharing, e-scooters, and Uber: Who are the shared mobility users and where do they live? *Sustainable Cities and Society*, 86, 104161.

NHTSA. (2017). *Automated driving systems 2.0: A vision for safety.* National Highway Traffic Safety Administration.

Nikiforiadis, A., Mitropoulos, L., Kopelias, P., Basbas, S., Stamatiadis, N., & Kroustali, S. (2022). Exploring mobility pattern changes between before, during and after COVID-19 lockdown periods for young adults. *Cities*, 125, 103662.

Oeschger, G., Carroll, P., & Caulfield, B. (2020). Micromobility and public transport integration: The current state of knowledge. *Transportation Research Part D: Transport and Environment*, 89, 102628.

Piplani, R., Piplani, R., & Boparai, A. (2020). Decision support systems for last mile delivery: A review of models and applications. *Computers & Industrial Engineering*, 148, 106797.

Pivtorak, H., Zhuk, M., Gits, I., & Galkin A. (2022). Shifting the population mobility of the Ukraine western region on the strength of the COVID-19 pandemic. *Archives of Transport*, 62(2), 146–160.

Przybylska, E., Kramarz, M., & Dohn, K. (2023). Analysis of stakeholder roles in balancing freight transport in the city logistics ecosystem. *Research in Transportation Business & Management*, 49, 101009.

Rao, C. R., & Vemuri, V. R. (2017). A comprehensive review of the last-mile problem: Models and methods in last-mile logistics. *Transportation Research Part D: Transport and Environment*, 53, 121–141.

Rietveld, P., & Daniel, V. (2004). Determinants of bicycle use: Do municipal policies matter?. *Transportation Research Part A: Policy and Practice*, 38(7), 531–550.

Rushton, A., Croucher, P., & Baker, P. (2017). *The handbook of logistics and distribution management: Understanding the supply chain.* Kogan Page Publishers.

Sakti, S., Zhang, L., & Thompson, R. G. (2023). Synchronization in synchromodality. *Transportation Research Part E: Logistics and Transportation Review*, 179, 103321.

Schultz, P. W. (2014). *Strategies for promoting proenvironmental behavior.* European Psychologist.

Seuring, S., & Gold, S. (2013). Sustainability management beyond corporate boundaries: From stakeholders to performance. *Journal of Cleaner Production*, 56, 1–6.

Shaheen, S., & Cohen, A. (2020). *Innovative mobility: Carsharing outlook carsharing market overview, analysis, and trends.* UC Berkeley Transportation Sustainability Research Center.

Shladover, S.E. (2018). Automated vehicle deployment: Should we wait for perfect?. *Transportation Research Part A: Policy and Practice*, 118, 131–142.

Shrivastava, A., Rizwan, A., Kumar, N. S., Saravanakumar, R., Dhanoa, I. S., Bhambri, P., & Singh, B. K. (2021). *VLSI implementation of green computing control unit on Zynq FPGA for green communication.* Wireless Communications and Mobile Computing.

Simchi-Levi, D., Kaminsky, P., & Simchi-Levi, E. (2019). *Designing and managing the supply chain: Concepts, strategies, and case studies.* McGraw-Hill Education.

Sodhi, M. S., & Tang, C. S. (2019). Research opportunities in supply chain finance. *Journal of Operations Management*, 65(5), 287–290.

Sustainable Development Goals. (2015). http://www.un.org.ua.

Teixeira, R., Leal, F., Leal, F., & Santos, B. S. (2017). Fleet management: A review of models for multi-objective vehicle routing problems. *Expert Systems with Applications*, 79, 59–82.

The Guardian. (2017). British airways: Chaos continues at heathrow.

Thompson, R., Garfin, D. R., & Silver, R. C. (2017). Evacuation from natural disasters: A systematic review of the literature. *Risk Analysis*, 37(4), 812–839.

Van den Berg, P., Van Den Berg, J., & Van den Hoed, R. (2017). Data governance for smart cities. In *Smart cities* (pp. 181–201). Springer.

Vulkan, N., Gullhav, S., & Malmström, M. (2019). Electrification of public transport in the Nordic countries: An explorative study of policy and progress. *Sustainability*, 11(17), 4649.

Wu, T., Djavadian, S., & Arentze, T. (2019). Integrating routing and job-shop scheduling for improving last-mile delivery operations. *Transportation Research Part E: Logistics and Transportation Review*, 130, 102–123.

Zheng, C., Wang, D., & Li, W. (2021). Developing resilient urban transportation systems in the era of global pandemics: Lessons from COVID-19. Cities, 111, 103035Top of Form

Zheng, S., Fan, D., & Li, X. (2017). Real-time urban traffic prediction with spatio-temporal correlations. *Transportation Research Part C: Emerging Technologies*, 81, 49–63.

Zheng, Y., Zhang, L., & Chen, W. (2018). IoT-based smart cities: A survey. In *2018 IEEE International conference on communications (ICC)* (pp. 1–6). IEEE.

Cairns, S., Sloman, L., Newson, C., Anable, J., Kirkbride, A., & Goodwin, P. (2012). Smarter choices: Assessing the potential to achieve traffic reduction using 'Soft Measures'. *Transport Reviews*, 22(5), 593–618.

13 Innovations in Sustainable Transportation for a Greener Tomorrow

Katarzyna Grondys

13.1 INTRODUCTION: BACKGROUND AND DRIVING FORCES

The significant role of transport, forwarding, and logistics services in the modern economy means that the contribution of these services also has a major impact on many different aspects of the business, social, and environmental environments. Progressive globalisation has contributed to an increase in the demand for the services of the TSL industry and is continuously causing its development, along with an intensive impact on public safety and health and the environment (Krawczyk, 2020). Therefore, the search for and design of innovative transport solutions that reduce the negative impact on the environment is part of the trends of a modern sustainable economy.

At the same time, the identification and creation of sustainable transport concepts and models is not a new process and accompanies every stage of civilisation development. Innovative solutions in transport have therefore been accompanying the development of the economy for some time; moreover, they often result from specific legal and environmental regulations. All pro-environmental measures are based on the use of state-of-the-art technology aimed at inhibiting the negative impact of transport (Moretti & Loprencipe, 2018). These solutions can be subdivided primarily into those used in freight and passenger transport and by broad system solutions. These divisions interlink and complement each other, creating synergies in the global transport system of tomorrow.

13.2 INNOVATIVE SYSTEM SOLUTIONS IN TRANSPORT

When analysing innovative system solutions, one observes numerous measures taken for this purpose both in individual economies and on a global scale (Starostka-Patyk et al., 2024). The so-called green concepts cover both freight and passenger transport and are regulated at the government level. Underpinning the solutions and tools

DOI: 10.1201/9781003475989-15

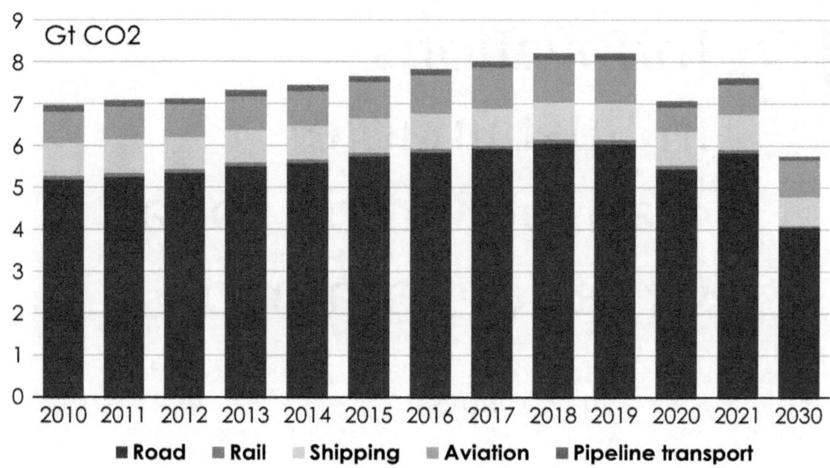

FIGURE 13.1 Global CO_2 emissions from transport (Global CO_2 emissions, 2022).

being designed is the concept of transported carbonisation, which aims to achieve zero carbon in freight transport. Current statistics indicate that carbon dioxide emissions from heavy vehicles only, i.e., lorries, buses, and coaches, account for around 25% of total carbon dioxide emissions from road transport and are on an increasing trend (Regulation (EU) 2019/1242…, 2019). Decarbonisation is the reduction or total exclusion of the production of carbon dioxide by humans. Therefore, it promotes environmental protection by stopping the destructive impact of carbon dioxide on the environment and its associated effects (Olojede, 2021).

Direct emissions from fossil fuel combustion in rail, road, sea, air, and pipeline transport are presented in Figure 13.1.

The largest contributor to carbon dioxide emissions is road transport, which emits 5 Gt on average per year, with the other modes of freight transport emitting around 10% of CO_2 on average. Due to regulations to reduce transport emissions, CO_2 reductions of up to a fifth are expected.

13.3 MODERN INFRASTRUCTURE

The overarching goal of a green future in transport is to prepare and upgrade infrastructures that co-create the main freight transport channels and can be effectively used in combination with innovative transport solutions. Innovation directions mainly focus on the expansion of inland transport infrastructure, the implementation of investments related to the construction of pipeline networks, and the encouragement of intermodal transport. Activities in the area of innovative infrastructure also include the opening and expansion of new routes for rail links, which from an environmental perspective are the most environmentally friendly means of transporting goods (OECD, 2020).

Although railways have the smallest share of greenhouse gas emissions from freight transport and the majority of rail networks operate with an electric overhead and live rail system, the railways still rely on diesel due to high infrastructure costs. In order to protect the environment, an innovative alternative to these cases is the use of electric trains powered by fuel cells. Within the European Union, almost half of the railway network is served exclusively by diesel trains, which have to store electricity themselves for this purpose (Gardner & Kries, 2020). As a result of the low battery capacity and high cost of batteries, battery power alone is not feasible. It is therefore optimal to use hydrogen cells to power the locomotive, especially on railway lines without electric traction. Additionally, hydrogen-powered trains could operate in hybrid mode (Rail Environment Policy…, 2021). On lines without an electric grid, hydrogen could be the source of energy, and when entering an electrified line, it would flow from electric voltage. An example of implementation is in Italy, where 20 trains have been equipped with battery hybrid technology. The train, made almost entirely of recyclable materials, can accommodate a maximum of 300 passengers. It has a special system that recommends an optimal speed to reduce energy consumption. Compared to diesel-only trains, hybrid trains can reduce fuel consumption by 50% (Kuta, 2023). In contrast, German rails began using passenger trains based on hydrogen technology for the first time in 2018. During operation, this train emits no CO_2, only water (Alstom CoradiaiLint, 2023).

Quite often in the context of green logistics, reference is also made to the use of inland shipping for urban transport services to relieve urban areas from pollution caused by road transport. A compilation of good practices for inland shipping in urban areas (Table 13.1) favours the reduction of car transport, protecting green zones in city centres from pollution by the last-mile transport.

The development of intelligent infrastructure and the transport innovations being implemented, e.g., smart parking, improve the mobility process, reduce congestion, and provide drivers with the necessary real-time traffic information. All of these contribute to an increase in the efficiency of transport processes and reduce greenhouse gas emissions from transport.

Solar roads, in which solar panels are integrated into the road surface, are also an innovation in sustainable transport. Such solutions allow the generation of clean energy while providing a sustainable transport infrastructure. An example project planned in Germany involves the construction of steel masts every 15 m along motorway shoulders, between which a structure will be installed to which photovoltaic panels will be attached (German highway…, 2023). This solution is consistent with the energy transition, which aims to use renewable energy sources. The energy produced from the solar panels is to be used to light the roads, to heat the roads when they are snowy, and to charge electric vehicles at stations along the motorway. Standard solar panels, whose use is growing year-on-year mainly among households and companies, are characterised by high efficiency, long service life, and low initial costs of use. An example of the application can be found in Switzerland, where the installation of solar panels on a retaining wall along a road is being tested (Lewis, 2023).

TABLE 13.1
Environmental Impact of Good Practices from Innovative Concepts in Inland Shipping (Janjevic & Ndiaye, 2014)

Type of Process Handled	Country of Implementation	Environmental Impact
Deliveries to local shops, restaurants, and hotels	Netherlands	Motor barge – reduction of emissions: particles (74%), CO_2 (27%), NO*x* (85%); electric barge – reduction of emissions: particles (98%), CO_2 (94%), NO*x* (100%);
Delivery of consignments	France	Reduction in tractor–trailer use by 15 per day, reduction in CO_2 emissions of 51 975 kg annually
Foating distribution centre	Netherlands	Reduction in tractor–trailer use by 10 per day; reduction in haulage work by 150,000 haulage units/year; reduction in fuel consumption by 12,000 litres/year
Deliveries to supermarkets	France	Shifting 450,000 km/year from car transport and reducing CO_2 emissions by 37% across the transport chain
Transport of food to shops	United Kingdom	Transfer of up to 350,000 tkm of freight work from road transport
Distribution of building materials on pallets	France	Annual decrease in car traffic by 2000 lorries; reduction of emissions by 220 t of CO_2 per year
Paper recycling	France	Annual decrease in automotive by 4,500 lorries, resulting in a reduction of 40% in fuel use and CO_2 emissions

An innovative alternative to traditional road infrastructure is pipeline transport, which is the use of lightweight capsules moving in pipelines directly from production companies to storage centres. The idea behind this transport is to design a pipeline system with a diameter that allows two Europallets to move simultaneously in a capsule weighing around 30 kg. The planned lifespan of the capsule should be around 30 months, while the transfer speed is estimated at around 40 km/h. Each capsule will be equipped with a safety system that will allow it to stop if an obstacle is detected. Between 30 and 50 capsules could replace a single European truck with a payload of up to 50 tonnes, and the fuel needed will be up to half that for transport by truck. This concept also considers the use of pneumatic propulsion for the system, which offers additional possibilities for the use of an air purification system and the use of carbon dioxide as a driving force for the pipeline. This solution will significantly reduce the level of heavy transport and therefore the associated pollution levels by up to 18% per year.

13.4 INNOVATIVE SOLUTIONS FOR FREIGHT TRANSPORT

Road transport is the most common mode of freight transport and, at the same time, the main source of pollutant emissions among all modes of transport; therefore, this mode of transport is focused on in this chapter. The general essence of greening activities in this area is to reduce emissions from the combustion of internal combustion engines used in freight transport. This causes a large share of transport innovation to focus mainly on solutions applied to the management of freight flows by road infrastructure. These solutions are conceptual and technological in nature. In the case of conceptual solutions, green innovation activities focus on the following (Becerra et al., 2021):

1. Packaging in transport – transport organisations that identify with an environmentally friendly approach to transport ensure that biodegradable packaging made from recycled and recyclable materials is used at an early stage in the planning of the transport of a particular commodity.
2. Optimising routes in line with the green supply chain concept: transported goods occupy as much available loading space as possible in order to minimise the number of so-called "empty runs" or raising the environmental awareness of end customers.
3. Selecting the right supplier, whose activities are part of a green supply chain, one of the criteria recently in force in the supplier selection process is the so-called environmental criterion, e.g., realisation of green competencies, green image, green product, pollution control, which prove the supplier's adaptation to environmentally safe operations.

The vast majority of innovations that have emerged or are emerging in road freight transport have a strictly technological dimension and include both solutions that have already been partially implemented in the transport business and those that remain at the testing stage.

Among the most advanced technological developments in road transport is the implementation of autonomous trucks. The most advanced technology of autonomous transport involves remote control of the vehicle by a computer system and does not require the driver's participation in the transport process (Mutabazi, 2023). Impersonal transport proves a significant reduction in pollutants' emission of pollutants compared to traditional means of transport, thus earning it the title of environmentally friendly freight transport. Impersonal vehicles are also preferred in terms of fuel consumption levels or safety aspects due to their greater ability to anticipate and react to traffic incidents (e.g., avoiding obstacles, ensuring the safety of pedestrians when trespassing, braking at an appropriate distance from traffic lights, controlling vehicle speed, etc.). It is estimated that by 2040, the level of fuel saved will increase by up to 10% (Berret et al., 2016). Furthermore, partially automated vehicles can result in fuel savings of up to 20%–30% (Zewe, 2022).

The development of autonomous transport is currently in the testing phase and is not yet widely used. The implementation of fully autonomous trucks is estimated to be around 2030. However, this requires changes to the road infrastructure and, above all, to regulations.

An extension of autonomous transport is the Autonomous Vehicle Platoon (AVP) or truck platooning, a concept of integrated truck convoys, whereby vehicles in a column adjust their speed to that of the vehicle at the head of the convoy (Syed & Abadin, 2020). The vehicles move at shorter distances between each other than in standard situations. This solution can be used on highways and expressways. Combined with the constant speed of the convoy, this allows for a reduction in fuel consumption, and thus a reduction in pollution resulting from fuel combustion and improved traffic safety. It is predicted that driving in a convoy can reduce the carbon dioxide emissions of the following vehicles by up to 16% and those of the convoy lead vehicles by 8%. The reduction in fuel consumption can be up to 7%. A convoy of this kind can be planned and include only fixed vehicles, or it can include a convoy of different vehicles (the real-time AVP) that decide in real time to join or separate from the convoy (the dynamic convoy) (Wu et al., 2022).

A similar concept is the road train, which is defined as a combination of vehicles consisting of a road tractor and a combination of trailers or semi-trailers connected to it. The use of a single tractor instead of several to make a particular delivery reduces costs associated with fuel consumption and also reduces environmentally harmful emissions. Australia has the longest and most heavy road trains, with a maximum length of more than 50 m and a weight of up to 200 tonnes (Bureau of Infrastructure…, 2022). Also, some European countries (Finland, Sweden, Germany, the Netherlands, and Denmark) allow trucks with a maximum length of 25.25 m. In most European Union countries, there is a maximum vehicle combination length limit of 18.75 m.

Another upgrade that has had a positive impact on the environment is the introduction of a new type of tractor trailers. The upgraded semi-trailers have been slimmed down, a double-decker freight transport system has been introduced, and technology has been used that allows energy to be generated during freight transport. The category of innovative green solutions in road transport therefore plays a significant role in the area of innovative trailers, where a distinction can be made (Brach, 2016):

- Mega semi-trailers – semi-trailers equipped with a lifting roof during loading, which makes it possible to transport goods up to 3 m high. Compared to a standard semi-trailer, a mega semi-trailer has larger dimensions and therefore allows more goods to be transported, reducing the number of journeys required for the required load size.
- Semi-trailer axles with electric drive – the operation of the first axle is based on its recovery of energy from the braking system on the basis of the so-called recuperation. The current generated in this way provides an alternative source of power for the vehicle, for example, providing temporary operation of cooling equipment solely via electricity. The solution reduces exhaust emissions and fuel consumption, reduces noise levels, or extends the service life of the components of the cooling equipment. This is complemented by a semi-trailer axle with an additional electric drive to drive the wheels of the semi-trailer. This solution relieves the tractor's

combustion engine and improves the dynamics of the vehicle combination (Kulikowska-Wielgus, 2018);

- Lean/lightweight semi-trailers and trailers are created through a unique method of machining sheet steel reminiscent of the Japanese art of origami by bending and folding paper. Innovative technology makes it possible to design much lighter trailers with the same carrying capacity as traditional trailers but requiring much less material for their construction (Rani et al., 2023). This increases the efficiency of the haulage process by reducing the weight of the kit and damage to road infrastructure. Less vehicle weight also reduces exhaust emissions. The slimmed trailer, compared to a traditional trailer, reduces the assembly of parts and materials to 10% and 70%, respectively, significantly simplifies the trailer manufacturing process and reduces the negative impact of this production on the environment, in addition to having more than 5% more payload capacity. This technology can be used to construct double-decker trailers.
- Double-decker semi-trailer – a semi-trailer with the ability to load pallets on two levels, allowing almost twice as many to be placed on the semi-trailer as on a normal semi-trailer. A second load level with a maximum load capacity of ten tons uses steel rails and aluminium beams to secure the load. This solution adds little to the trailer's own weight and therefore has no significant impact on its load capacity limitations. At the same time, it allows for much better management of loading space and reduces the number of trips when transporting goods on unstacked pallets and is completely weatherproof;
- Refurbishing old semi-trailers and restoring used semi-trailers serve as a green alternative to investing in new trailers. Using current greenhouse gas emission conversion factors, it has been calculated that the production of a new curtain-sided semi-trailer has a carbon footprint of 18.6 tonnes. In comparison, the refurbishment of a used curtain-sided semi-trailer, due to the materials and energy used, results in emissions of around 2.7 tonnes of carbon dioxide (DSV actions circular economy…, 2023).

Creating trailers that help to reduce emissions, using e-mobility solutions, e.g., electric axles indirectly powering generators. Sustainability measures such as reducing the empty weight of trailers and using alternative materials for their manufacture are of great importance.

13.5 ALTERNATIVE FUELS AND PROPULSION SYSTEMS FOR ROAD TRANSPORT

Until recently, both freight and passenger transport by road was mainly made possible by traditional types of fuel, such as petrol or diesel. However, their combustion contributes significantly to pollutant emissions in road transport. To reduce harmful exhaust emissions, European-wide emission standards have been in force in most European countries for more than a dozen years. Currently, the EURO 6 standard is in force for new vehicles entering the market, for which the nitrogen oxide emission value is a maximum of 400 mg/kWh and the particulate emission level is 10 mg/kWh (Commission Staff Working Document…, 2022). Furthermore, vehicles in this combustion class are also more economical. Legislation and climate change are forcing the automotive industry to be constantly active in the search for innovative solutions. One way to reduce exhaust emissions is to use alternative fuels and propulsion systems, which are increasingly popular among both transport companies and the private customer market. The fuel types that are most commonly used to reduce emissions into the environment are as follows (Unnasch & Goyal, 2017):

- Liquefied petroleum gas (LPG) – directly extracted from some natural gas and oil wells, comprising propane, butane, and ethane in winter. It is used as a propulsion fuel for cars with petrol engines;
- Natural gas (CNG, LNG) – more than 90% of its composition is methane. It can exist in two forms: compressed (CNG) and liquid (LNG). It is used as a propellant for gasoline and diesel cars.

Innovative types of propulsion in road transport include electric propulsion, hydrogen propulsion, and hybrid propulsion.

Battery electric vehicles (BEVs), powered by battery energy, have a significant place in the concept of sustainable transport. An electric vehicle (EV) uses an electric motor or motors powered by a battery pack, which is recharged at a dedicated charging point or on the home electricity grid (Fara et al., 2021). It is a solution to reduce greenhouse gas emissions while moving away from fossil fuels. Electric motors provide an environmentally friendly means of transport that do not emit pollutants into the air (Bhambri et al., 2023). Electric vehicles also provide quiet engine operation, are easy to install, and have zero tailpipe emissions. A major concern is the availability of vehicle charging stations, especially in small towns and cities (cited as the main reason inhibiting the purchase of electric vehicles worldwide by 34% of respondents), the short range of possible kilometres (33%), and whether the initial purchase costs are high (27%) (Leading reasons inhibiting…, 2024). The rate of availability of public EV chargers per kilometre of paved main road in selected countries in 2022 is presented in Figure 13.2.

China dominates when it comes to the number of electric vehicle chargers per kilometre of road. It has 1.76 million free and fast chargers, the largest public charging infrastructure in the world. The United States only ranks fifth in terms of the indicator studied, but in terms of the number of chargers, the country is already second in the world (around 200,000 units). The solutions used in the EV charging process and their features are described in Figure 13.3.

A not-so-optimistic aspect of electric vehicles is presented by research conducted at the University of Leeds. According to them, in addition to the limited availability of public chargers, the main reason inhibiting the purchase of electric vehicles worldwide (34% of respondents) is also the low mileage range. Furthermore, an electric vehicle puts more than twice as much stress on the roads as a petrol vehicle and almost twice as much as a diesel vehicle. Moreover, this factor increases with the weight of the vehicle (Lorimer, 2023). Current BEVs are mainly designed for short-haul routes of up to 600 km per day (Siddiqi & Edmondson, 2023). The level

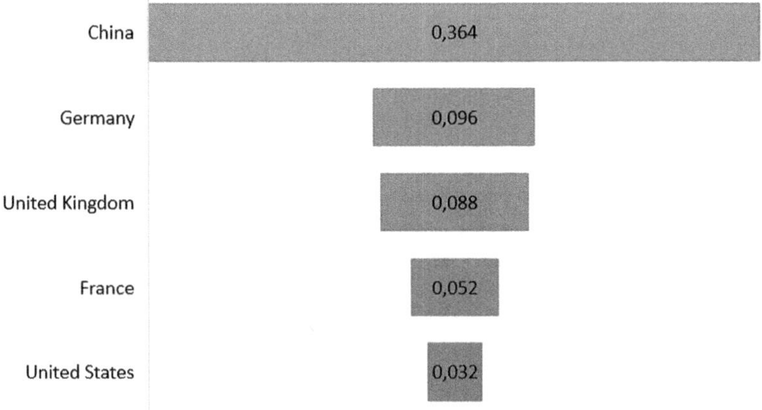

FIGURE 13.2 Electric vehicle charger availability rate in selected countries around the world in 2022 (Number of publicly available…, 2024).

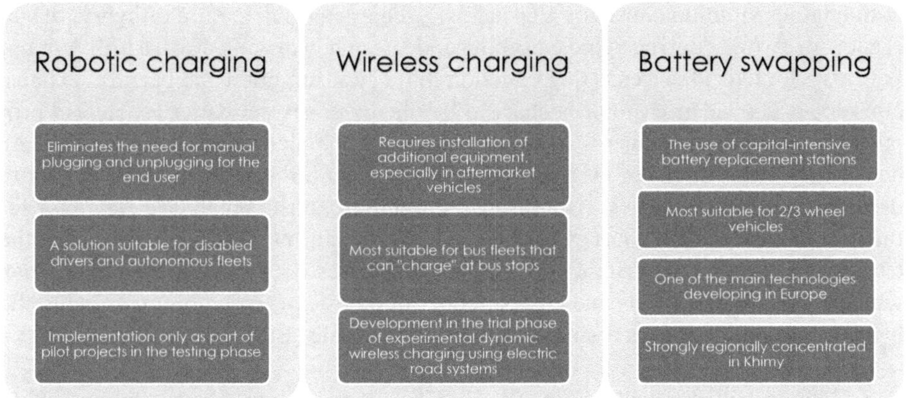

FIGURE 13.3 Solutions used in the electric vehicle charging process.

of engine efficiency of zero-emission trucks according to their weight is shown in Figure 13.4. It is the highest for the heaviest electrically charged vehicles and the lowest for fuel cell vehicles weighing less than 15 tonnes.

One of the latest ways of counteracting harmful emissions is the use of hydrogen propulsion, including heavy goods vehicles. The hydrogen-powered tractor–trailer prototype has an engine output similar to that of diesel trucks with a maximum range of 500 km. The vehicle is equipped with an electric motor and two sets of fuel cells that generate electricity from hydrogen. Using hydrogen to generate emission-free electricity, hydrogen fuel cell technology significantly expands the scope of sustainable mobility.

Another innovative solution is the hybrid drive system, which combines an internal combustion engine and an electric motor. HEVs (hybrid electric vehicles) can use

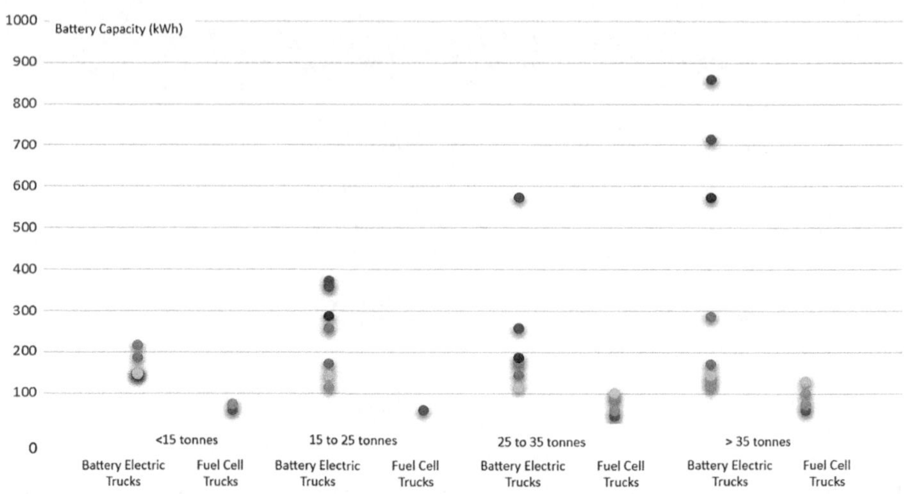

FIGURE 13.4 Battery capacity by truck weight (Shazan S., 2023).

both engines simultaneously or alternately. A characteristic feature of hybrid drives is energy recovery during vehicle braking and also when driving down a hill. A major advantage of hybrid drives is the reduction in fuel consumption and harmful exhaust emissions. The electric motor is charged by means of several batteries. Hybrid propulsion is used in passenger vehicles, heavy goods vehicles, and urban transport. As heavy-duty vehicles make far fewer stops, even downhill gradients of up to 1% are used to charge the batteries. As a result, the vehicle can run on an electric motor for the vast majority of the journey during long-distance transport. This means that the truck can cover even a distance of several kilometres on the electric engine alone, which reduces noise levels and completely eliminates exhaust emissions. Due to the hybrid technology used, a distinction can also be made (Burt, 2020):

- Plug-in Hybrid Electric Vehicle (PHEV) – these vehicles have a larger battery pack that allows them to travel longer distances at higher speeds; in addition, the electric and petrol engines can operate independently.
- Fuel cell electric vehicle (FCEV) – an all-electric vehicle that, in order to charge a large battery pack, uses oxygen from the air and hydrogen from an onboard tank to generate electricity in a device called a fuel cell.
- The extended-range electric vehicle (E-REV) – a vehicle with a much larger battery that uses an internal combustion engine to run a generator, which in turn provides power to the electric motor and recharges the battery when it discharges.

Table 13.2 presents a summary of innovative powertrain technologies and fuel types with their characteristics.

Challenges facing the electrification of trucks concerning mainly the battery and the need for its frequent recharging are somewhat dismissing the concept of

TABLE 13.2
Characteristics of Innovative Technologies in Powertrains and Fuels (Hacker & Helms, 2019)

Type of Drive Systems	Fuels	Maximum Range	Pollutant Emissions	The Weaknesses
Diesel drive system fuels	Fossil diesel	>1,000 km	High	High levels of pollution and noise production, limited scope for further development while remaining cost-competitive
	Renewable diesel fuels such as biodiesel, hydrotreated vegetable oils (HVO)	>1,000 km	A significant reduction	The availability of fuels is limited, since they are based on vegetable oils and waste streams
	Methanol	>1,000 km	The reduction of greenhouse gas emissions	Lack of infrastructure for its distribution, low energy efficiency, risk due to its toxicity
	Dimethyl ether (DME)	>1,000 km	No soot formation during combustion, making it easier to reduce NO*x*	Lack of infrastructure for its distribution and limited availability of vehicles with the appropriate installation
	Electrofuels (e-fuels)	>1,000 km	The reduction of CO_2; potential to balance the power system and store energy over longer times	Significantly higher power requirements compared to direct electricity consumption, high production cost
Liquefied natural gas and biogas	Liquefied natural gas (LNG) and liquefied biogas (LBG)		The reduction of greenhouse gas emissions	Because they are fossil fuels, they require complex station technology, carry the risk of methane leakage, and have limited availability of stations

(Continued)

TABLE 13.2 (CONTINUED)
Characteristics of Innovative Technologies in Powertrains and Fuels (Hacker & Helms, 2019)

Type of Drive Systems	Fuels	Maximum Range	Pollutant Emissions	The Weaknesses
Hydrogen and fuel cells	Hydrogengas	>800 km	No emissions during using vehicle	Lack of infrastructure for its distribution, low energy efficiency, limited availability of stations especially for heavy duty vehicles
Battery electric drive systems	Pure battery electric	<400 km	No emissions during use of vehicle	Requires very high battery charging power for heavy-duty vehicles travelling long distances; limited network of charging points

implementing fully autonomous vehicles in favour of hydrogen cell-based technologies. Although they too are not without their drawbacks, many industry experts consider hydrogen vehicles to be more promising than electric vehicles (Road transport in Poland 2023, 2023). The prospects of implementing the described technologies in transport were considered in the TLP and Spot Data surveys (Road Transport in Poland 2023). According to them, almost half of the representatives of the TSL industry believe the described technologies will become widespread within 5–15 years (Figure 13.5). The technology that, according to those surveyed, will find

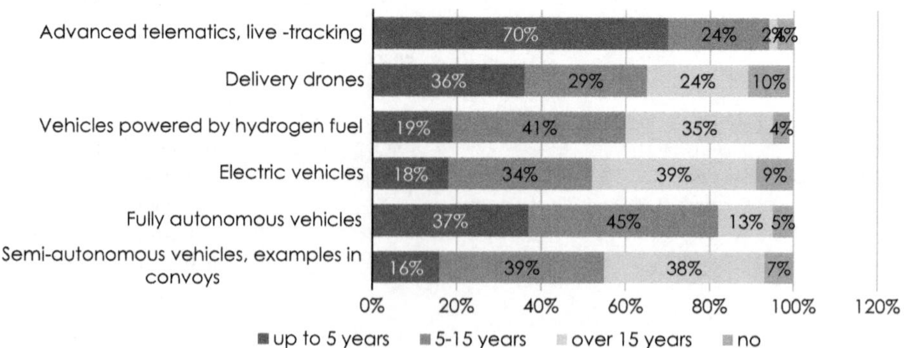

FIGURE 13.5 Forecasts for the deployment of innovative technologies in transport (Road Transport in Poland 2023).

widespread market use the fastest, within the next five years, is developed telematics and live-tracking.

13.6 PUBLIC TRANSPORT SOLUTIONS

An important element in the implementation of the concept of sustainable development is public transport, which is formed by a set of interconnected and organised activities that aim to provide passenger traffic services within a given area (Nordin et al., 2020). It is one of the most important factors in the spatial development of urban areas. Its task is to ensure that residents can use the various facilities or institutions and to reduce the number of motor vehicles, especially in city centres. An increase in the use of public transport brings with it a reduction in emissions and congestion. This effect is also actively reinforced by applying or testing innovative solutions that already exist in freight transport. In order to improve the sustainable transport system in the city, autonomous delivery vehicles are also intended to fully replace traditional modes of transport in urban areas. The aim is to reduce carbon dioxide emissions, lower noise levels, and reduce traffic congestion in cities.

Urban transport sustainability relies on alternative fuels and energy sources. Biofuels, hydrogen fuel cells, and electric vehicles are examples of low-carbon transport options that effectively implement a sustainability strategy. For this reason, the use of hybrid or electric buses and passenger vehicles is becoming increasingly popular. Electric vehicles are increasingly powered by renewable energy sources such as solar and wind power, further reducing greenhouse gas emissions.

Shared mobility and innovative means of transport are also working towards sustainable transport. These include bike sharing, car sharing, or so-called micromobility options such as electric bicycles and electric scooters (Fanchao & Gonçalo, 2022). Many transportation companies like Uber, for example, offer electric cars and ride-sharing services to provide environmentally friendly transport in cities. Mobile applications for integrating these modes of transport, on the other hand, make it possible to share real-time information and optimise routes in urban areas. Such solutions not only reduce congestion but also minimise emissions from transport systems. What is more, new innovative concepts are constantly being developed in cities all over the world in order to move towards sustainable transport. These aim to design energy-efficient, cost-effective, and high-speed public transport systems (Lopez, 2023), which are intended to provide an alternative to current transport strategies in the future. An example of this is the Siemens prototype autonomous capsules, which, based on the phenomenon of magnetic levitation, can transport passengers along a special overground track (Siemens Xcelerator…, 2023).

The transport sector is in a constant state of flux, pursuing the goal of a fully sustainable global transport system. Challenges, including the reduction of pollution from transport activities, congestion phenomena, and limited resources, are constantly creating new opportunities for innovation and change. For example, the deployment and use of low-carbon transport options effectively reduce greenhouse gas emissions, and the resulting improvement in mobility has a positive impact on economic growth.

With the technological solutions introduced, the implementation of innovative changes and concepts, and the creation of corresponding regulations, there is an opportunity to make a real impact on the environment. Innovative transport solutions that reduce emissions and improve mobility are key to the decarbonisation of the transport sector and a sustainable green future for transport operations.

REFERENCES

Alstom CoradiaiLint – the world's 1st hydrogen powered passenger train (2023).https://www.alstom.com/solutions/rolling-stock/alstom-coradia-ilint-worlds-1st-hydrogen-powered-passenger-train

Becerra P., Mula J., Sanchis R. (2021). Green supply chain quantitative models for sustainable inventory management: A review, *Journal of Cleaner Production*, 328, doi.org/10.1016/j.jclepro.2021.129544.

Berret M., Mogge F., Schlick T., Fellhauer E., Söndermann Ch., Schmidt M. (2016). Global automotive supplier study 2016, Being prepared for uncertainties, Lazard & Co. GmbH and Roland Berger GmbH, https://www.rolandberger.com/publications/publication_pdf/roland_berger_global_automotive_supplier_2016_final.pdf

Bhambri P., Singh S., Jain S., Dhanoa S. I. (2023). Plants recognition using leaf image pattern analysis with focus on advanced smart computing technologies, in AIP Conference Proceedings 2916, 020003.

Brach J. (2016). The ways of cutting of CO2 Emissions in the road freight transport, Zeszyt yNaukoweUniwersytetuGdańskiego. EkonomikaTransportuiLogistyka: Challenges of transport development, *Enterprise-Technology Policy*, 59, 283—297.

Bureau of Infrastructure and Transport Research Economics Rail Trainline (2022). Statistical report, department of infrastructure, Transport, Regional Development and Communications Canberra, Australia.

Burt M., PHEV, BEV, FCEV? Hybrid, electric and fuel cells explained (2020).https://mag.toyota.co.uk/phev-bev-fcev-powertrains-explained/

Commission Staff Working Document Impact Assessment Report Annex 5–8 (2022). Accompanying the proposal for a regulation of the European Parliament and of the council on type-approval of motor vehicles and of engines and of systems, components, and separate technical units intended for such vehicles, with respect to their emissions and battery durability (Euro 7) and repealing Regulations (EC) No 715/2007 and (EC) No 595/2009, Brussels.

DSV actions circular economy plan to refurbish 1,100 trailers (2023).https://www.globaltrailermag.com/2023/01/20/dsv-actions-circular-economy-plan-to-refurbish-1100-trailers/

Edmondson J., Holland A., Jeffs J., Siddiqi S. (2023). Electric vehicles: Land Sea & Air 2024–2044, https://www.idtechex.com/en/research-report/electric-vehicles-land-sea-and-air-2024-2044/962

Fanchao L., Gonçalo C. (2022). Electric carsharing and micromobility: A literature review on their usage pattern, demand, and potential impacts, *International Journal of Sustainable Transportation*, 16(3), 269–286, doi:10.1080/15568318.2020.1861394

Fara A., Ambikapathy A., Thangavel S., Logavani K., Arun Prasad G. (2021). Battery electric vehicles (BEVs), in Patel N., Bhoi A. K., Padmanaban S., Holm-Nielsen J. B., Eds. *Electric vehicles*, Springer, Singapore, doi:10.3390/systems10060250

Gardner N., Kries S. (2020). The low emissions railway, https://www.europebyrail.eu/the-low-emissions-railway/

German highway PV could generate up to 200 TWh a year (2023).https://www.pv-magazine.com/2023/04/11/german-highway-pv-could-generate-up-to-200-twh-a-year/

Global CO2 emissions from transport by sub-sector in the Net Zero Scenario 2000-2030(2020). https://www.iea.org/data-and-statistics/charts/global-co2-emissions-from-transport-by-sub-sector-in-the-net-zero-scenario-2000-2030

Gustavsson M., Hacker F., Helms H. (2019). Overview of ERS concepts and complementary technologies, Report from CollERS: Overview of ERS concepts and complementary technologies.

Janjevic M., Ndiaye A. B. (2014).Inland waterways transport for city logistics: A review of experiences and the role of local public authorities, *WIT Transactions on The Built Environment*, 138, doi:10.2495/UT140241.

Krawczyk P. (2020). Innovations in image creation and management – a case study of an entity from the TSL industry, *ZeszytyNaukowe. Organizacja i Zarządzanie / Politechnika Śląska*, 149, 351–359, doi: 10.29119/1641-3466.2020.149.30.

Kulikowska-Wielgus A. (2018). This innovative solution will allow for considerable savings. Axles of a semi-trailer generate electrical energy, https://www.globaltrailermag.com/2018/10/12/saf-holland-responds-to-electrification-trend/.

Kuta S. (2023). Europe's First Battery-Powered Trains Are Here. The tribrid trains now running in Italy can switch between battery power, electricity and diesel, https://www.smithsonianmag.com/smart-news/first-battery-powered-trains-europe-180982468/

Leading reasons inhibiting the purchase of electric vehicles according to consumers worldwide as of March 2022, Statista database 2024.

Lewis M. (2023). Switzerland put vertical solar panels on a roadside retaining wall, https://electrek.co/2023/12/01/switzerland-put-vertical-solar-panels-on-a-roadside-retaining-wall/

Lopez L. (2023). Sustainable transportation solutions: A path to a greener future, https://go-freight.io/sustainable-transportation-solutions-a-path-to-a-greener-future/.

Lorimer K. (2023). EV pothole threat prompts call for concrete roads,https://www.constructionnews.co.uk/sustainability/ev-pothole-threat-prompts-call-for-concrete-roads-30-06-2023/.

MorettiL.,LoprencipeG. (2018). Climate change and transport infrastructures: State of the art, *Sustainability,* 10, doi.org/10.3390/su10114098.

Mutabazi P. (2023). Self-drive/autonomous trucks and their impact on logistics, https://www.linkedin.com/pulse/self-driveautonomous-trucks-impact-logistics-patrick-mutabazi/.

Nordin V., Kharitoshkin N., Czerwińska-Lubszczyk A. (2020). Rating of the activities of urban transport policy in the context of sustainable development, *New Trends in Production Engineering*, 3(1), doi: 10.2478/ntpe-2020--0030.

Number of publicly available electric vehicle chargers (EVSE) in 2022, by major country and type, Statista database 2024.

OECD. (2020). *Transport bridging divides, OECD urban studies*, OECD Publishing, Paris, doi.org/10.1787/55ae1fd8-en.

OlojedeO A. (2021). Transport decarbonisation in South Africa: A case for active transport, *Scientific Journal of Silesian University of Technology. Series Transport*, 110, 125–142. ISSN: 0209–3324, https://doi.org/10.20858/sjsutst.2021.110.11.

Rail Environment Policy Statement On Track for a Cleaner, Greener Railway, Department for Transport, London 2021, https://assets.publishing.service.gov.uk/media/60eee7498fa8f50c7f08ae4b/rail-environment-policy-statement.pdf

Rani, S., Kaur, J., & Bhambri, P. (2023). Technology and gender violence: Victimization model, consequences and measures, in *Communication Technology and Gender Violence* (Vol. 1, pp. 1–19). Springer.

Regulation (EU) 2019/1242 of the European Parliament and of the Council of 20 June 2019 setting CO2 emission performance standards for new heavy-duty vehicles and amending Regulations (EC) No 595/2009 and (EU) 2018/956 of the European Parliament and of the Council and Council Directive 96/53/EC, PE/60/2019/REV/1.

Road transport in Poland 2023. The report was prepared by SpotData on behalf of and in cooperation with the employers' association, Transport and Logistics Poland, SpotData's Analysis Centre, Warsaw 2023.

Shazan S. (2023). Zero emission truck tech: Future of hauling, https://www.idtechex.com/en/research-report/electric-and-fuel-cell-trucks-2024-2044-markets-technologies-and-forecasts/971

Siddiqi S., Edmondson J. (2023). Electric and fuel cell trucks 2024–2044: Markets, *Technologies, and Forecasts*, www.IDTechEx.com/eTrucks.

Siemens Xcelerator speeds development of maglev system to beat global traffic (2023). https://newsroom.sw.siemens.com/pl-PL/skytran-simcenter-magnet/

Starostka-Patyk M., Bajdor P. Białas J. (2024). Green logistics performance Index as a benchmarking tool for EU countries environmental sustainability, *Ecological Indicators,* 158(3), doi: 10.1016/j.ecolind.2023.111396

Syed M., Abadin Z. (2020). An architectural design pattern for autonomous vehicle platoon. In Proceedings of European Conference on Pattern Languages of Programs (EuroPLoP 2020) Virtual Event, Germany, doi.org/10.1145/3424771.3424813

Unnasch S., Goyal L. (2017). Life cycle analysis of LPG transportation fuels under the Californian LCFS, Life Cycle Associates Report LCA.8103.177.2017, Prepared for WPGA.

Wu B., Wu Q., Ying Z. (2022). GAP-MM: 5G-enabled real-time autonomous vehicle platoon membership management based on blockchain, *Security and Communication Networks*, doi: 10.1155/2022/7567994.

Zewe A. (2022). On the road to cleaner, greener, and faster driving, *MIT News Office*, https://news.mit.edu/2022/ai-autonomous-driving-idle-0517

Part III

Challenges, Risks, and Ethical Considerations

14 Environmental Challenges and Technological Solutions

Rachna Rana and Pankaj Bhambri

14.1 INTRODUCTION

The significance of the environment in solving ecological, community, and money-making issues is increasingly acknowledged. here is knowledge of this in the domains of concern and insurance policy, as well as in society at large. Since the inception of its Framework Programme for Research and high-Tech evolution (FP), the European Union (EU) has provided substantial financing for environmental research.

The primary focus of these programs was on biodiversity surveys, improving our knowledge of ecosystem functioning and architecture, evaluating ecosystem services, and determining how vulnerable these ecosystems were to external stresses. The Sixth Framework Programme (FP6; 2002–2006) places research within a social–ecological framework, demonstrating the increasing consolidation of social group and public opinion discipline (Anderson and Mammides, 2020).

The use of environment-based solutions (EBS) to protect human welfare, improve biodiversity, and mitigate and adapt to climate change is becoming more and more common. But not every environment has received the same amount of attention since the advent of EBS. One such neglected habitat that is declining in number and quality is pools and collections of pools within a landscape.

Since the European Union (EU) is thought to have had a significant impact on the conceptualization and internationalization of EBS, it is looked for potential and barriers to the adoption of pools and pool's group as EBS by conducting a quantitative content analysis of 38 EU legislations. Although the researchers concentrate their research on pools and pool spaces, they also draw some conclusions about the expression of the EU argumentation setting for EBS generally due to their widespread occurrence in many landspaces and myriad of advantages.

They have discovered that EBS associated with current and prospective Natura 2000 areas, as well as with biospheres safeguarded by the EU's Water Framework Directive (WFD) and Birds and Habitats Directive (BHD), are strengthened by EU policies. However, the implementation of ecosystems as EBS may be hindered if they are not covered by these already accepted regulations, as is the case with the huge bulk of land and land-spaces (Biggs et al., 2016).

DOI: 10.1201/9781003475989-17

In many nations, renting agricultural land is a regular practice. Renting land gives farmers more options when it comes to the volume of their output. While research analyzing these effects has shown conflicting findings, land rental may have a variety of implications on farmland management (Rana et al., 2024). Maintaining the highest quality of farmland is essential to achieving the goal of feeding an expanding population via sustainable agriculture. Making decisions on farming investments is crucial. The intent of this investigation is to better interpret the weather conditions that purposefulness farmers' conclusion on what to invest in on the property they rent. The researchers conducted a survey using a questionnaire, and then they looked at 34 factors using multiple linear regression (Blicharska et al., 2016).

$R^2 = 0.22$ indicates that while the factors in our model are relevant in justifying investments, a considerable portion of the change remains inexplicable. This finding suggests that choosing investments is a difficult process. The unexplained variance may have been partially explained by our variables, but non-economic constituents influencing farmers' investment judgment, so much reliance or measure, may also have played a role. Furthermore, choices on investments might differ not only across farmers but even within an individual farmer's portfolio. Legislators find it difficult to create policies aimed at farmers who rent property since the options about how to handle leased land are so complicated (Blicharska et al., 2016).

The use and expansion of environment-based solutions (EBS) might be aided by the information gained from successful case studies. In order to obtain a general understanding of EBS practices and their contribution to decrease the negative effects of environmental condition modification and earthy disasters, this chapter analyzed 547 case illustrations. Compared to the remainder of the world, where they are below depicted, the bulk (60%) of case surveys are situated on the Continent. Out of 547 case surveys, 33% utilized green solutions, followed by assorted (27%), blue (10%), and composite (31%). Of these EBS initiatives, ecosystems found in rivers and lakes (24%) and urban areas (24%) accounted for around half (48%) of the total. Of which 92% of the case surveys were operationalized at the national (50%) and watershed (46%) scales of intervention, whilst just 4% were enforced at the landscape scale of measurement.

In addition, the findings demonstrated that 63% of EBS have been applied to address earthly disasters, environmental condition change, and diversity loss, with the other 37% focusing on socioeconomic issues (such as social justice, economic development, inequality, and cohesiveness). Approximately 88% of EBS implementations were facilitated by national policies, with the other 12% coming from municipal and regional frameworks. The majority of the examined examples supported the strategic goals for biodiversity and Sustainable Development Goals 13, 15, and 6. The co-benefits of EBS were also illustrated by case studies: 64% of them were environmental (such as enhancing biodiversity, air and water quality, and chemical element storehouse), 27% were social, and 9% were economic. By bridging the content space between science, policymakers, and professionals, this chemical process of case studies can help the approval and grading up of EBS for disaster risk reduction and environmental condition modification adjustment, as well as gain their preference in decision-making routines (Boothby, 1999).

14.2 ENVIRONMENTAL CHALLENGES

Environmental problems are caused by disruptions in the natural cycle of the environment. Current ecological issues can range from weather condition modifications and contamination to ecological degradation and resource depletion. These types of ecological issues can be caused by human activities or natural occurrences. They can range in scale from local to domestic or international. To address ecological issues, we can promote green innovations, reduce, reuse, and recycle items, conserve water and energy, avoid single-use plastics, and recycle waste to preserve raw materials.

The environment plays a vital role in sustaining life on Earth. However, as the population has grown, so has the need for food, clothes, fuel, and housing. This has put a lot of strain on earthly generator, leading to biological impurity, resource depletion, biodiversity loss, etc. (Singh and Bhambri, 2023). These problems are influencing the equilibrium of the Earth's system. Environment degradation is another term for environmental issues. It refers to the degradation of resources such as air, water, soil, biosphere, natural habitat destruction, wildlife elimination, and contamination. These problems need to be self-addressed globally in order to reduce their consequence and safeguard the environment for future generations. Environmental issues can be divided into two main categories: pollution and solid waste management (Briggs et al., 2019).

14.2.1 Pollution

Pollution is any unwanted modification in the bodily material or biologic features of the surroundings, such as air, soil, water, etc. Substances that cause pollution are called "pollutants". Wind contamination is the pollution of the air with vicious gases, dust, and fumes, mainly caused by the burning of fossil fuels, cars, industry, and foundries. Air pollution can negatively affect people's health, weaken crops, and cause plant death. Water pollution is the presence of harmful substances in natural bodies of water such as lakes, rivers, and oceans. Other pollutants include soil, noise, and radioactive pollution. The waste is either incinerated to reduce the quantity or disposed of in a hygienic landfill.

14.2.2 Solid

Solid garbage management entails the collection, disposal, and recycling of various non-liquid items such as paper, plastic, and organic garbage in order to reduce environmental impact and promote sustainability. Proper processing can save natural resources, lessen pollution, and reduce harm to the environment and public health.

14.2.3 Chemicals Used in Agriculture

The chemicals used in agriculture, such as fertilizers and pesticides, seriously harm the ecosystem. When they get into bodies of water, they can pollute the water, which harms aquatic ecosystems and people's health. Overuse of chemicals can lead to the loss of beneficial organisms, degraded soil, and nutritional imbalances. They can cause harm to non-target species and disturb the ecology. These substances may produce atmospheric gases that contribute to global warming. To solve these concerns, we must adopt sustainable agriculture practices and reduce our reliance on pesticides.

14.2.4 Worldwide Heating and Atmospheric Phenomenon

Worldwide heating and atmospheric phenomenon consequence, the somatic perception of the world and its surface, is said to have expanded due to global warming. The principal cause of temperature rise is the increase of atmospheric greenhouse gases, and this has brought about a dramatic warming of the planet (Bhambri and Khang, 2024). The atmospheric phenomenon effect is the name given to this warming of the planet and atmosphere. Gases from industry, fossil fuels, vehicle exhaust, and other sources are discharged into the atmosphere.

14.2.5 Radioactive

Radioactive waste from nuclear power plants and other industrial activities is a major environmental issue. Inadequate storage and disposal practices may result in soil and groundwater contamination, endangering human and ecological health over the long run.

14.2.6 UV Radiation

Since radioactive waste persists as a threat for thousands of years, managing it is challenging. Safe storage and disposal solutions are critical for preventing environmental damage and potential calamities like nuclear accidents or spills. To address this environmental issue, public awareness and rigorous laws are required. The destruction of the stratosphere's ozone layer is another significant concern. The ozone layer in the upper atmosphere absorbs sunlight and damaging UV radiation. The stratosphere maintains a balance between ozone generation and destruction. The growing consumption of ozone-depleting chlorofluorocarbons (CFCs) has disturbed this delicate balance. Ozone holes, which allow UV light to pass through, were created as a result of this. UV rays can result in DNA alterations, cataracts, skin cancer, and other conditions.

14.2.7 Deterioration Caused by Improper Use and Establishment of Resources

Natural resource degradation can be caused by inefficient resource utilization. Excessive cultivation, deforestation, unrestrained grazing, and inadequate methods of irrigation have all exacerbated soil erosion and desertification. Over the years, an inhospitable environment is created when these arid regions combine. The issue we face today is growing urbanization.

14.2.8 Water Collection and Soil Salinity

Water collection and soil salinity rose with the Green Revolution, resulting in irrigation problems and the formation of salt crystals on the surface. Water harvesting and soil salinity are also factors that influence yield.

14.2.9 Forested Lands

Forested lands become new forests as a result of deforestation. Agriculture replaced woodland areas due to an increase in population. This results in habitat loss, endangering many species and biodiversity. It also contributes to climate change since trees serve as carbon sinks. Deforestation affects both local and global ecosystems by causing soil to erode and altering the moisture cycle (Bhambri and Rani, 2024a). To lessen these biological challenges and protect the world and its trees, sustainable vegetation administration and restoration are needed (Debele et al., 2023).

Environmental issues are concerns that genuinely jeopardize the well-being of the planet and its ecosystem. Several disputes consider: merchandise of noxious matters into the situation, such as atmospheric pollution from commercial enterprise breathe and transportation fumes, water impurity from industrial and cultivation overflow, and marine pollution in the body of water.

Overpopulation is one of the most serious environmental issues, as the world's population has tripled in the last six decades. This places a strain on the ecosystem due to accumulating food supply, real estate development, and condition.

Because of atmospheric greenhouse gases, the earth's temperature and climate are rising, resulting in extreme weather events and sea level rise. Pollution, habitat destruction, overexploitation, and the introduction of exotic species are the main causes of species extinction. Large-scale deforestation for agriculture, logging, and urban growth causes habitat loss, decreased carbon storage, and changes in local and global climate (Faivre et al., 2017).

There are possible solutions to environmental problems:

- Debar single-use plastic.
- Publicity of fresh and low-cost produce.
- Keep property farming.
- Cut down nutrient body processes.
- Industrial plants show more woody plant growth and go material.
- Boost green concept.
- Prevention of water and energy.
- Use useful merchandise instead of usable products.
- Reprocess waste matter to save earthly resources.
- Assistance of local and geographical area affable preparation.

14.3 ENVIRONMENTAL CHANGE MITIGATION

Decreasing the sum of energy-housing atmospheric phenomenon gases incoming the standard atmosphere is the first step to edge condition modification. Some of this will reduce atmospheric phenomenon gas emissions from key sources such as industry, workplaces, vehicles, and powerfulness plants. Soil, forest, and sea play an important role in the solution because they absorb and store these gases. To reduce and prevent emissions, we need to change all aspects of our lives, including the way we produce energy, grow food, travel, live, and use things. There are problems here both locally and globally.

Thanks to the EU and its aggressive climate change policies in recent decades, EU emissions will be reduced by more than 31% in 2022 compared to 1990 levels. This is mainly due to an increase in the use of renewable energy sources and a decrease in the use of fossil fuels rich in carbon dioxide.

Getting the scores also changed the procession in vigor, skillfulness, and knowledge accommodation of the efficiency. More assertive references have also been set, such as a 55% net decrement from 1990 levels by 2030 and the goal of achieving condition non-participation by 2050. This definite quantity importantly reduces emissions when switching from fossil fuels to clean, exhaustible energy sources to get these aims. To achieve a balance between releasing greenhouse gases into the atmosphere and absorbing and storing them in soil, seas, and forests, we must also stop deforestation, manage the land responsibly, and restore nature.

The EU is only responsible for 6% of global emissions, so it is powerless to act alone (Gonzalez-Ollauri et al., 2021). Any means to curb climate change requires global cooperation. International cooperation in combating environmental modification is ensured by the United Nations Framework Convention on Climate Change (UNFCCC) and the Paris Agreement. In the European Union, greenhouse gas emissions decreased by 2% in 2022 compared to 2021 levels, according to the latest "Trends and Forecasts". The study warns, however, that despite progress in reducing carbon dioxide emissions, renewable energy, and energy efficiency, the EU #039 still has to meet aggressive climate and energy goals that must act quickly (Hatziiordanou et al., 2019).

14.4 ENVIRONMENT-BASED TECHNOLOGICAL SOLUTIONS

There are the following technological solutions for saving the environment.

14.4.1 Ocean Cleanup

Trash is a major problem, but it is especially common in our rivers and oceans. Particularly concerning is plastic contamination. Sea-bin is one ingenious way to approach the problem. It is a basic yet very useful device that functions similarly to a drain (Bhambri and Rani, 2024b). Trash and pollutants are collected within by the water as it washes over the top. Sea-bin is then discharged in a manner that is more eco-friendly. Other technologies are being developed in addition to sea-bin to assist in addressing the issue of plastic garbage in the ocean. Using self-propelled micro-robots that are programmed to break down hazardous plastic trash and aid in the preservation of the ocean is one such example.

14.4.2 Sunlight Panels

Utilizing renewable energy sources is necessary for environmental conservation, and even a tiny bit of effort may have an impact. Solar panel material-related technologies are growing. We can now utilize sunshine-powered glass in our houses and businesses, as well as commute on solar-paneled roads. Clearly, a lot of area is needed for roads in the developed world. Researchers are now working to maximize this space using technology. By harnessing the energy inherent in them, these solar panel highways may be used to generate heat and improve safety in addition to supplying power to the national grid. Installing solar glass panels in lieu of regular windows would make adding solar panels to your house much simpler.

14.4.3 Carbon-based Computing

Numerous well-known businesses have been researching applications for carbon-friendly computing. This reasoning is applied at their data centers, where a significant amount of work is completed during periods when renewable energy is most abundant. Less energy from fossil fuels is utilized, and more work is done when the sun and wind are shining. It's clear that adopting this approach may help organizations lessen their environmental impact, and the more often they do so, the better.

14.4.4 The Material of the Future

Although it isn't really a technical improvement in and of itself, this might be used in many other technologies in the future, perhaps even ones that have the ability to repair whole planets. One substance that was initially found by the University of Manchester is the material of the future. It is said to be translucent, conductive, flexible, and constructed of a thin coating of graphite, among other qualities. This implies that it may be applied to a wide range of fields, including photovoltaics and water filtration.

14.4.5 Electric Cars

Automobiles and other motorized vehicles have long been a major environmental issue. We depend on automobiles to get around, transport our belongings, commute to work, vacation, and many other activities, yet conventional fossil fuel-powered cars harm the environment. Electric cars are the way of the future; therefore, we should start using them as soon as possible.

14.4.6 Grown in a Lab

According to food research, cattle are accountable for 14.5% of atmospheric phenomenon gas emissions worldwide. Although many people are moving to lab-grown or non-meat choices, many will still consume meat because there are so many plant-based options available. Meat created in a lab is one solution. Because we wouldn't be raising animals for slaughter and all the consequences that go along with manufacturing and shipping, we would therefore have less of an influence on the environment. Researchers have also produced lab-grown meats that are developed to resemble genuine meat in terms of flavor and texture. 3D printed food is another solution to this problem. It could be possible in the future to print cakes, pizza, and other delectable delicacies that are on par with your current favorites.

14.4.7 Carbon-sucking Structures

Regretfully, not enough trees exist to absorb all of the CO_2 that humans emit. Fortunately, scientists and researchers have created a few methods for CO_2 absorption. An intriguing solution has been devised by a business named Eco-Logic Studio in the shape of an exterior shell for buildings made of algae. With the use of this technique, pollutants are drawn in, CO_2 from the algae is captured, and then photosynthesized oxygen is released back into the environment. After that, the algae may be used as fertilizer Air-Powered.

14.4.8 Energy

A technique that uses moisture in the air to create energy has been developed by the University of Massachusetts Amherst. In essence, it uses electrically conductive protein nanowires to create electricity out of thin air. This readily available, sustainable energy source may become increasingly prevalent in the future.

14.4.9 Innovative Glass Recycles

DB A machine designed by New Zealand breweries turns beer bottles into sand. The idea is that the generated sand will serve as a substitute for removing sand from nearby beaches and may be utilized to feed the building sector. Furthermore, after witnessing recycling in action, individuals will be more likely to recycle in the future.

14.4.10 Using Hydroponics

An intriguing and environmentally friendly agricultural method is hydroponics. It's quite easy to use; just combine plants with fish tanks. Fish feces is a natural source of nutrients that may be utilized as fertilizer. The plants receive nutrition by means of a hydroponics system, which also filters out nitrogen waste from the water before returning it to the fish tank.

14.5 SUSTAINABLE DEVELOPMENT GOALS

Clean Earth is a large non-profit environmental health organization. It works with authorities, communities, and business leaders to find and implement solutions that eliminate hazardous exposures, protect public health and restore ecosystems, pollution, and the UN for Sustainable Development (Potthoff and Dramstad, 2023).

14.5.1 Achieving 11 of the 17 SDGs Depends on Addressing Pollution

Environmental condition restraint measures assist in accomplishing the sustainable development goals of the United Nations, which includes measures that directly and indirectly contribute to a clean environment. One particularly important pollution goal is 3.9.2: Deaths from environmental pollution: to reduce disease and wellness induced by hazardous substances and pollution of air, water, and dirt. Impurity control improves access to clean water and sanitation services (SDG 6), builds sustainable cities and communities (SDG 11), supports sustainable economic growth (SDG 8, 9, 12, 17), and protects land and water (SDG 2, 14, 15).

Additionally, it advances world health (SDG 3). The 2030 Schedule for Sustainable Development, which was ratified by all UN member states in 2015, is a collective roadmap aimed at achieving prosperity for the group and the entire planet now and in the future. The 17 Sustainable Development Goals (SDGs) are the main focus, inspiring every newly established and functioning administrative district to take urgent action in matters of global commercial significance.

They infer that concluding impoverishment and other privation needs combined activeness, in addition to boosting worldly maturation, combating environmental condition modification, securing our seafaring and forests, rising well-being and pedagogy, and reducing differences. The 2030 Agenda for Sustainable Development Goals are the work of the UN Division of Social and Economic Affairs, in collaboration with other nations. At the June 1992 Earth Summit in Rio de Janeiro, Brazil, more than 178 countries adopted Schedule 21, an all-encompassing program of action to form a global partnership for sustainable development to improve people's lives and safeguard the status quo. The Millennium Declaration was irresistibly adopted in September 2000 at the UN business office in New York. As a result of the elevation, eight Millennium Development Goals were formed to cut down utmost poorness by 2015. The elevation on Sustainable Development in South Africa (2002) led to the proceedings of the Metropolis Statement and System of Human Action on Sustainable Development, which strongly emphasized planetary seriousness to situation protective cover and poorness obliteration. Based on Agenda 21 and the Millennium Statement, these written documents included "The Future We Want" and quote: Final document of the United Nations Conference on Sustainable Development (Rio+20) held in June 2012 in Rio de Janeiro, Brazil. Among other things, it was decided to create a high-level UN political forum for sustainable evolution and to start the organic evolution process with sustainable development goals substantiated based on the Millennium Development Goals.

The Rio+20 determinations also included other sustainable development initiatives, such as programs for funding for improvement, small earth processing nations, and other sectors in the future. In 2013, the General Assembly defined an open working group of 30 members to set up a report on the scores of sustainable development. The General Assembly began negotiations on the post-2015 development program in January 2015. At the UN Sustainable Development Summit in September 2015, the process culminated with the adoption of the 2030 Agenda for Sustainable Development, which focuses on 17 Sustainable Development Goals.

With the adoption of numerous essential agreements, 2015 became an accidental year for multilateralism and global policymaking: in March 2015: Sendai Framework for Disaster Risk Reduction; in July 2015 Addis Ababa Action Agenda on Financing for Financing; in September 2015, Changing the world with inside the UN Sustainable Development elevation; the 2030 movement application for sustainable improvement, which incorporates 17 sustainable improvement goals, was confirmed with inside the new Paris climate agreement (December 2015). Currently, thematic subjects of the Sustainable Development Goals (DSDG) department are the World Wide Sustainable Development Report (GSDR), Partnerships, Small Island Developing States, Water, Energy, Climate, Oceans, Urbanization, Transport, Science, and Technology.

The branch additionally offers essential aid and ability building on these issues. In addition to influencing sustainability goals and communiqué activities, the DSDG has a decisive function in evaluating the systematic implementation of Agenda 2030 and #039 at the United Nations. All parties to the 2030 Agenda need to strongly commit to the implementation of the Worldwide goals so that the SDGs are broadly recognized. DSDG aims to facilitate this communication.

14.5.2 2022 Sustainable Development Goals Report 07.07.2022

Based on contemporary estimates and available data, the record of the Sustainable Development Goals 2022 offers an overview of the progress made with inside the implementation of the 2030 Agenda for Sustainable Development. It tracks worldwide and nearby development in the direction of the 17 goals using comprehensive assessments of complete exams of precise signs for each goal. The report argues that the 2030 Agenda for Sustainable Development and human life are under extreme risk from a series of interconnected and sequential disasters. The report highlights the seriousness and scale of the challenges we face. All the Sustainable Development Goals (SDGs) are affected by the convergence of crises, specifically as a result of COVID-19, climate change, and wars. These crises have further implications for health, education, food and nutrition, environment, and peace and security (Reguero et al., 2022).

14.5.3 2023 Sustainability Goals Report

The special issue, which offers an honest assessment of the SDGs based on the latest facts and assessments, is a call to action. The study highlights the enormous potential for success in applying the significant political will and existing technologies, resources, and knowledge, but also notes the shortcomings and calls on the world to redouble its efforts. Working together, the worldwide community can accelerate progress toward the Sustainable Development Goals and create a more promising future for all.

The document argues that the consequences of the COVID-19 pandemic and the war in Ukraine, the consequences of the climate disaster, and the volatility of the worldwide economy have exposed shortcomings and prevented the achievement of the goals. The study also warns that while the lack of progress is worldwide, the poorest and most vulnerable people on the planet are suffering, leading to unprecedented worldwide challenges. It also identifies areas that require continuous care to prevent the Sustainable Development Goals and accomplish substantive advancement for groups and the situation by 2030.

14.5.4 The 2023 Worldwide Sustainable Development Report (GSDR)

The 2023 World Wide Sustainable Development Report (GSDR) titled "Times of Crisis, Times of Change: Science to Accelerate Change towards Sustainability" and notes that, given the critical midpoint of 2030, incremental and piecemeal changes will not be enough to achieve the SDGs over the remaining seven years. Active mobilization of political leadership and a strong desire for science-based reforms are essential for the 2030 Agenda and its implementation. Worldwide, this must be achieved without leaving any nation, community, or individual behind. The study is a call to welcome the changes needed to accelerate the SDGs and development.

14.6 REDUCTION OF DISASTER RISK (DRR)

Disasters that are "natural" only exist in the sense that they are natural dangers. Through an ethic of prevention, the disaster risk reduction (DRR) movement aims to decrease the harm caused by earthly disasters such as disruptions, torrents, droughts, and atmosphere. Natural dangers are typically followed by disasters. The magnitude of a disaster is determined by the extent to which a danger affects the environment and society. The decisions we make for our environment and our lives will determine how big of an influence it will have. These decisions affect our food production practices, the location and design of our houses, the structure of our government, the operation of our financial system, and even the curriculum in our educational institutions. We become more susceptible to calamities – or more impervious to them – with every choice we make and action we take.

14.6.1 Reducing Disaster Risk Involves Making Decisions

The idea and process of lowering catastrophe risks by methodically analyzing and minimizing the causes of disasters are known as disaster risk reduction. Disaster risk reduction includes lowering exposure to risks, reducing the vulnerability of people and property, managing land and the environment wisely, and enhancing readiness and early warning for unfavorable occurrences.

14.6.2 Everyone Has a Stake in Reducing the Danger of Disaster

Although disaster risk reduction (DRR) is a component of sustainable development, it also encompasses fields like catastrophe management, mitigation, and preparedness. Development initiatives must lower the risk of disasters in order to be sustainable. However, poor development strategies would raise the likelihood of disasters as well as their costs. DRR, therefore, encompasses all facets of society, government, and the business and professional sectors.

14.7 CONCLUSION

In summary, technology is influencing how we not only use the environment but also contribute to its preservation. Science and research have made great strides toward meeting our current environmental demands and situations. In this research, we evaluated the potential and constraints for pools and pool spaces to be implemented as EBS under the EU policy framework. Although there is strong evidence in favor of EBS overall, there are very few chances pertaining to pools and landscapes. Legally binding objectives and safeguards, initiatives to enhance biodiversity management and monitoring, and funding for EBS work together to support WFD-compliant protected areas, prospective and existing Natura 2000 sites, and carbon-rich places. The majority of pools and pool spaces do not fit into these groups. The EU's goal of zero pollution may present some chances for pools and poolscapes.

This includes managing activities at the poolscape scale, such as lowering pollution in agriculture, as well as the construction and restoration of pools and poolscapes, such as wastewater treatment pools. Similarly, pools and pool spaces may benefit from the goal of improving climate change adaptation and mitigation through EBS, such as through Natural Water Retention Measures (NWRM), particularly if they are regarded as (part of) wetlands. We contend, however, that implementation routes are less clear in these situations and rely far more on the aspirations and consciousness of individual landowners and lower-level governing bodies. In this regard, research projects – especially those included in Horizon Europe Missions – may be crucial to the open rationalization of EBS as they combine research efforts with funding, policymaking, and stakeholder involvement. Furthermore, our analysis highlights certain areas that require further investigation. Using "opportunities" and "limitations", we build on earlier research on the variables that facilitate and hinder the implementation of EBS by offering a conceptual tool to better characterize the possible combined impacts of these elements.

Subsequent research endeavors might potentially extend the application of the idea of possibilities and restrictions to non-pool surfaces (EBS) beyond pools and pool-scapes. This would facilitate the identification of potential recurrent patterns and improve the concept's analytical purchase via further refinement. To determine what factors might have a significant impact on the future of pools and pool-scapes, such study may also follow the path taken by European policies as they relate to and interact with local policies. To the best of our knowledge, no prior research has outlined in such detail the space that a supranational framework affords for the implementation of EBS. The next stage would be a thorough review of the actual implementation on the ground, as many choices still rest with lower-level public and commercial players, even if our study offers preliminary insights into the EU policy framework's support for EBS. In particular, this calls into question the following for pools and pool-spaces: (a) the use of wetland-focused measurements for pools; (b) the understanding of non-economic advantages in reality; and (c) the consideration of non-priority BHD habitats and generally non-BHD habitats as EBS. Last but not least, given the EU's role as an advocate for EBS even outside of its borders, future studies ought to assess the effects of the EU's emphasis on the management and restoration of carbon-rich areas as EBS, paying particular attention to the issue of land rights to these areas and the communities that may be impacted by EBS projects.

REFERENCES

Anderson, E., & Mammides, C. (2020). Changes in land-cover within high nature value farmlands inside and outside natura 2000 sites in Europe: A preliminary assessment. *Ambio*, 49(12), 1958–1971. https://doi.org/10.1007/s13280-020-01330-y

Bhambri, P., & Khang, A. (2024). Biosensor applications and principles of agricultural and aquacultural sectors. In A. Khang (Ed.), *Agriculture and Aquaculture Applications of Biosensors and Bioelectronics* (pp. 1–17). IGI Global. https://doi.org/10.4018/979-8-3693-2069–3.ch001

Bhambri, P., & Rani, S. (2024a). Ethical issues for climate change and mental health. In D. Samanta & M. Garg (Eds.), *Impact of Climate Change on Mental Health and Well-Being* (pp. 178–198). IGI Global. https://doi.org/10.4018/979-8-3693-2177–5.ch012

Bhambri, P., & Rani, S. (2024b). Challenges, Opportunities, and the Future of Industrial Engineering with IoT and AI. In *Integration of AI-Based Manufacturing and Industrial Engineering Systems with the Internet of Things* (pp. 1–18). CRC Press.

Biggs, J., Von Fumetti, S., & Kelly-Quinn, M. (2016). The importance of small waterbodies for biodiversity and ecosystem services: Implications for policy makers. *Hydrobiologia*, 793(1), 3–39. https://doi.org/10.1007/s10750-016-3007–0

Blicharska, M., Andersson, J., Bergsten, J., Bjelke, U., Hilding-Rydevik, T., & Johansson, F. (2016). Effects of management intensity, function and vegetation on the biodiversity in urban pools. *Urban Forestry & Urban Greening*, 20, 103–112. https://doi.org/10.1016/j.ufug.2016.08.012

Boothby, J. (1999). Framing a strategy for pool landscape conservation: Aims, objectives and issues. *Landscape Research*, 24(1), 67–83. https://doi.org/10.1080/01426399908706551

Briggs, A. J., Pryke, J. S., Samways, M. J., & Conlong, D. E. (2019). Complementarity among dragonflies across a poolscape in a rural landscape mosaic. *Insect Conservation and Diversity*, 12(3), 241–250. https://doi.org/10.1111/icad.12339

Debele, S. E., Leo, L. S., Kumar, P., Sahani, J., Ommer, J., Bucchignani, E., Vranić, S., Kalas, M., Amirzada, Z., Pavlova, I., Shah, M. A., Gonzalez-Ollauri, A., & Di Sabatino, S. (2023). Nature-based solutions can help reduce the impact of natural hazards: A global analysis of EBS case studies. *Science of The Total Environment*, 902, 165824. https://doi.org/10.1016/j.scitotenv.2023.165824

Faivre, N., Fritz, M., Freitas, T., De Boissezon, B., & Vandewoestijne, S. (2017). Nature-based solutions in the EU: Innovating with nature to address social, economic and environmental challenges. *Environmental Research*, 159, 509–518. https://doi.org/10.1016/j.envres.2017.08.032

Gonzalez-Ollauri, A., Munro, K., Mickovski, S. B., Thomson, C. S., & Emmanuel, R. (2021). The 'Rocket framework': A novel framework to define key performance indicators for nature-based solutions against shallow landslides and erosion. *Frontiers in Earth Science*, 9. https://doi.org/10.3389/feart.2021.676059

Hatziiordanou, L., Fitoka, E., Hadjicharalampous, E., Votsi, N., Palaskas, D., & Malak, D. (2019). Indicators for mapping and assessment of ecosystem condition and of the ecosystem service habitat maintenance in support of the EU biodiversity strategy to 2020. *One Ecosystem*, 4. https://doi.org/10.3897/oneeco.4.e32704

Potthoff, K., & Dramstad, W. E. (2023). Management of rented farmland in Norway: Factors impacting on tenants' decisions to make investments. *Land Use Policy*, 135, 106941. https://doi.org/10.1016/j.landusepol.2023.106941

Rana, R., Bhambri, P., & Chhabra, Y. (2024). Evolution and the future of industrial engineering with the IoT and AI. In *Integration of AI-Based Manufacturing and Industrial Engineering Systems with the Internet of Things* (pp. 19–37). CRC Press.

Reguero, B. G., Renaud, F. G., Van Zanten, B., Cohen-Shacham, E., Beck, M. W., Di Sabatino, S., & Jongman, B. (2022). Editorial: Nature-based solutions for natural hazards and climate change. *Frontiers in Environmental Science*, 10. https://doi.org/10.3389/fenvs.2022.1101919

Singh, G., & Bhambri, P. (2023). Simulation analysis of AODV and DSDV routing protocols for secure and reliable service in mobile adhoc networks (MANETs). In *Integration of AI-Based Manufacturing and Industrial Engineering Systems with the Internet of Things* (pp. 205–216). CRC Press.

15 Eco-Ethics in the Digital Age

Tackling Environmental Challenges through Technology

J. Vigneshwari, P. Senthamizh Pavai, L. Maria Suganthi, and Pankaj Bhambri

15.1 INTRODUCTION

In the contemporary world, technology has evolved into a pivotal instrument for achieving sustainability on a global scale. Recognizing the inadequacy of current economic development patterns, people worldwide acknowledge the critical role of technology, education, and training in steering society toward a more sustainable future. The paradigm of sustainability represents a visionary approach, wherein environmental, social, and economic aspects are harmonized to pursue development while enhancing the quality of life. The recognition of the need for sustainable practices arises from the realization that existing economic trajectories are not conducive to the long-term health of the planet. In confronting issues like climate change, resource depletion, and environmental degradation, societies increasingly agree that technology, along with well-informed education and training, is essential for tackling these challenges and promoting sustainability. In this context, sustainability encapsulates a holistic perspective on the future, where the three pillars of environmental responsibility, social equity, and economic viability are carefully balanced. It goes beyond the traditional model of development that solely focuses on economic growth, emphasizing the interconnectedness of ecological health, societal well-being, and economic prosperity.

15.1.1 EDUCATION AND SUSTAINABILITY

Education plays a pivotal role in this shift toward sustainability. It serves as a conduit for disseminating knowledge about the environmental impact of human activities and the role of technology in mitigating or exacerbating these effects. By fostering

DOI: 10.1201/9781003475989-18

eco-friendly attitudes and practices, education becomes a catalyst for creating a populace that is conscious of the delicate balance between human development and environmental preservation. Furthermore, technological advancements contribute significantly to sustainable development. Innovations in renewable energy, waste management, and resource efficiency are reshaping industries and communities. Sustainable technologies, often referred to as "green tech", encompass a range of solutions aimed at minimizing environmental impact, conserving resources, and promoting a circular economy. Solar and wind power, among other renewable energy technologies, have risen in importance as cleaner substitutes for fossil fuels, thereby cutting greenhouse gas emissions and combating climate change. Furthermore, improvements in energy storage systems bolster the dependability of renewable energy sources, tackling intermittency concerns and encouraging their broad acceptance. Efficient waste management technologies, including smart waste systems and recycling innovations, are crucial for minimizing the environmental footprint of human consumption (Rana et al., 2024). These technologies optimize waste collection, sorting, and processing, diverting materials from landfills and contributing to the creation of a more sustainable, circular economy. Technological solutions also play a pivotal role in agriculture, water management, and biodiversity conservation. Precision farming, enabled by digital technologies, enhances agricultural productivity while minimizing resource inputs.

Moreover, the digital age has ushered in transformative approaches to education, enabling the widespread dissemination of knowledge on sustainability. Online platforms engage learning experiences and foster a deeper understanding of environmental challenges and sustainable solutions.

15.1.2 The 3 P's of Environmental Management

The 3 P's represent key principles and aspects that are essential for effective environmental stewardship. These principles are often used as a framework for understanding and addressing environmental issues. The 3 P's are:

15.1.2.1 People

The "People" aspect of environmental management emphasizes the role of individuals, communities, and stakeholders in environmental conservation. This includes

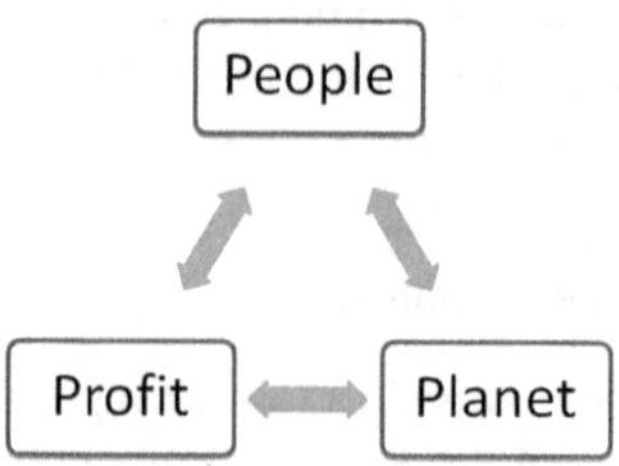

FIGURE 15.1 3 P's of environmental stewardship.

raising awareness, promoting environmental education, and engaging the public in sustainable practices. In the context of the 3 P's, People highlights the importance of human involvement and responsibility in protecting and preserving the environment, as shown in Figure 15.1.

15.1.2.2 Planet

The "Planet" component focuses on the natural environment itself. This involves understanding ecosystems, biodiversity, and the overall health of the planet. Conservation efforts, sustainable resource management, and initiatives to reduce pollution and environmental degradation fall under the Planet category. The goal is to ensure the long-term health and resilience of the Earth's ecosystems and natural resources.

15.1.2.3 Profit

The "Profit" dimension recognizes the economic aspect of environmental management. It emphasizes the need for sustainable business practices that consider environmental impact and promote long-term profitability. Companies and organizations are encouraged to adopt environmentally responsible policies, sustainable resource use, and eco-friendly innovations. The Profit element underscores the notion that economic prosperity should not be achieved at the cost of environmental degradation but rather through practices that are economically viable and environmentally sustainable.

The 3 P's of environmental management provide a holistic approach that integrates social, ecological, and economic considerations. This framework is often associated with the concept of the "triple bottom line", which seeks to balance social, environmental, and economic outcomes for a more sustainable and responsible approach to business and development.

15.2 GREEN TECHNOLOGY

Green technology represents a pivotal approach to developing and implementing products, equipment, and systems that prioritize the conservation of the natural environment and resources. Its overarching goal is to curtail and mitigate the adverse effects of human activities on the planet, fostering sustainability and ecological harmony (Bhambri and Khang, 2024a). Through continuous innovation, green technology encompasses a spectrum of solutions, incorporating sustainable elements such as solar and wind power, energy-efficient technologies, waste reduction measures, and the adoption of eco-friendly materials. By integrating these practices, it seeks to create a synergy between technological progress and environmental preservation.

This multifaceted approach addresses pressing environmental challenges. Green technology not only reduces carbon footprints but also promotes responsible consumption and production patterns. The emphasis on sustainable practices in manufacturing, energy production, and daily living reflects a commitment to long-term environmental health. As societies increasingly recognize the importance of ecological balance, the adoption and advancement of green technology become integral to creating a resilient and environmentally conscious future where human activities coexist harmoniously with the planet's ecosystems.

15.2.1 Benefits of Green Technology

Green technology brings a host of benefits, fostering a more sustainable and environmentally conscious world. Notably, it reduces environmental impact by curbing pollution and lowering greenhouse gas emissions, contributing significantly to the fight against climate change. Benefits of Green Technology are displayed in Figure 15.2. The emphasis on reusable energy sources and optimal energy utilization not only diminishes dependence on finite fossil fuels but also results in long-term cost savings. Job creation is stimulated in the burgeoning green technology sector, supporting economic growth and innovation. Beyond economic advantages, green technology enhances public health by reducing air and water pollution, while its focus on resource conservation helps mitigate waste and preserve natural resources. Embracing green technology also positions countries and businesses at the forefront of global competitiveness, providing a pathway toward a resilient and sustainable future that harmonizes technological progress with environmental stewardship.

15.3 ECO-ETHICS: AN ETHICAL PRINCIPLE

"Eco-ethics" refers to a branch of ethics that focuses on the ethical principles and values related to ecological and environmental issues. It involves examining the moral dimensions of human interactions with the natural world and addressing questions about how individuals, societies, and institutions should behave in relation to the environment.

Eco-ethics often encompasses a wide range of topics, which includes various components as shown in Figure 15.3.

Eco-ethics promotes a comprehensive and interconnected viewpoint that acknowledges the mutual reliance between human well-being and the well-being of the planet. It seeks to provide ethical guidance for decision-making in areas such

FIGURE 15.2 Benefits of green technology.

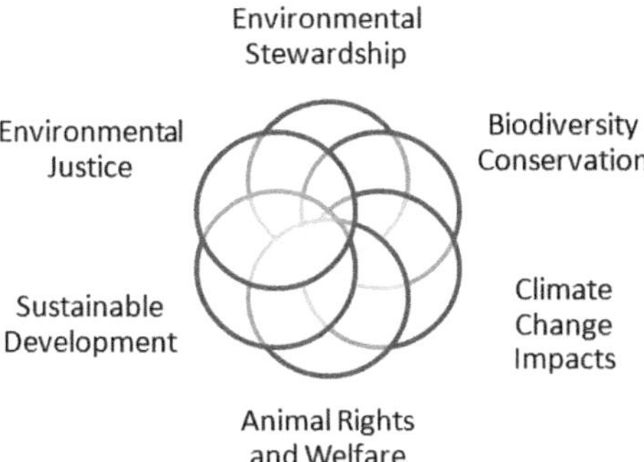

FIGURE 15.3 Ranges of eco-ethics.

as policy development, resource management, and individual behavior to promote a more sustainable and harmonious relationship between humans and the environment. In essence, embracing eco-ethics leads to harmonious coexistence between humans and the environment, fostering a more sustainable, just, and ethically responsible relationship with the natural world (Bhambri and Khang, 2024b). The benefits extend to ecological health, societal well-being, and the preservation of the planet for future generations.

15.3.1 Green Good Deeds

"Green good deeds" refer to environmentally friendly actions and practices that individuals can adopt in their daily lives to contribute positively to the planet. These actions aim to reduce the ecological footprint and promote sustainability. Some examples of green good deeds are given below:

- Reduce, reuse, recycle

Minimize waste by recycling materials, reusing items, and reducing unnecessary consumption.

- Preservation of water

Use water efficiently, fix leaks promptly, and consider installing water-saving devices.

- Sustainable energy practices

Switch off lights and electronic devices when they are not in use, opt for energy-saving appliances, and explore renewable energy options.

- Eco-friendly transportation

Opt for walking, biking, carpooling, or using public transportation to reduce carbon emissions.

- Sustainable shopping

Support environmentally conscious products, choose items with minimal packaging, and consider secondhand or locally produced goods.

- Plant trees

Contribute to reforestation efforts by planting trees or supporting organizations involved in tree-planting initiatives.

- Green gardening

Use organic and eco-friendly gardening practices, such as composting and avoiding harmful pesticides.

- Mindful consumption

Make informed choices about the products you buy, considering their environmental impact and ethical production practices.

- Educate and advocate

Raise awareness about environmental issues, share information, and advocate for sustainable practices in your community and workplace.

- Waste reduction

Minimize single-use plastics, participate in local recycling programs, and choose products with minimal packaging.

- Energy-efficient home practices

Implement energy-saving measures in your home, such as proper insulation, energy-efficient lighting, and programmable thermostats.

- Participate in clean camps

Participate in or coordinate community cleanup initiatives to eliminate litter and waste from public areas.

- Support conservation organizations

Contribute to or volunteer with organizations working to protect and preserve the environment.

- Environmental education

Stay informed about environmental issues and educate others to foster a collective commitment to sustainability.

Green good deeds are small yet impactful actions that, when adopted collectively, can contribute to a more sustainable and environmentally conscious world. By integrating these habits into everyday routines, individuals significantly contribute to fostering a healthier planet for present and future generations.

15.3.2 Recent Eco-friendly Innovations

Tackling environmental issues using technology involves leveraging innovative solutions to address challenges such as pollution, climate change, resource depletion, and habitat destruction. Some of the recent innovations help us tackle environmental problems using technology.

- **Reusable energy sources**

Solar and wind power – advancements in solar and wind technologies have made renewable energy sources more efficient and cost-effective, contributing to a reduction in reliance on fossil fuels.

Smart grids – implementing smart grid technologies enhances the efficiency of energy distribution and consumption, optimizing the use of renewable energy.

Reusable energy sources encompass hydropower, geothermal, biomass, and ocean energy. Tapping into these sustainable reservoirs presents a greener and eco-conscious substitute for conventional energy, diminishing dependency on fossil fuels and alleviating the effects of climate change.

- **Environmental monitoring**

Satellite technology – satellites help monitor deforestation, track wildlife, and observe changes in the Earth's climate, providing valuable data for environmental conservation.

Remote sensing – remote sensing technologies, including drones and sensors, assist in monitoring air and water quality, biodiversity, and illegal activities such as poaching and logging.

- **Bakey's edible cutlery**

It offers a sustainable substitute for plastic utensils. Available in three distinct flavors – plain, savory, and sweet – this eco-friendly alternative is entirely natural and will decompose naturally if not consumed.

- **Waste management**

Smart bins and IoT devices – deploying smart waste management systems with sensors and Internet of Things (IoT) devices to optimize collection routes, monitor waste levels, and reduce unnecessary pickups.

Waste-to-energy technologies – implementing technologies that convert waste into energy, such as incineration or anaerobic digestion, helps reduce landfill dependence.

- **Water conservation**

Sensor technologies – using sensors to monitor water quality, detect leaks, and optimize irrigation in agriculture to reduce water wastage.

Desalination technologies – progress in desalination technologies contributes to the sustainable management of water resources, particularly in areas confronting water scarcity.

- **Circular economy practices**

Product lifecycle tracking – implementing blockchain technology to trace the lifecycle of products, promoting transparency and accountability in supply chains.

3D printing with recycled material – utilizing 3D printing technology to create new products from recycled materials, reducing the demand for new raw materials.

- **Sustainable transportation solutions**

Electric vehicles (EVs), high-speed rail systems, and sustainable aviation fuels are emerging as environmentally friendly alternatives, reducing carbon emissions in the transportation sector.

Anti-pollutant tires – often referred to as eco-friendly or green tires – are designed to reduce the environmental impact associated with tire manufacturing, usage, and disposal. These tires incorporate various features and technologies aimed at minimizing pollution and promoting sustainability. One of the primary features of anti-pollutant tires is low rolling resistance. Lower rolling resistance reduces the energy required to keep the tires moving, improving fuel efficiency and decreasing the amount of CO_2 emissions from vehicles.

- **Afforestation and reforestation**

Drone planting – drones equipped with seed-dispersal technology can be used for large-scale afforestation efforts, particularly in areas that are difficult to access.

Precision forestry – using data analytics and sensors for precision forestry to optimize tree planting, harvesting, and biodiversity conservation.

- **Climate modeling and prediction**

Supercomputing – high-performance computing facilitates sophisticated climate modeling, enhancing scientists' comprehension of climate trends, forecasting extreme weather occurrences, and devising adaptation strategies.

Machine learning for climate data analysis – these algorithms possess the capability to scrutinize extensive datasets, pinpointing patterns and trends within climate data. This facilitates more precise predictions and informed decision-making processes.

- **Carbon capture and storage (CCS)**

Direct air capture – technologies for directly capturing carbon dioxide and helping mitigate greenhouse gas emissions, contributing to climate change mitigation efforts.

- **Robotics surveillance**

Robotic surveillance – using robots for surveillance and protection of endangered species, monitoring ecosystems, and patrolling protected areas to combat illegal activities.

- **A smart reusable notebook**

It combines traditional note-taking with innovative technology, providing an environmentally friendly and efficient way to capture and organize information. The core feature of a smart reusable notebook is its pages, which are made from a material that allows writing with a specialized pen and erasing multiple times without causing damage. This feature promotes sustainability by reducing paper waste.

- **Eco-friendly fibers**

Many reusable dress materials utilize eco-friendly fibers such as organic cotton, Tencel, bamboo, or recycled polyester. These materials are produced using environmentally conscious methods and often have a lower ecological footprint (Bhambri and Khang, 2024c). Reusable dress materials come in versatile designs that can be adapted to various fashion styles and preferences. This versatility encourages users to create different looks using the same material, extending the life of the garment.

- **Green hydrogen**

Green hydrogen, produced using renewable energy sources, notably through electrolysis, stands in contrast to conventional methods reliant on fossil fuels, which often lead to carbon emissions. Recognized for its environmental friendliness, green hydrogen serves as a pivotal element in the shift toward a low-carbon or carbon-neutral economy. The key distinguishing factor of green hydrogen is the integration of renewable energy sources in the production process. This ensures that the overall carbon footprint is minimized, making green hydrogen a sustainable alternative to conventional hydrogen production.

- **Hydroponics: A technique of growing plants without soil**

The process called photosynthesis makes the plants grow. They utilize sunlight and chlorophyll, a chemical found inside the leaves, to transform carbon dioxide from the air and water into glucose and oxygen.

The absence of "soil" in this discussion serves as evidence that plants can thrive without it. They require water and nutrients for growth, both of which can be obtained by immersing their roots in a nutrient-rich solution without soil. This fundamental concept underlies hydroponics.

These innovations collectively contribute to a more sustainable and environmentally conscious future. By combining these technological solutions with sustainable practices and policy initiatives, it is possible to make significant strides in addressing environmental challenges and fostering a more sustainable future.

15.4 ECO-ETHICS: APPS AND SOFTWARE

Numerous mobile apps and software tools are available to support and encourage green good deeds, helping individuals adopt environmentally friendly practices in their daily lives. A few are mentioned below. These apps and software tools empower individuals to incorporate green good deeds into their daily routines, making it easier to lead more environmentally conscious lives. Whether it is reducing waste, conserving energy, or supporting sustainable practices, technology plays a significant role in promoting a greener lifestyle.

Joule Bug – a mobile application that gamifies sustainability. Users accumulate points by participating in a range of eco-friendly activities and challenges, transforming the adoption of green habits into an enjoyable and interactive experience.

Ecosia – a search engine application that utilizes its advertisement revenue to support tree-planting initiatives globally. Through everyday internet searches, users actively contribute to environmental conservation endeavors.

Good On You – an application designed to assist users in making ethical and sustainable fashion decisions. It offers insights into brands' environmental and ethical practices, empowering users to make informed choices.

Too Good to Go – This app connects users with local restaurants and stores to rescue surplus food that would otherwise go to waste, reducing food waste and supporting sustainability.

Oroeco – tracks your carbon footprint based on your daily activities and purchases. It provides personalized tips to reduce your environmental impact and allows users to set and track sustainability goals.

Happy Cow – for those adopting a plant-based diet or looking for sustainable dining options, Happy Cow helps users find vegetarian, vegan, and environmentally friendly restaurants worldwide.

Recycle Nation – this app helps users locate nearby recycling centers and provides information on how to recycle specific items, making recycling more accessible and convenient.

Forest – Forest is a productivity app that encourages users to stay focused by growing virtual trees. For each virtual tree planted, the app developers plant a real tree through partnerships with tree-planting organizations.

Share the Meal – developed by the United Nations World Food Program, Share the Meal allows users to donate meals to children in need with a simple tap on their smartphones.

Green Choice – assists users in making sustainable food choices by offering insights into the environmental impact of various food products. It enables users to shop for groceries that reflect their values and sustainability preferences.

Water Minder – a hydration tracking app, which serves as a tool for users to monitor their daily water intake. Raising awareness of individual consumption patterns encourages responsible water usage and promotes water conservation efforts. Through a simple interface and reminders, users develop habits aligned with sustainable practices, contributing to a more environmentally conscious lifestyle (Bhambri, 2024). The app's emphasis on personal accountability empowers users to understand and manage their hydration needs while fostering a broader appreciation for the importance of water conservation in everyday life.

Mobile apps and software are powerful tools for fostering green good deeds and sustainability. From tracking carbon footprints to promoting ethical shopping, these applications empower individuals to make eco-friendly choices in their daily lives. Whether planting virtual trees or rescuing surplus food, technology provides accessible and engaging avenues for environmental action. These tools not only raise awareness but also transform sustainable practices into habits. As users collectively adopt greener lifestyles, the impact extends beyond individual actions, contributing to a more environmentally conscious and sustainable global community. Embracing technology for green initiatives is a pivotal step toward a healthier planet for current and future generations.

15.5 CONCLUSION

In conclusion, the pivotal fusion of technology, education, and training emerges as an indispensable force guiding society toward sustainability. Embracing a paradigm that views environmental, social, and economic factors as interwoven pillars of development is paramount for attaining a harmonious and enhanced quality of life for both present and forthcoming generations. The nexus of informed education and the implementation of sustainable technologies catalyzes charting a course toward a future in which societal development harmoniously aligns with ecological well-being, social equity, and economic prosperity. By fostering a deep understanding of the intricate connections between these pillars, societies can transcend conventional paradigms and cultivate a collective consciousness that prioritizes holistic development.

In this envisioned future, education not only imparts knowledge but also instills a sense of responsibility towards the environment and society. Concurrently, sustainable technologies become the instruments through which societies enact positive change, forging a path where progress is synonymous with environmental stewardship, social inclusivity, and economic resilience. Through this integrated approach, a sustainable future becomes not merely an aspiration but a tangible and achievable reality for the well-being of individuals and the planet alike.

REFERENCES

Akhil Jabbar Meerja, M. B. (2022). *Emerging Technologies and Applications for a Smart and Sustainable World.* Bentham Science Publishers.

Beder, S. (1994). The role of technology in sustainable development. *IEEE Technology and Society Magazine, 13*(4), 14–19.

Bhambri, P. (2024). Wallets and transactions. In D. Darwish (Ed.), *Decentralizing the Online Experience With Web3 Technologies* (pp. 90–106). IGI Global. DOI: 10.4018/979-8-3693-1532–3.ch004

Bhambri, P. & Khang, A. (2024a). Managing and monitoring patient's healthcare using AI and IoT technologies. In A. Khang (Ed.), *Driving Smart Medical Diagnosis Through AI-Powered Technologies and Applications* (pp. 1–23). IGI Global. https://doi.org/10.4018/979-8-3693-3679–3.ch001

Bhambri, P. & Khang, A. (2024b). AI-Integrated biosensors and bioelectronics for healthcare. In A. Khang (Ed.), *AI-Driven Innovations in Digital Healthcare: Emerging Trends, Challenges, and Applications* (pp. 82–96). IGI Global. https://doi.org/10.4018/979-8-3693-3218–4.ch004

Bhambri, P. & Khang, A. (2024c). Machine learning advancements in e-Health: Transforming digital healthcare. In A. Khang (Ed.), *Medical Robotics and AI-Assisted Diagnostics for a High-Tech Healthcare Industry* (pp. 174–194). IGI Global. https://doi.org/10.4018/979-8-3693-2105–8.ch012

Bowen, K. J., Cradock-Henry, N. A., Koch, F., Patterson, J., Häyhä, T., Vogt, J., & Barbi, F. (2017). Implementing the "Sustainable Development Goals": Towards addressing three key governance challenges—collective action, trade-offs, and accountability. *Current Opinion in Environmental Sustainability, 26*, 90–96.

Cramer, J., &Zegveld, W. C. (1991). The future role of technology in environmental management. *Futures, 23*(5), 451–468.

Findler, F., Schönherr, N., Lozano, R., Reider, D., & Martinuzzi, A. (2019). The impacts of higher education institutions on sustainable development: A review and conceptualization. *International Journal of Sustainability in Higher Education, 20*(1), 23–38.

Khan, H., Dong, Y., Nuţă, F. M., & Khan, I. (2023). Eco-innovations, green growth, and environmental taxes in EU countries: A panel quantile regression approach. *Environmental Science and Pollution Research, 30*(49), 108005–108022.

Mitlin, D. (1992). Sustainable development: A guide to the literature. *Environment and Urbanization, 4*(1), 111–124.

Parris, T. M., & Kates, R. W. (2003). Characterizing and measuring sustainable development. *Annual Review of Environment and Resources, 28*(1), 559–586.

Rana, R., Bhambri, P., & Chhabra, Y. (2024). Evolution and the future of industrial engineering with the IoT and AI. In P. Bhambri, S. Rani, V. Emilia Balas, A. A. Elngar (Eds.), *Integration of AI-Based Manufacturing and Industrial Engineering Systems with the Internet of Things* (pp. 19–37). CRC Press.

Sætra, H. S. (2023). *Technology and Sustainable Development: The Promise and Pitfalls of Techno-Solutionism*. Routledge.

Tracey, B., & Florian, K. (Eds.). (2016). *Educational Research and Innovation Governing Education in a Complex World*. OECD Publishing.

Webliography

https://www.jagranjosh.com/current-affairs/sustainable-development-and-india-1503408725- 1

Technologies for Sustainable Development. (n.d.). https://www.eli.org/international- programs/technologies-sustainable-development

https://www.sciencedirect.com/science/article/abs/pii/S0959652622042111

https://www.frontiersin.org/articles/10.3389/fenvs.2023.1174827/full

https://openyls.law.yale.edu/bitstream/handle/20.500.13051/1019/Scientific_Innovation_and_Environmental_Protection___Some_Ethical_Considerations.pdf?sequence=2&isAllowed=y

https://www.un.org/development/desa/dpad/wp-content/uploads/sites/45/CDP-excerpt-2013–3.pdf

http://www.ecoeet.com/Environmental-Technological-Innovations-and-the-Sustainability-of-their-Development,162708,0,2.html

https://news.climate.columbia.edu/2015/10/19/sustainability-requires-technology-ethics-and-political-will/

Important Topics: Science & Technology in Sustainable Development. (n.d.). http://www.iasplanner.com/civilservices/important-topics/science-technology-in-sustainable- development

16 Technological Risks and Resilience

Tomasz Lis and Aleksandra Ptak

16.1 INTRODUCTION

Human functioning, both privately and professionally, is related to the constant pursuit of goals using available means. Implementing activities is not only about overcoming obstacles, solving problems, and dealing with adversities, which are their natural elements. It is also about dealing with problems that arise from the tools used, the approach to them, their features, conditions, and methods of use. The tool dimension of achieving goals is related to the technology used. The concept of technology does not only refer to the tools used (Badri et al., 2012). It also includes methods of their use, or generally knowledge and skills in using all available factors helpful in carrying out specific tasks.

Elements of available technology that are tailored to the implementation of specific activities are not only attributes that help improve work. Like everything that occurs in people's surroundings and what it interacts with on a daily basis, use, or even simply coexist with, carries threats. The aforementioned knowledge about the ability to use, understanding the essence, features, conditions, and circumstances – coexistence, which in the tool aspect can be called use – must be inseparably related not only to the aspect of positive effects but also to limiting/eliminating threats and dangers.

Each activity can be considered in terms of time and expected effects at the operational and strategic levels. Operational effects occur within a short period of time after the action, almost immediately. Strategic effects are visible over a longer period of time. The dimension determines the company itself, but also the conditions in which it operates, which include, among others: legal regulations, behavior of market participants, market events, and technological development (Zalutska et al., 2021). In this respect, the threats that occur during the operation of the enterprise in a dynamically changing free market economy also have to be considered. It should be noted that due to the widespread use and even increasing dependence of organizations on information and communication technologies, digital threats are becoming more and more important. Due to the above, the aspect of management, which focuses on this type of threat and building an appropriate security policy tailored to the challenges and realities, is becoming more and more important (Ziemba, 2017). Importantly, given the speed of ICT development and its importance, security

DOI: 10.1201/9781003475989-19

policy should look into the future. This makes security management a very complex undertaking, but one that has a direct impact on the organization's ability not only to succeed but also to function normally.

The use of ICT is an important area of human activity. The development of this technology, especially the Internet, means that human functioning, both in the personal and professional spheres, is increasingly considered from the point of view of virtuality. This provides market participants with new and very extensive opportunities, but it is also a source of threats that increase the complexity of protective measures and is also a serious ethical problem. ICT is not an isolated area of technology application in enterprises. One of its most important features is its widespread use. ICT supports human information activity, which is the basis of everything that is planned, organized, and undertaken and is also an effect that, through informational impact on others, generates market information subject to management (using ICT) – an information feedback loop (Tohidi, 2011).

The areas of use of ICT, as well as information/digital technology in general, which allow achieving certain benefits but are also a source of threats, include the use of artificial intelligence, communication, sustainable development, robotics and automation, and data management. The most important goals of this work include the identification and analysis of threats occurring in enterprise management in connection with the use of modern technological solutions, especially in terms of IT/digital solutions. A natural element is the confrontation of threats with benefits, which results from their simultaneous occurrence. Another goal is to analyze the possibilities and conditions for counteracting threats and possibly correcting the effects. The development of this work is related to the literature analysis carried out by the authors and qualitative research based on observation. Literature analysis and observations were conducted in parallel with the authors' scientific and research work – for a period of approximately 20 years.

16.2 TECHNOLOGY IN BUSINESS MANAGEMENT

Since the beginning of our existence, humans have been using available means to facilitate and improve the activities they perform. This applies to functioning both privately and professionally. The most important reasons include the following (Chingning & Ping, 2012):

- Increasing the effectiveness of human activity/work as the basic unit but also of the structures they co-create, e.g., enterprises;
- Improving/optimizing interpersonal and inter-entity communication (between elements of the same organizational structure, as well as market participants with whom relationships exist) – achieving the expected best results requires collaboration and cooperation between many units;
- Individual/human development, but also organizational development;
- Increasing the "area" constituting the source of acquired data and information, their quantity, while reducing the time of analysis and conclusions;

- Increasing the security of the organization's application infrastructure, and therefore of the collected resources, such as data, information, and knowledge;
- Optimization of enterprise management – in every dimension, including human resources, occupational safety, data, functional areas, processes, time, finances, etc.;
- Optimization of resource management – identification, acquisition, disposal, and use, including identification and activation of relationships between resources;
- Increasing safety in the organization – safety in terms of people, but also resources (efficiency, service life, level of use – devices, tools, transport, etc.);
- Increasing awareness and the level of use of aspects of corporate social responsibility in achieving goals;
- Blurring the limitations related to the physicality of functioning – limitations of time and place that have an activating effect on the phenomenon of globalization;
- Creating new opportunities for work and work organization, i.e., running a business, including gainful activity.

The use of elements of available technology is closely related to adapting functioning to emerging challenges. This is called adapting to the dynamically changing reality, which is particularly important in relation to economic and market activity. It can be noted that technological development is both the effect and the causative factor of dynamism (Bhambri, 2024). This applies especially to those areas of technology that are related to the broadly understood flow/movement of people, the resources they use, and the goods produced. But, in the opinion of the authors of this paper, the most important thing is data, information, and knowledge. By eliminating the limitations of time and place, which have traditionally been factors slowing down human activity and the scope of human functioning, dynamism naturally increases (Martincevic & Kozina, 2018). Thanks to the development of technology, activity is increasingly considered in the physical (real) and information (virtual) sense. Proximity in the physical dimension is influenced by technology in terms of transport and in the information dimension, ICT, especially the Internet.

Increasing the effectiveness of human activity/work as a result of the technology used results from the relationship between possibilities (abilities) and effects. The most important "productive" assets of a human being include the ability to think logically, associate, and plan. This also includes awareness of one's own limitations that determine the discrepancy between possible effects in terms of thoughts and plans and their physical implementation. At the same time, as a result of everyday experience, acquiring information and knowledge, the awareness of the possibilities of using available means of work, as well as the ability to create new means of work, increases. Physical activity becomes more effective and thus closer to achieving plans. Human development applies not only to the physical area but also to the mental dimension – analysis and reasoning. In this respect, the greatest breakthrough occurred with the development of information and communication technology: computers, information systems, the spread of information and communication, and the Internet. It also had a positive impact on the development of physical activity, including the creation of new means of work and the organization of work itself. This is undoubtedly another positive effect of involving technology in management (Bankole et al., 2011)

The availability and involvement of an increasing amount of work, resources, and employees force development in the field of management. Increasing efficiency is related to the optimization of the implementation of basic and classic management functions: planning, organizing, controlling, motivating, and coordinating – relating to every dimension of the organization's activity. Technological development considered in the context of its impact on efficiency in the area of management refers primarily to the development of ICT. This is due to the fact that management is primarily the acquisition, disposal, and dissemination of data and information as well as the formation of knowledge on an individual and group basis (Gashaw, 2010). This translates into better use of existing resources, as well as identifying and engaging new ones, and, importantly, identifying and using new market relationships in every possible aspect of their existence.

The currently observed technological development in management serves is primarily to optimize the following areas: information and knowledge management, including communication, identification and acquisition, dissemination, analysis and inference, and finally decision-making processes (Sanghamitra & Malaya, 2011). Nowadays, technological development is at such a level that it allows the implementation of many complex projects. At the same time, the value of the individual, both in the sense of a person and the structure created – an enterprise – is losing importance. An individual is not able to use all possibilities. Even with the use of modern technologies, the person's individual "productive" abilities are limited. In this regard, relationships have become more important. Their shaping and organization are some of the most important challenges in the dynamically changing economic reality in the conditions of globalization. Technology, especially ICT, allows optimization of this area of management (Rani et al., 2024). What is particularly important here is the effectiveness of information management between all interacting elements, which depends largely on the course of the communication process. Business relationships involving technology allow an increase in the area of active search and acquisition of information (Qureshi & Abro, 2016). The technology itself has a positive impact on the duration and efficiency of analyses, even on very large data and information resources.

The use of modern technologies, including ICT, allows for the activation of activities based on the identification, acquisition, dissemination, analysis, and use of information. The effects of this state of affairs include the organization's ability of the organization to take into account the social factor in the developed and adopted strategies, including, in the dimension of environmental safety, employees and people living in the areas of physical impact and functioning of the enterprise and its business activities. This is an issue related to a sustainable approach to enterprise management, the assumptions of which are based on environmental protection and the standard of living of local communities and employees themselves (Đordević et al., 2021).

Technological development and the use of modern management solutions, along with the increasing importance of business relationships, make it necessary to pay special attention to security issues. This concerns, on the one hand, the physical safety of employees, material assets at the disposal of the enterprise, local communities, and the environment, and, on the other hand, stored and distributed data, information, and knowledge (Kopytko, et al., 2021).

Security relations with the technological solutions used in the ICT dimension have become particularly important in connection with the phenomenon of globalization and the global COVID-19 pandemic. Globalization is the ability to conduct activities, and strictly business activities, without the importance of geographical distances. On the one hand, the company can make its product offer available anywhere in the world, and on the other hand, employ and/or cooperate with entities located anywhere in the world. As a result of the COVID-19 pandemic, the principles of human functioning have changed globally. A very large number of employees had to switch to remote work, and children and teenagers had remote classes. It is currently difficult to imagine that this would be possible without the availability of appropriate ICT technological solutions (Lee et al., 2021). The number of sick people and deaths would have been much higher, which would certainly have negatively affected the world economy: countries, companies, and people, to an even greater extent than before. Remote work, already known, then became particularly important.

A new technological solution used in enterprises, various types of institutions, and generally by private individuals, about which practical and theoretical discussions are currently taking place, is artificial intelligence. It allows specific activities to be performed without the need to involve a human, or where a human takes a supervisory role, defines goals, and sets limiting conditions (Raisch & Krakowski, 2024). In 2023, the authors of this work conducted qualitative research, which included interviews with employees of two IT companies. During these interviews, questions were asked about the use of solutions based on artificial intelligence. It turned out that in both cases such solutions are used to support decision-makers in decision-making processes. Interestingly, however, it was found that employees of these companies conducted experiments on the possibilities of engaging artificial intelligence in the implementation of other activities. They were based on creating fragments of code to be used in the process of creating larger and more complex software. The artificial intelligence systems used for this purpose coped with the problem very well. The employees described the effects of their work by saying "the code created by artificial intelligence allowed the implementation of planned activities. It was readable and easy to merge with other man-made pieces". The conclusion, also presented by employees, was that artificial intelligence can successfully perform the same activities that programmers with a little experience do. It should be assumed that the use of modern technological solutions in enterprise management in the near future will largely involve the use of artificial intelligence tools. However, this raises many concerns and questions about the importance of people in the functioning of enterprises in the future. Will human work become redundant? Will humans only perform control activities? To what extent can we rely on artificial intelligence? Will solutions of this type result in the disappearance of certain professions? And finally, will humans become dependent on technology to an extent that limits their ability to exist without it? These are just some of the questions and problems we will face in the near future, which result from technological development in the functioning of humans and enterprises (Habbal et al., 2024).

16.3 TECHNOLOGY AS A SOURCE OF THREATS IN THE ENTERPRISE

Every activity is associated with threats. These threats may concern both the actions themselves and the person as the main causative factor. The tools, methods, solutions, and technology in general are intended to improve and increase the effectiveness of achieving goals. At the same time, they are intended to have a positive impact on increasing the safety level of their operator/user. However, as Kisiel notes,

> safety and organizational effectiveness are often perceived antagonistically to each other. (...) The disconnect between efficiency and safety is even more visible when it comes to the costs of ensuring safety, as safety expenses are perceived as reducing the efficiency and profitability of the organization.
>
> **(Kisiel, 2017)**

According to the authors, companies are constantly investing in modern solutions. They do this because they are interested in obtaining maximum profits and an appropriate competitive position. They treat the costs incurred for safety issues as a factor that reduces profitability.

Modern technological solutions can now not only increase work efficiency, but at the same time have a positive impact on increasing the level of safety of employees and the organization (Badri et al., 2018).

> The introduction of industrial robots and IoT systems not only reduces the risk of accidents, but also allows for quick response in the event of sudden threats to employee health. Combined with appropriate procedures and employee education, modern technologies create the work environment in which everyone can feel safe and comfortable. It is not only an investment in the development of the industry, but also an investment in the health and well-being of people who are the heart of every company.
>
> **(Nowe ... , 2023).**

The impact of introducing and using modern technological solutions on the level of safety of employees, enterprises, societies, and the environment is an area of research conducted by many entities and organizations. In the years 2009–2011, the FESTOS project was initiated, financed from EU funds (Foresight of evolving security threats posed by emerging technologies). Its aim was to identify, assess, and understand the threats to security posed by the improper use of modern technologies. The project also focused on identifying ways to counteract these threats. Eighty technologies in the fields of robotics, ICT, new materials, nanotechnology, and biotechnology were subjected to research. It should be noted that the research was not only related to the occurrence of natural hazards accompanying the normal use of technology. Attention was also drawn to the threats posed by the possibility of intentional use of new technological solutions by third parties to disrupt the functioning of countries' economies, individual enterprises, corporations, people, and societies. Taking into account that every novelty, including in the technological area, is the result of existing and acquired knowledge, attention was paid to the problem of free spread and access to knowledge. The most important threats included the following areas: communication and broadly understood exchange of information, access to military technologies, and improper use of technologies developed to facilitate functioning and ensure security. One of the conclusions indicated that, taking into account the risks, in the future the issues of benefits and dangers coexisting should be considered simultaneously (Zagrożenia … 2023).

The economic development we are currently experiencing is related to the "fourth industrial revolution". What distinguishes it, compared to the previous ones, is that the driving factor is information (Prisecaru, 2016), which, due to its form and method of dissemination, is called digital. In such a reality,

> digitalization and information and communication technologies (ICT), apart from, among others, artificial intelligence, advanced data analytics, robotics, automation, autonomous vehicles, drones, smart devices, 3D printers, human-machine interfaces, Internet of Things, big data, cyberphysical systems, advanced technological sensors, cloud computing, quantum computers, e-commerce, or electronic waste, are becoming more and more popular and ubiquitous.
>
> **(Bezpieczeństwo, 2019)**

Contemporary technological development is focused on broadly understood digital technologies. This concerns the use of data and information to increase knowledge, which in turn aims to optimize and increase the efficiency of activity. It should be noted that data, information, and knowledge are factors that support employees but at the same time directly supply the work of machines or robots. The latter operate increasingly independently, using, among other things, artificial intelligence. Widespread computerization forces us to produce ever more efficient devices, which have to be smaller and smaller. Taking into account the presented directions of development of digital technologies, it is indicated that the most important areas related to the safety of employees and enterprises include digitization and ICT, automation and robotization, as well as nanotechnology (Bezpieczeństwo, 2019).

Digitization and the widespread use of ICT are undoubtedly sources of many benefits in the area of not only organizational efficiency but also security. However, they are often sources of threats that affect both employees and entire organizations. Interestingly, it can be pointed out that the vast majority of benefits have a corresponding disruption. One of such threats is taking over responsibilities traditionally performed by people. Robots and automation based on artificial intelligence are cheaper labor factors for decision-makers. They allow for the performance of difficult and accident-prone duties. However, they are sources of frustration and problems that negatively affect the human psyche, which is one of the most important elements of security. Another case in which technological development is characterized by the duality of effects in positive and negative directions is the possibility of remote work (Strelicz, 2021). Greater freedom of work, flexibility, and no need to travel to the workplace are accompanied by the blurring of the boundaries between work and private life, a sense of social isolation and disturbances in interpersonal relationships, or a reduction in the ergonomics of the workplace. Mobile devices are increasingly used to monitor the health and fatigue levels of employees. At the same time, they cause social isolation, a sense of constant control and surveillance, and a loss of privacy. The increasing virtualization of professional life may also lead to the occurrence of harassment, aggression, and online attacks. The employee whose primary work element is the Internet is exposed to access to a large amount of negative information (events on a global scale). These have a negative impact on their psyche, limiting efficiency and effectiveness. The negative effects of the use of modern technologies also include addiction to technology, information overload, increased risk of losing sensitive and strategic data and information, exposure to electromagnetic radiation, and a sedentary lifestyle (Oldroyd & Morris, 2012).

One of the most important challenges regarding information management using modern technologies in enterprises is the security of data, information, and knowledge (Soomoro et al., 2016). Companies' databases contain information about persons and cooperating entities that must be protected against leakage due to legal regulations. Other information influences decisions made and competitive strategies implemented. Their loss may cause disruptions in normal functioning. The work "The use of new technologies carries new threats" presents the results of research conducted among enterprise employees regarding information security in companies and public institutions. Based on the results obtained, it was found that managers are more afraid of negative actions by their own employees than of hackers when it comes to information leaks. This is due to, among other things, the belief in irresponsibility when using social networking sites – 45% of management staff. Based on the research conducted, it was found that only 40% of entities use various types of encryption techniques to protect data and information. The result is an increased likelihood of data loss as a result of the use of mobile devices, social media, and cloud computing. It has been found that more and more organizations are choosing to use cloud solutions. Among them, approximately 38% admitted that they do not take any action to increase safety. Data and information security issues are widely discussed in the media. They are also the subject of inquiry and discussion by scientists and practitioners. At the same time, it was noted that, according to employees, approximately 63% of enterprises do not have a formally developed or implemented security architecture. In the case of employees of approximately 70% of organizations, the security system used does not meet their existing needs. The most important reasons for the lack of an appropriate approach to the security of IT systems were identified by respondents as too low financial outlays – approximately 62% of enterprises. This is consistent with the approach described earlier, where security spending is often treated as unnecessary and a drag on profits. Another reason turned out to be an insufficient number of qualified employees – approximately 43%

of organizations. In the case of approximately 20% of enterprises, the lack of support from management staff was considered to be the reason. This seems to indicate that, despite the "widely known seriousness" of the problem, decision-makers seem to be limiting the costs of action, including hiring appropriate staff (Korzystanie, 2023). This is confirmed by the results of observations and interviews conducted by the authors of this work. During interviews, decision-makers often say that they know about these types of threats and their negative effects, but there seems to be a belief that "these problems do not and will not occur in my case".

Nowadays, it seems that one of the most important technological solutions is artificial intelligence. It allows enterprises to obtain a number of benefits. However, as Ślusarczyk notes, its use, especially on a wider scale, is accompanied by certain threats. The author includes, above all, the increase in unemployment and the spread of information in the environment that is of little truth or even completely untrue (Ślusarczyk, 2021). According to the authors of this work, the disadvantages of using this type of solution also include a reduction in job security, a resulting reduction in employee efficiency, and a possible long-term reduction in the number of employees in specific professions (whose work may be replaced by artificial intelligence in the future). Reducing the number of specialists in specific fields may in the future have a very negative impact not only on the functioning of enterprises but also on societies. Considering potential negative effects in terms of possible cataclysms or crisis events with a wide impact is now justified. This was influenced by the emergence of the global COVID-19 pandemic, the effects of which have been and still are felt throughout the world, which has never happened before and whose probability was considered low. This resulted in virtually no level of preparation of enterprises and institutions. This situation should and must draw the attention of those responsible for security issues to events with a low probability of occurrence, but characterized by a wide range and spectrum of negative impact (Reed, 2021). According to the authors of this work, the use of artificial intelligence should be classified as such a potential threat.

The need to include this type of solution among those that may cause significant damage confirms the increasing popularity of their use in organizations. As research has shown (Nowak, 2024), in 2023, compared to the previous year, the level of their use in enterprises increased by approximately 400%. The report presented by Netskope states that currently approximately 10% of employees in enterprises in various industries use at least one generative artificial intelligence application per month. This percentage in the previous year was approximately 2%. The most frequently used application was ChatGPT – 7% of the share. The report also draws attention to the fact that, according to forecasts, the percentage of active users will increase significantly in the coming year. The reason for this state of affairs is that new areas will be identified in enterprises in which these tools can be used.

The potentially negative impact on the safety of using solutions based on artificial intelligence does not only apply to the situation of their conscious use by humans. Due to the increasing scope of their use, the spectrum of threats increases. In his work, Lampart refers, among others, to the dangers arising from the development of autonomous means of transport based on AI. The author notes that despite the existence of a number of legal provisions on this issue in national and EU regulations, there is a noticeable lack of one that would solve the issue of liability for artificial intelligence activities. It also states that the need for change is included in the EU White Paper on AI:

> Under the Product Liability Directive, a manufacturer is liable for damage caused by a defective product. However, in the case of AI-based systems such as autonomous cars, it may be difficult to prove that the product is defective, the damage caused or the causal relationship between them. Furthermore, it is not entirely clear how and to what extent the Product Liability Directive applies to certain types of defects, for example those resulting from deficiencies related to the cybersecurity of a product. (…). Artificial intelligence technologies embedded in products and services may create new security risks for users. For example, due to an error in object recognition technology, the autonomous car may incorrectly identify an object on the road and cause an accident resulting in bodily injuries and the property damage. As with threats to fundamental rights, these threats may be due to flaws in the design of AI technologies, related to data availability and quality issues, or other issues arising from machine learning. While some of these risks are not limited to products and services that rely on AI, the use of AI may exacerbate them (…). If security threats materialize, the lack of clear requirements and the above-mentioned features of AI technology make it difficult to trace potentially problematic decisions made using AI systems. This, in turn, may make it difficult for injured parties to obtain compensation under applicable EU and national liability legislation.
>
> **(Lampart, 2023)**

The problem identified and described by the author will become more and more relevant due to the previously indicated growth rate in the use of AI technology. Without appropriate preparation, including legal provisions and procedures in the event of an occurrence, it is difficult to determine the potential consequences.

Technologies used by humans are often the source of benefits and threats. It should be emphasized that the benefits are directly related to the purpose of use, while the threats are a side effect resulting from improper use, the human factor, defects, and/or a combination of many simultaneously occurring factors – which are difficult to predict. With the development of technology and man himself, the group of threats increases, which are not just a side effect. These are activities performed by humans with the active participation of available technologies, which are intentionally and consciously used to cause a threat to other people, entities, institutions, enterprises, or even countries. Elements of ICT technology play a special role in this respect. They enable and improve the processes of theft of information resources, sensitive data, and even money. At a time when information resources constitute a strategic resource of man-made structures (institutions, enterprises), their loss causes disruptions that may have a real impact on the continuity and ability to function (Harris & Patten, 2014). The phenomenon of simultaneous occurrence of threats, especially in relation to ICT, is so common that often before and during the development of a specific technology, the expected and assumed benefits and possible threats are analyzed at the same time.

This type of situation currently occurs, among others, in the case of quantum computing technology. Taking into account computing power, there is a belief that its use, including in management, may, like artificial intelligence, contribute to a change in the approach to business. It is stated that the speed and computing power "will impact many fields of science and aspects of everyday life – from particle research, through weather forecast accuracy, traffic optimization and financial modeling, to cybersecurity". At the same time, it is indicated that as a result of the development of quantum computer technology, cybercriminals will be able to gain access to it. As a result, similar to the scale of positive effects, they will be faced with a tool that opens up new, previously unknown opportunities. The scale of the problem is so large that, according to Kutyłowski,

> financing work on technologies that may lead to the blocking of ICT systems is at least incomprehensible (…). The success of work on the construction of a quantum computer capable of breaking systems based on RSA and DLP on a production scale would lead to huge disruptions in economic transactions, a breakdown in the protection of many systems and a return to paper trading mechanisms from several dozen years ago.
>
> **(Kutyłowski, 2024)**

At the same time, it should be noted that the threat forecasted by the author seems to be exaggerated at the moment. Technological development, especially ICT, has from almost the very beginning necessitated the development and implementation of tools to protect against inappropriate use. This is undoubtedly the case with quantum technology. Taking into account the indicated revolutionary nature of the technology, it is necessary that the sequence of events does not include the sequence: technology – positive and negative effects – corrective actions (introduction of security solutions) in the area of security. It should be based on the simultaneous development of technology and safe ways/methods of its use. Actions of this type are called preventive (Flowerday & Tuyikeze, 2016).

16.4 TECHNOLOGICAL DEVELOPMENT AND ORGANIZATIONAL SECURITY

The issue of security is a very important aspect of human functioning. As the level of safety increases, the efficiency and effectiveness of the activities carried out increase (Dziuba et al., 2020). Taking into account the physical dimension of the effects of events, safety can be considered in physical and mental terms. Based on the materiality of factors affecting human (employee) safety, they can be considered from the material and non-material perspectives. The issue of security from the point of view of enterprise management is associated with the concepts of occupational health and safety, security systems, and increasingly often the security of data, information, and information/IT systems. Health and safety can be defined as:

- Total legal standards and research, organizational and technical measures aimed at creating such working conditions for the employee that he can perform work in a productive manner, without exposing him to an unjustified risk of accident or occupational disease and excessive physical and mental strain.

- The set of rules and regulations relating to occupational health and safety, and an independent field concerned with determining appropriate working conditions. Occupational health and safety should be understood as both protecting the employee's life and health and ensuring ergonomic conditions at the workplace. The scope of issues related to occupational health and safety also includes concepts related to occupational psychology, occupational medicine and technical safety.

(Pasoń, 2021)

Analyzing the definitions of occupational health and safety, it can be concluded that the area of interest covers everything related to the employee's physical and mental safety. This includes the use of any technology that is implemented to achieve individual and team goals. An interesting area is the security in the mental and intangible dimensions related to the use of ICT. An employee who uses various channels for disseminating information at work is exposed to information overload. Receiving negative information from the outside world burdens the mind, which is accompanied by an increase in stress levels. In each case, work efficiency decreases.

The occupational safety system is "specific procedures enabling the use of ready-made tools in the enterprise to support managers in controlling and reducing unnecessary costs" (Bąk & Kapusta, 2015). The term unnecessary costs mentioned in the definition can be replaced by the concept of risk. This is due to the fact that unnecessary costs are related to the occurrence of disruptions, which always result in losses, including financial ones. In this respect, controlling risk means eliminating negative effects (Adler et al., 1999). The concept of the safety system has a much more extensive meaning than occupational health and safety. It covers all aspects of the enterprise's operation in which disruptions occur and/or may occur. It is related to every technology used.

Taking into account the importance of intangible resources: data, information, and knowledge, the element of the organization's security system is the information security system. It covers everything, including the technologies used, related to information management (including data and knowledge). The information security system deals with security in the intangible dimension and applies to both information processed in a traditional way and using ICT tools. As Myśko and Młodzik note,

> information processing in conditions of constant search for ways to reduce the uncertainty of decisions made, and therefore limit the risk of losses and thus strive to increase the chances of success, means that information resources are constantly exposed to unauthorized access, modification, or even partial or total loss. Actions of this type can be performed both from outside the individual and from inside the organization, both consciously and unconsciously.
>
> **(Myślo & Młodzik, 2014)**

Due to the fact that the basic form of information processing is digital, the information security system is primarily an IT security system.

Counteracting disruptions in the functioning of enterprises in the era of rapid technological development involves the introduction of two types of activities. The first are actions aimed at correcting/neutralizing the negative effects of events; the second are actions aimed at reducing and/or completely eliminating the possibility of undesirable events. The first are referred to as corrective actions and concern what has already happened, and the second are preventive actions, which concern what is forecast/expected, but as such has not yet occurred.

The indicated activities are often considered in the context of quality management, which is related to the pursuit of the best effect. Everything that is implemented in the company is assessed in terms of the possibility of introducing improvements. The goal of the approach should be the pursuit of perfection.

> Corrective actions are an example of reactive actions, they are undertaken as a result of detecting a non-compliance (error, failure, incident, accident, etc.) (a non-compliance has occurred, then corrective actions are proposed in relation to it). (…) They involve introducing changes that will prevent recurrence of non-compliance. For this purpose, the causes of non-compliance are examined. The effectiveness of the entire process depends on this analysis. It is then determined how the causes can be removed. The next step is to plan and implement actions that will eliminate the causes of non-compliance or at least significantly reduce them. The introduction of corrective actions has to be verified – so it should be checked whether, as a result of the changes introduced, the non-compliance cannot recur.
>
> **(Knop & Mielczarek, 2015)**

Preventive actions refer to situations that have not yet occurred. It is necessary to identify possible negative situations and develop an approach that will reduce their likelihood. For this purpose, it is necessary to recognize and understand the factors, relationships, and circumstances that may contribute to their occurrence.

Both corrective and preventive actions refer to the optimization of future functioning. Their use is related to risk management in enterprises. The basic goal is to prevent the occurrence of threats while limiting the negative effects of the events that have occurred. As a result, the probability of disruptions is reduced, and at the same time, procedures and corrective methods are developed and implemented to limit the effects of events that occurred as a result of the previously identified risks. As Serafin rightly notes,

> risk management is a management technique that rationalizes decision-making in conditions of uncertainty. This approach is considered the most effective when operating in conditions of uncertainty, it allows for reducing various types of security, which has a beneficial impact on the economic calculation of activities.
>
> **(Serafin, 2013)**

Taking into account the uncertainty and dynamic changes accompanying the use of technology, risk management is a natural approach in the functioning of enterprises. The risk of disruption increases with the level, scale, and intensity of the technologies used. It is also influenced by the knowledge, competences and skills of employees, and the form of organizational culture.

16.5 CONCLUSIONS

The development of man, the structures he co-creates, and in the work in which he participates are closely related to technological development. On the one hand, new technologies cause human development, and on the other hand, human development influences the development and implementation of new technologies. So it is a self-propelling mechanism. Information and communication technology plays a particular role here. It helped to eliminate the traditional limitations of time and place. In the information dimension, it can "reach" almost immediately to any place on earth. In this way, it can cooperate with people physically residing in various places around the world, reach virtually unlimited numbers of potential customers, and acquire information and knowledge needed for development and building a competitive position.

In addition to the benefits, the use of each technology is accompanied by certain risks. Their occurrence always has a negative impact on the user (who may be a person, a department, or an enterprise). This simultaneously results in the development and/or improvement of technology. Modern times are called the digital era, which is related to the dynamic development of digital technologies. It increases the efficiency of enterprises, which at the same time become very sensitive to digital disruptions. Burglaries, unauthorized access to data and information, and their theft are just some of the potential threats.

The dangers resulting from the development of digital technology are also related to artificial intelligence and quantum computers. AI creates a new type of ethical and legal threat. Very fast computers will allow for more effective breaking and bypassing of existing security measures.

Awareness of the simultaneous occurrence of benefits and disruptions should result in taking both factors into account. It is a mistake to develop and/or implement technologies without simultaneously preparing to reduce the likelihood of threats occurring and developing procedures and methods of conduct in the event of their occurrence. The effects are then much worse and more difficult to eliminate. This area of the enterprise's operation is related to risk management.

REFERENCES

Adler T. R., Leonard J. G., Nordgren R. K. (1999). Improving risk management: Moving from risk elimination to risk avoidance, *Information and Software Technology*, 41(1), 29–34.

Badri A., Boudreau-Trudel B., Souissi A. S. (2018). Occupational health and safety in the industry 4.0 era: A cause for major concern?, *Safety Science*, 109, 403–411.

Badri A., Gbodossou A., Nadeau S. (2012). Occupational health and safety risks: Towards the integration into project management, *Safety Science*, 50(2), 190–198.

Bąk P., Kapusta M. (2015). Rola bezpieczeństwa w zarządzaniu przedsiębiorstwem, Zeszyty Naukowe Uniwersytetu Szczecińskiego nr 855, *Finanse, Rynki Finansowe, Ubezpieczenia* 74(2), 18.

Bankole F. O., Shirazi F., Brown I. (2011). Investigating the impact of ICT investments on humanDevelopment, *The Electronic Journal on Information Systems in Developing Countries*, 48(8), 1–14.

Bezpieczeństwo i zdrowie w centrum przyszłości pracy (2019). *Międzynarodowa Organizacja Pracy,* CIOP-PIB, 5.

Bhambri, P. (2024). NFTs: Transforming digital ownership in the Web 3 Era, in D. Darwish (ed.), *Decentralizing the Online Experience With Web3 Technologies* (pp. 151–167). IGI Global. DOI: 10.4018/979-8-3693-1532–3.ch007

Chingning W., Ping Z. (2012). The evolution of social commerce: The people, management, technology, and information dimensions, *Communications of the Association for Information Systems*, 31, 106-–120.

Đordevi´c A., Đordevi´c Y., Arsovski S., Stefanovi´c N., Shamina L., Pavlovi´c A. (2021). The impact of ICT support and the EFQM criteria on sustainable business excellence in higher education institutions, *Sustainability*, 13, 1–22.

DziubaSz. T., Ingaldi M., Zhuravskaya M. (2020). Employees' job satisfaction and their work performance as elements influencing work safety, *CzOTO*, 2(1), 18–25.

Flowerday S. V., Tuyikeze T. (2016). Information security policy development and implementation: The what, how and who, *Computers & Security*, 61, 169–182.

Gashaw K. (2010). Knowledge management: An information science perspective, *International Journal of Information Management*, 30, 416–421.

Habbal A., Ali M. K., Abuzaraida M. A. (2024). Artificial Intelligence trust, risk and security Management (AI TRiSM): Frameworks, applications, challenges and future research directions, *Expert Systems with Applications*, 240, 233–254.

Harris M. A., Patten K. P. (2014). Mobile device security considerations for small- and medium-sized enterprise business mobility, *Information Management & Computer Security*, 22(1), 97–110.

Kisiel S. (2017). Efektywność organizacji a jej bezpieczeństwo, *Digital & More*, available on the website: https://digitalandmore.pl/efektywnosc-organizacji-a-jej-bezpieczenstwo/ (access: 18.12.2023)

Knop K., Mielczarek K. (2015). Aspekty doskonalenia procesu produkcyjnego, Zeszyty Naukowe, *Quality Production Improvement*, 1(2), 76.

Kopytko M., Fleychuk M., Veresklia M., Petryshyn N., Kalynovskyy A. (2021). Management of security activities at innovative-active enterprises, *Business: Theory and Practice*, 22(2),299–308.

Korzystanie z nowych technologii niesie nowe zagrożenia, forsal.pl, available on the website: Korzystanie z nowych technologii niesie nowe zagrożenia - Forsal.pl (access: 19.12.2023).

Kutyłowski M. (2024). Cyber (nie)bezpieczeństwo a kryptografia kwantowa, *Informatyka i Bezpieczeństwo*, available on the website: https://portal.pti.org.pl/wp-content/uploads /2022/07/7.-Cyberbezpieczenstwo-a-kryptografia-kwantowa_Domena_1-2022.pdf (access: 20.01.2024).

Lampart M. (2023). *Rozwiązania generatywnej sztucznej inteligencji – zagrożenia i aspekty prawne* (pp. 40–41), Polska Agencja Rozwoju Przedsiębiorczości, Warszawa.

Lee Y. Ch., Malcein L. A., Kim S. C. (2021). Information and Communications Technology (ICT) Usage during COVID-19: Motivating Factors and Implications, *International Journal of Environmental Research and Public Health*, 18, 1–14.

Martincevic I., Kozina G. (2018). The impact of new technology adoptation in bussines, 36th International Scientific Conference on Economic and Social Development – "Building Resilient Society" - Zagreb, 14–15 December 2018, 842–847.

Myśko A., Młodzik E. (2014). Bezpieczeństwo informacji – dylematy związane z realizacją obowiązku prowadzenia audytu wewnętrznego w jednostkach sektora finansów publicznych, Zeszyty Naukowe Uniwersytetu Szczecińskiego 833, *Finanse, Rynki Finansowe, Ubezpieczenia* 72, 108.

Nowak K. (2024). Sztuczna inteligencja szturmem zdobyła firmy. Wykorzystanie AI wzrosło o ponad … 400%, *forsal.pl*, available on the website: https://forsal.pl/lifestyle/technologie/artykuly/9403479,sztuczna-inteligencja-szturmem-zdobyla-firmy-wykorzystanie-ai-wzroslo.html (access: 20.01.2024).

Oldroyd J. B., Morris S. S. (2012). Catching falling stars: a human resource response to social capital's detrimental effect of information overload on star employees, *Academy of Management Review*, 37(3), 396–418.

Pasoń J. (2021). *Bezpieczeństwo i higiena pracy biurowej – podręcznik* (p. 7), Ekonomik, Warszawa.

Prisecaru P. (2016). Challenges of the fourth industrial revolution, *Knowledge Horizons – Economics*, 8(1), 57–62.

Qureshi Z. H., Abro M. M. Q. (2016). Efficient use of ICT in administration, *International Journal of Economics, Commerce and Management*, IV(10), 540–548.

Raisch S., Krakowski S. (2024). Artificial intelligence and management: The automation-augmentation paradox, available on the website: https://www.researchgate.net/profile/Sebastian-Krakowski/publication/339184283_Artificial_Intelligence_and_Management_The_Automation-Augmentation_Paradox/links/5c46af8a299bf1cdb92a57b0/Artificial-Intelligence-and-Management-The-Automation-Augmentation-Paradox.pdf (access: 23.01.2024).

Rani S., Bhambri P., Kaur J., Sangwan Y. S. (2024). Exploring the application domains of ML-based facial emotion recognition systems: Framework, techniques and challenges. In Second International Conference on Computing and Communication Networks (ICCCN 2022) (pp. 090008-1–090008–9). AIP Conference Proceedings 2919. https://doi.org/10.1063/5.0184852

Reed J. H. (2021). Operational and strategic change during temporary turbulence: evidence from the COVID-19 pandemic, *Operations Management Research*, 15, 589–607.

Sanghamitra M., Malaya K. N. (2011). Optimization model in human resource management for job allocation in ICT project, *International Journal of the Computer, the Internet and Management,* 19(3), 21–26.

Serafin R. (2013). Koncepcja systemu adaptacyjnego zarządzania ryzykiem dostaw w procesach produkcyjnych, *Zarządzanie Przedsiębiorstwem / Polskie Towarzystwo Zarządzania Produkcją*, 3, 46.

Strelicz A. (2021). Risks and threats in cyberspace – The key to success in digitization, *Journal of Physics: Conference Series*, 1935(2021), 1–7.

Soomoro Z., Shah M., Ahmed J. (2016). Information security management needs more holistic approach: A literature review. International *Journal of Information Management*, 36(2), 215–225.

Ślusarczyk Z. (2021). Znaczenie sztucznej inteligencji w działalności przedsiębiorstw. Podstawowe informacje, *Zarządzanie Innowacyjne w Gospodarce i Biznesie,* 2(23), 50.

Tohidi H. (2011). Human resources management main role in information technology project management, *Procedia Computer Science*, 3, 925–929.

Zagrożenia dla bezpieczeństwa ze strony nowych technologii, *CORDIS*, available on the website: https://cordis.europa.eu/article/id/85979-security-risks-of-emerging-technologies/pl (access: 19.12.2023)

Zalutska K., Petrushka K., Myshchyshyn O., Danylovych O. (2021). Strategic management of the innovative activity of the enterprise, *Journal of Optimization in Industrial Engineering*, Special issue, 22, 95–103.

Ziemba E. (2017). The contribution of ICT adoption to the sustainable information society, *Journal of Computer Information Systems*, 7, 1–8.

17 Ethics in the Tech Age
Resilience Strategies for Mitigating Technological Risks

Helena Fidlerová and Dejan Mirčetić

17.1 INTRODUCTION

The idea of the importance of ethics in the invention and use of technologies was introduced in the 20th century when American mathematician and philosopher, the founder of cybernetics, Norbert Wiener introduced the concept of computer ethics and of the need to understand technology as a stimulus to the flourishing of humanity and to building a better world.

Globalization of economic life is dependent on the technical infrastructure of information technology. The social and environmental changes are often the result of the interaction of technology, economy, and social institutions in specific and differentiated contexts.

A significant challenge in today's turbulent environment is that risks occur not "from outside", as a result of fatal forces of nature, but "from within" human societies, based on and as a result of human decision-making and changes in the new social context as part of the post-industrial society known as concept of risk society. This is not only as a result of partial deficiencies, malfunctions and dysfunctions, and unintended consequences, but also directly in the systemic nature of industrial modernization and technological revolution in which is contained the growth of productive forces, the integration of markets and cultural symbols connected to the market, and the inherent risk of destruction and threat to the natural foundations of life.

Nowadays, it is necessary to monitor and predict the impact of using technology, which on the one hand can improve the life of society and its users andwhile on the other we must ensure that this technology does not endanger the environment, which is the basis of society's existence. Therefore, for the evaluation of used and planned technologies, it is necessary to use the indicated indicators taking into account economic, environmental, and social aspects of sustainability model: enterprenurship, society, and the environment (Ginters & Revathy, 2021).

The main objective of this chapter is to improve awareness about the possibilities of using sustainable strategy and strategic management methods to minimize

 DOI: 10.1201/9781003475989-20

technological risks regarding sustainability. The presented results are part of the research project "KEGA 025STU-4/2023 Building the Modular Laboratory for the Development of Management Systems Auditing Skills".

17.2 RISK MANAGEMENT AND SUSTAINABILITY

Risk management is often understood as the major driving force of sustainable development and corporate social responsibility, especially in the industry (Anderson, 2005; Schulte & Hallstedt, 2018; Veselovská et al., 2020; Nikolaou et al, 2013; Kurowski & Huk, 2021; Ingaldi & Klimecka-Tatar, 2020). Risk management is essential to carry out an efficient risk analysis (Bolaños et al., 2019).

Although the need for and application of risk management has developed gradually in numerous fields to satisfy various needs, it is the adoption of a consistent risk management process within the system framework that ensures coherent and effective risk management. The general system approach detailed in the international standard ISO 31000 defines and offers guidance on the correct application of the principles and guidelines for the management of any form of risk, while it is ensured in a systematic and transparent manner for all processes and organizations (ISO 31000).

17.2.1 Sustainability Risk Management

Sustainability risk management (SRM) can be distinguished from a strategic point of view on international, national (regional), corporate, and individual levels. When implementing any principles it is essential to anchor the top-down principle so that hierarchy and continuity are preserved. The basic philosophy during implementation should be "think globally, act locally".

Dan R. Anderson and Kenneth E. Anderson (2009) considers sustainability risk as being part of emerging risks and sees it as a critical risk of this century. The integration of sustainable perspectives into risk management systems has been first established. It can be concluded that there are limited systematic methods available for identifying and managing sustainability risks. Moreover, existing solutions exhibit notable limitations (Schulte, 2018).

Risk is considered as something impermanent, and indeterminate, which is related to the course of the case and which disturbs purposeful behavior. Risk, uncertainty, and indeterminacy are part of human actions in relation to the environment.

Risk is associated with an event, human activity, uncertainty of the environment, or limitations in the system. An important factor in understanding risk is regarding its relevance and extent. Like any other quantity, risk must be measured in order to assess its significance. Mathematical–statistical methods use probability theory for the estimation of risk. The purpose of this approach is to reduce the uncertainty related to possible outcomes of different processes. Accordingly, if the processes and activates in modern-day business operations were deterministic the need for statistics and other related areas of mathematics like machine learning, data science and artificial intelligence will be diminished. These methods and approaches have only one task which is related to unveiling the mystery – what will be the outcome of some processes bearing in mind different possible outcomes and risks related to external factors, especially product demand spikes and surges and generally market demand for company's products (Mircetic et al., 2016, 2022; Rostami-Tabar & Mircetic, 2023). Therefore, it could be argued that the main task of modern analytical risks is mitigating the risks and providing decision support guidelines for steering via risks.

The absolute value of the risk is determined as the variance of the achieved result from the expected one (plan, target, etc.). The smaller the standard deviation, the narrower the distribution of probability over a given time period and the smaller the risk in its absolute understanding. The relative value of risk is understood as the variation of possible outcomes from the expected outcome. The most suitable characteristics of risk measurement are the probability over a given time period with which the risk may occur and the degree of its significance.

From previous experiences of considering the probability of occurrence, we can assume that in practice there usually occur smaller risks (problems), howeverbig risks (big problems) occur rarely. Serious risks are few, but their consequences are significant. Every business entity (organization) must draw up a scale of frequency, importance of possible risk, and the scale of their consequences and determine their impact on the activity of the organization and on the costs or for losses resulting from occurring risks and risky situations. On this basis, measures can be implemented to eliminate risks and calculate the associated financial costs (Singh et al., 2004). The organizations must pay special attention on drawing all the conclusions about the potential risks from the previous experiences, since this can be misleading in designing the strategies for mitigating risks which never before occurred in the past experiences. These risks are the most significant driver of costs and business breakouts, which usually need business reengineering of all, or several main processes (Nikoličić et al., 2021; Todorović et al., 2018; Maslarić et al., 2016). The researches in the area of sustainability and risk resilience highlight the importance of system approach for solving the risk disruptions and bottlenecks (Bojić et al., 2023; Maslarić et al., 2024). Still there are no definitive guides on how to holistically encompass this area.

The main problem with dealing with risk disruption emerges from established common practice to heavily relyon past experience as the main source of potential risks. As an example of this, COVID-19 comes in place. COVID-19 caused a total halt of the global supply chain, which resulted in shortages of consumer goods and eventually inflation of product prices. One of the reasons for such a big disruption was that in the last 30 years predominant supply chain design strategy was in the concept of lean logistics and past experiences. Lean logistics design philosophy totally disregarded big risks of product shortages and availability of material sourcing options, since it was not the risk occurrence in the past decades. Maslarić and colleagues proposed in 2013 a general supply chain risk management model (SCRM) that demonstrates balance between efficiency and resilience to overcome these types of risks (Figure 17.1).

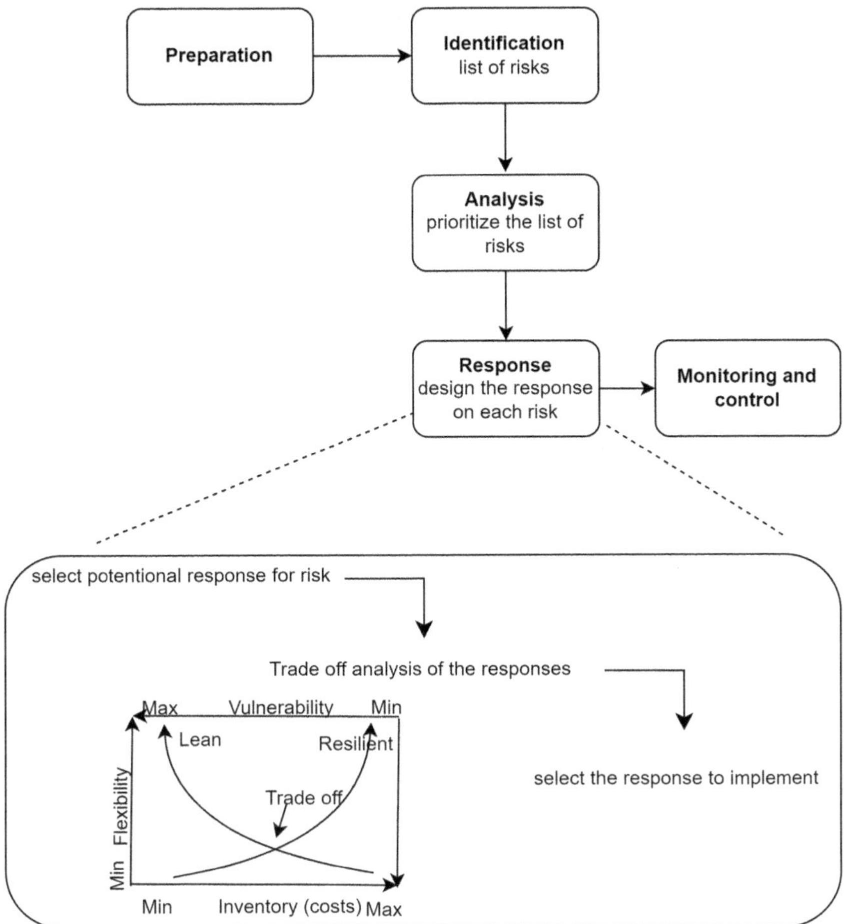

FIGURE 17.1 Framework for balance between lean and resilience within SCRM model.

The figure demonstrates the methodology for several steps for mitigating risks embedded in the above-mentioned SCRM framework. Definition of risk management is provided by ISO 31 000, where the risk management is defined as coordinated activates with the aim of controlling and directing the organization's risks. Sustainability risk management considers risks arising due to environment and corporate social responsibility (CSR), therefore the activity of each organization should be analyzed considering the impact on the environment, community, and stakeholders (Singh et al., 2005). Actions can be considered sustainable if the chance that the threat will be needed by the next generation is limited by a constant (Krysiak, 2009). Undesirable is a situation where the future generation would rather prefer a situation without the implementation of the action than the situation if the action was implemented.

Recent studies have defined sustainable risk in the following way: Czajkowska & Ingaldi (2021) highlighted the risks arising from environmental or social justice issues. Numerous examples illustrate how issues may impact the livelihood of enterprises negatively and positively. Sustainability risk is interpreted and defined in different ways, from partial to attempts at generalization.

The strategic intent is manage risks considering sustainable development, identify risks, quantify their severity, then determine potential sources of risk, initiate risk reduction and prevention measures, monitor and report methods necessary to ensure sustainable development and address identified threats.

The work (Krysiak, 2009) points out that it is often reasonable to expect a certain risk of harm to future individuals, because otherwise the potential increase in the well-being of individuals would be endangered and limited, since it could only happen trying to maintain the status quo without change. In reality, probably in the context of available information, a situation would arise when individuals decided to accept a potential risk in order to achieve a higher expected standard of living.

17.2.2 Strategic Connection of Ethics, Technology, and Risk Management

The World Commission on the Ethics of Scientific Knowledge and Technology (COMEST) is a recognized working group under UNESCO that ensures the connection of ethics, technology, and possible risks at the international level. An important step in this area has been the general conference of UNESCO (2021), when the first global instrument in this area was introduced, which concerns general recommendations for the Ethics of Artificial Intelligence (2021).

The given document is a guideline with the aim of maximizing the benefits and eliminating the risks associated with the design and usage and application of artificial intelligence (AI). The general ethics recommendations on the area of artificial intelligence is human-centered, based on existing human rights standards. The Recommendation on the Ethics of Artificial Intelligence includes two tools: an ethical impact assessment and a preparedness assessment methodology. The ethical impact assessment is intended to help assess the impact that datasets or algorithms may have on society. The proposed readiness assessment methodology has the task of setting regulatory and control mechanisms to facilitate the implementation of artificial intelligence. In April 2021, the European Commission submitted to the European Union a proposal for a regulatory framework on artificial intelligence – the Proposal for a Regulation of the European Parliament and of the Council with harmonization of application rules and ethics for artificial intelligence. Through this, the European Union aims to create legislative guidelines for the development, usage, and application of AI, as well as to facilitate and increase the level of investment and innovation in this area in accordance with the creation of a single market for the safe and reliable application of artificial intelligence systems.

Artificial intelligence is defined in various ways. Most definitions describe it as software (software model) trained on a dataset to perform specific tasks, such as recognizing certain patterns, etc. An independent expert group of the European Commission provided the following definition: Artificial intelligence is defined as a system that exhibit intelligent and reasonable behavior by making decisions based on the analysis of their environment – with a certain degree of autonomy – to achieve specific goals. We are aware that there are obvious differences and therefore can be found to work in virtual environments such as virtual assistants and avatars, photo analysis software, Internet search engines, voice recognition systems, and facial recognition (computer vision systems). Additionally, it can be combined with hardware systems (e.g., highly advanced robots, cobots, autonomous vehicles, drones, etc.).

One of the leaders in ethical directions regarding the usage of artificial intelligence and risks related to this technology is the Republic of Serbia with its guidelines for the Ethical Development, Application, and Deployment of Reliable and Responsible Artificial Intelligence (Republic of Serbia, 2022). The aforementioned ethical guidelines formulate different potential risk and mitigation strategies. Their aim is to ensure that science, especially in the field of artificial intelligence, can develop and advance without compromising the central role of humans in all processes that affect them, whether they are directly or indirectly involved. Furthermore, special attention devoted towards future artificial intelligence systems must also be in harmony with the welfare of animals. The guidelines aim to encompass the broadest range of participants in the artificial intelligence ecosystem to establish a horizontal approach to rule application. The guidelines apply to the following entities:

- Individuals involved in the development and/or implementation of artificial intelligence systems
- Individuals who apply artificial intelligence systems, especially in their work involving interactions with others (e.g., market participants)
- Individuals who use artificial intelligence systems and are affected by them
- Directly affected (e.g., using systems for accessing some public service)
- Indirectly affected (e.g., part of a group researching rare diseases, whose medical data are processed as part of the strategy of the Republic of Serbia for better national health)
- General public, in the broadest sense

17.2.3 Ethical Guidelines and High-Risk Artificial Intelligence Systems

A high-risk system is one that tends to directly or indirectly violate the principles and conditions ethics. Under the ethical guidelines, a high-risk system is considered part of a product's security system or its own security system, and a third party must comply with standards for the deployment of artificial intelligence systems. High-risk systems including biometric identification and classification, management and operation, education, and professional training are development critical infrastructure, employment, workforce management and access to self-employment, healthcare – particularly, access to many public and social services as well as essential private, criminal prosecution, management of human migrations, judiciary, and democratic processes considered to be areas of implementation.

High-risk systems also include algorithms and artificial intelligence systems for platforms such as social networks. As critical systems from the point of view of influence, we can also include systems supported by employing artificial intelligence on extensively utilized social platforms, like social networks, which serve a significant user population. These systems make numerous decisions about what content is presented to public based on data analysis, predictive behavior, setting, preferences, and set goals. This significantly affects decisions and behavior, actions, opinions, preferences, creativity, and emotions on a conscious and unconscious level. Socially, it can create social bubbles, deceive or polarize important social and other problems, ultimately affecting democracy and sustainability.

In order to mitigate risk of misusing artificial intelligence, the following steps are proposed: intervention (mediation, control, participation) and supervision; technical reliability and safety, privacy considerations, transparency in operations, fairness in AI systems, safeguarding of data and effective data management; alongside a commitment to diversity, prevention of discrimination, and promotion of social and environmental well-being (Republic of Serbia, 2022).

17.2.4 Risk Assessment

Assessing risk represents a key process that is an integral part of the strategic planning and operational management of organizations, projects, and various activities. It serves as a systematic approach for identification, assessment of risks, as well as elimination of possible risks that could negatively affect the intended goals. The main goal is to provide support in decision-making for a comprehensive understanding of threats and vulnerabilities, which facilitates more informed decision-making.

The risk assessment process takes place through several interrelated steps. In the beginning, the crucial task is to identify the risks. This means a thorough examination of all aspects of a system, project, or operation to determine potential threats. Risks are diverse and can include financial, operational, technical, legal, and environmental dimensions.

The object of risk analysis of technical systems and technologies must be in their life cycle: risks in the design and construction stage, risks in the production stage, risks in the use stage, risks during maintenance and repairs, risks during the recovery of parts of the system or during its deterioration (scrapping, destruction).

Once identified, risks are analyzed to assess their likelihood and potential impact. This involves evaluating both the likelihood of a risk occurring and the potential consequences. The methods used for analysis can be both qualitative and quantitative and offer a different view of the risk area.

The next step is risk evaluation, where the results are analyzed and compared with risk criteria. This process helps in determining the significance of each risk and prioritizing them based on their potential impact. Once priorities are established, organizations develop strategies to address risks, which may include avoidance, reduction, transfer, or acceptance with contingency plans in place.

17.2.5 The Life Cycle of Risk Assessment

The life cycle of risk assessment does not end, but continuous monitoring, improvement and regular checks are necessary. This ongoing process ensures that new potential risks are identified in time and eliminated or their consequences reduced.

Effective risk assessment brings many benefits. It allows decision-makers to see potential challenges, allowing them to make informed decisions. Bromiley et al. (2014) emphasizes the importance of understanding how risk is defined depending on individuals and groups within the organization, drawing attention to possible biases in the assessment and challenges in the implementation of risk management.

Effective resource allocation is facilitated by identifying and prioritizing risks, ensuring that resources are directed to managing the most significant threats. In addition, risk assessment helps with regulatory compliance, especially in industries that are subject to standards mandating systematic risk assessment. In essence, risk assessment is not just a procedural formality; it is a dynamic and integral aspect of organizational resilience. By systematically analyzing and addressing potential risks, organizations can improve their ability to navigate uncertainties and thus support a safer and more sustainable future.

Risk treatment according to international standard ISO 3100, discusses the possibility of mitigating the risk by deciding on whether the organization should begin or continue the activities that possibly lead towards the risk; accepting any possible increase in risk when trying to take advantage of opportunities; possible removal of the source of risk, analysis, and determination of change in probability and consequences, the possibility of spreading the risk with another party; and retaining risk for an informed decision (Figure 17.2).

The risk management framework is created as a system of related components that provide the basis and necessary measures for the designing, monitoring, implementation, control, and continuous improvement of their organizations.

The risk management framework is intricately woven into the fabric of the organization, seamlessly integrated within its strategic, tactical, and operational policies and practices (ISO 31 000). In crafting a resilient risk management approach, the framework, intricately aligned with the Plan, Do, Check, Act (PDCA) cycle as emphasized by Chojnacka-Komorowska & Kochaniec (2019), navigates through essential components. Governance and policy set the organizational tone, outlining a mission and showcasing unwavering commitment. Program design proposes a global framework for seamless risk management, while implementation brings it to life, establishing a robust program. Monitoring and evaluation vigilantly oversee system structures and performance, while continuous improvement actively seeks avenues for enhancement. This comprehensive framework, deeply rooted in the PDCA cycle, is a dynamic roadmap ensuring governance, adaptability, and continual refinement in the face of evolving risks.

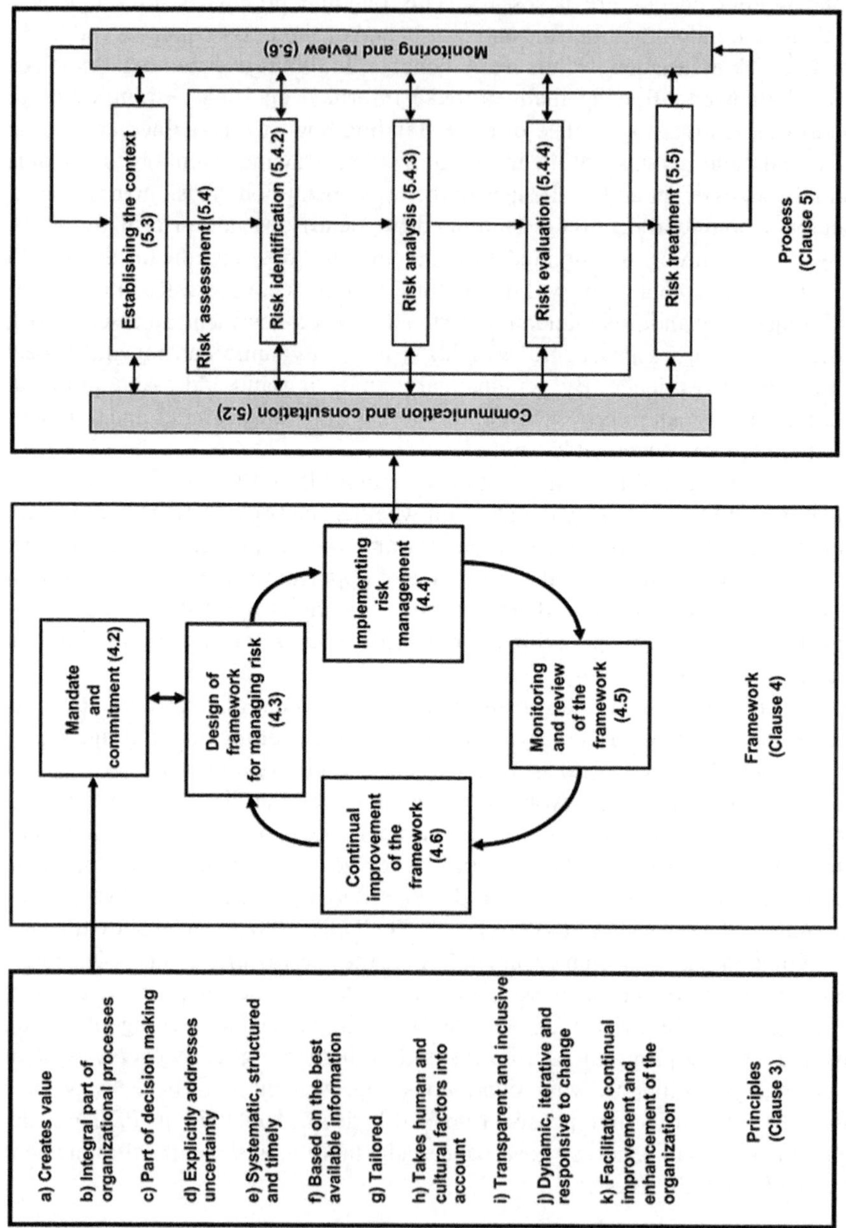

FIGURE 17.2 The synergy among risk management principles, frameworks, and processes (ISO 31 000).

17.3 METHODS FOR MANAGING THE SUSTAINABILITY RISK

Krysiak (2009) emphasizes that sustainability has become a major framework for evaluating long-term decisions on a political- and organization-level scale. Sustainable decision-making means to be responsible for eliminating present and future sustainability risks for all stakeholders and the environment. In this section, some methods for professional multicriterial decision-making considering risks analysis and identification of causes of problems are discussed.

When choosing a suitable method for risk management, we need to consider the possibilities of the method and the ability to distinguish details at a given level, express information in a suitable form (matrix, number, graph), scope of use of the method, objectivity of the results (objective or subjective), and the difficulty of the method (economic, time, personnel-qualification, technical requirements).

When distinguishing between methods, the approach of the method to the description of the causal connection may also be different: either inductive (bottom-up procedure) prediction of the probability in future by predicting consequences of a possible adverse event (mathematical models) or deductive-based on events that have already happened (statistical data on system failures).

In the section of the chapter are introduced selected methods of strategic management and quality management that can be implemented for decision-making and risk management.

17.3.1 Analysis of Strengths, Weaknesses, Opportunities, and Threats

One of the important tools of strategic management for decision-making that can be used is the analysis of Strengths, Weaknesses, Opportunities, and Threats (SWOT). SWOT is often used as an expert method by including several experts into the interdisciplinary working group.

It is possible to compare the criteria in the context of sustainable development goals, analyze them, and determine the sustainability strategy, to determine the hierarchy and importance of established criteria in terms of risk. It is appropriate to use expert methods of multi-criteria decision-making, for example, the analytical hierarchical method by Saaty (1987). Multicriterial analysis is useful for incorporating sustainability criteria regarding ecological, social, and economic implications (Fidlerová et al., 2016; Senthil et al., 2018; Veira et al., 2020).

As presented in the study (Paulíková et al., 2021) SWOT analysis has been used as a tool to assess many aspects considering impact of robotization in industry. We agree that SWOT analysis provides an indication of the feasibility of robotics in the workplace, helps to find the optimal solution, and represents new areas of research for future.

Furthermore, SWOT analysis has been used successfully for risk assessment and strengths for future strategic decisions considering new technologies such as the green vehicles market for a green transport considering sustainable logistics (Klimecká et al., 2021).

The main limitation of the use of SWOT analysis method is that it is possible to analyze only a limited number of individual criteria, divided into four defined dimensions (Strengths, Weaknesses, Opportunities, Threats) (Paulíková et al., 2021).

17.3.2 Method Five Whys

The five whys method is based on the principle ofmultiple iterations of "why" to identify the root cause. First, the identified problem is posed, and the assessor asks, "Why?" Once an answer is obtained the assessor asks "Why?" again to understand the reason of the identified circumstance of the problem. Usually, after repeating more (five) times, we might find the root cause of the problem to propose risk mitigation strategies. It is often used in industrial organizations (Skurkova & Prajová, 2022).

17.3.3 The Failure Mode and Effect Analysis

The Failure Mode and Effect Analysis (FMEA) has been used since the 1970s to the 20th century, when the National Aeronautics and Space Administration (NASA) brought an method to find out the reliability of the FMEA for the Apollo project. That was the first time critical groups of possible errors and their consequences were defined (Ignáczová, 2016). We can conclude that FMEA is with a modification still used nowadays, especially in industry (Renosori et al., 2023; Shafiee et al. 2019).

A cause-and-effect diagram (CE diagram), also known as an Ishikawa diagram, can be used as part of an FMEA to display risks and possible causes. The main axis is a root of the problem and the branches are visualization of the effects that are caused. When creating this diagrams, it is necessary to define first the root causes of the problem (mostly divided into five branches – material, equipment, methods, people, nature, etc.).

Another analytical tool, the Pareto diagram, can be used in strategic management for identification, which prioritizes and guides using the 80/20 principle. The most important issues (20%) should be given the most attention because they cause 80% of the consequences.

The FMEA analysis aims to identify and quantitatively assessing the risk of possible errors, as well as their consequences. The method is based on the analytical determination of cause-and-effect relationships of potential defects and identification of the criticality factor (risk). The FMEA method can be used for the prediction of errors, risks, nonconformity (Doshi & Desai, 2017) and predictive maintenance (Mesarošová et al., 2022).

In FMEA, it is appropriate to include environmental aspects in the analysis of failure modes and effects because is a technique that allows errors or problems to be identified and helps to minimize them (Czajkowska & Ingaldi, 2021).

The FMEA method was developed due to technological changes and innovations in production, product, and environment, and a modification of the method was created under the title Failure Mode and Effect Critical Analysis (FMECA). FMECA is a logical extension of FMEA that examines consequence of failure and failure criticality, with each identified failure rated for the combined impact of its probability, outcome, and severity (Sakly et al., 2022).

The authors (Nguyen et al., 2013) claim that the main strength of FMECA is that it is simple to use, howeverit offers quantitative evaluation using a sophisticated combination of three parameters: frequency, severity, and detectability. Moreover, the FMECA can be helpful for identifying the most important critical events as support for decisions.

17.3.4 Fault Tree Analysis

Fault Tree Analysis (further FTA) is popular and often used technique, technology widely used in the evaluation of the reliability of various systems. (Kabir, 2017; Tao et al., 2010; Shafiee et al., 2019). This graphical and mathematical tool is used to analyze the risk of machine or system failure. It is a top-down approach that seeks to reverse-analyze the root causes of potential failure. FTA creates a system of logic error models that uses Boolean logic gates (AND, OR ...) to describe how simple events can be combined and lead to an accident the information shown in the fault tree depends on the resolution it is need to solve the problem.

FTA helps to visualize, by means of a graphic model called fault trees, the course of how failure spreads through the system and how failures of individual parts in the form of small faults can lead to failures of the entire system. The task is to set up risk management so that component failures do not turn into a failure of the entire system, which enables the implementation of preventive internal controls to minimize risks.

17.3.5 Checklist

Checklists are a proven method for verifying the correct execution of risk management process. The procedure is as follows: first a list of all identified risks is created, then each defined risk is checked, and the relevant task is completed.

Its principles are as follows: a simple comparison of information contained in the fact checklist based on a simple distinction – yes/no, was/wasn't, did/didn't – can be stated which condition, property whether the activity exhibits risk characteristics that need to be addressed and/or precede. The main advantage of this method is its simplicity and versatility, as it can be used for any project.

17.3.6 Methods "What if" and Structured "What-if"

Following the checklist, it is possible to use the "what if" method in the risk assessment process, which can be easily understood and implemented. The evaluator gathers a group of selected experts from various fields on the given issue, who are sufficiently familiar with the process that is the subject of the risk analysis. The team identifies various "what if" scenarios to brainstorm possible risks.

The disadvantage of this method is that its use and documentation are not sufficiently structured to guarantee continuity. However this shortcoming can be overcome by combining the method of creative questioning with a systematic procedure according to a diagram or checklist. This analysis is particularly suitable in the pre-project phase of planning new technologies.

The modification of this method is the structured what-if technique (SWIFT) (Card et al., 2012). This method is suitable for use due to its flexibility in identifying risk at a high level. the advantage is that the mentioned method can be used in both ways only on its own or as part of an FMEA.

REFERENCES

Anderson, D. R. (2005). *Corporate Survival: The Critical Importance of Sustainability Risk Management*. Lincoln: iUniverse.

Anderson, D. R., & Anderson, K. E. (2009). Sustainability risk management. *Risk Management and Insurance Review*, 12(1), 25–38. https://doi.org/10.1111/j.1540–6296.2009.01152.x.

Bojic, S., Maslaric, M., Mircetic, D., Nikolicic, S., & Todorovic, S. (2023). Simulation and genetic algorithm-based approach for multi-objective optimization of production planning: A case study in industry. *Advances in Production Engineering & Management*, 18(2), 250–262.

Bolaños, E. R. L., Barriga, G., & Noriega, E. (2019). Risk management and anti-bribery: An operational approach from the perspective of ISO 31000 and ISO 37001. *Revista Universidad Empresa*, 21, 79–118. https://doi:10.12804/revistas.urosario.edu.co/empresa

Card, A. J., Ward, J. R., & Clarkson, P. J. (2012). Beyond FMEA: The structured what-if technique (SWIFT). *Journal of Healthcare Risk Management*, 31(4), 23–29. https://doi.org/10.1002/jhrm.20101

Chojnacka-Komorowska, A., & Kochaniec, S. (2019). Improving the quality control process using the PDCA cycle. *PraceNaukoweUniwersytetuEkonomicznego We Wrocławiu*, 63(4), 69–80. doi:10.15611/pn.2019.4.06; oai:dbc.wroc.pl:73412

Czajkowska, A., & Ingaldi, M. (2021). Structural failures risk analysis as a tool supporting corporate responsibility. *Journal of Risk and Financial Management*, 14(4), 187. https://doi.org/10.3390/jrfm14040187

Doshi, J., & Desai, D. (2017). Application of failure mode & effect analysis (FMEA) for continuous quality improvement – Multiple case studies in automobile S.M.E.s. *International Journal for Quality Research*, 11(2), 345–360. https://doi.org/10.18421/IJQR11.02–07

Fidlerová, H., Jurík, L., & Sakál, P. (2016). Application of AHP in the process of sustainable packaging in a company. *Production Management and Engineering Sciences*, 585(591), 585–591.

Ginters, E., & Revathy, J. C. (2021). Hidden and latent factors' influence on digital technology sustainability development. *Mathematics*, 9(21), 2801. https://doi.org/10.3390/math9212801

Ignáczová, K. (2016). FMEA (Failure mode and effects analysis) and proposal of risk minimizing in storage processes for automotive client. *Acta Logistica*, 3(1), 15–18.

Ingaldi, M., & Klimecka-Tatar, D. (2020). People's attitude to energy from hydrogen—from the point of view of modern energy technologies and social responsibility. *Energies*, 13(24), 6495. https://doi.org/10.3390/en13246495

ISO 31000:2009. Risk management — Principles and guidelines.

Junkes, M. B., Tereso, A. P., & Afonso, P. S. L. P. (2015). The importance of risk assessment in the context of investment project management: A case study. *Procedia Computer Science*, 64, 902–910. https://doi.org/10.1016/j.procs.2015.08.606

Kabir, S. (2017). An overview of fault tree analysis and its application in model-based dependability analysis. *Expert Systems with Applications*, 77, 114–135. https://doi.org/10.1016/j.eswa.2017.01.058

Krysiak, F. C. (2009). Risk management as a tool for sustainability. *Journal of Business Ethics*, 85(3), 483–492. https://doi.org/10.1007/s10551-009-0217–7

Kuchenbrandt, D., & Eyssel, F. (2012). The mental simulation of a human-robot interaction: Positive effects on attitudes and anxiety toward robots. In 2012 IEEE RO-MAN: The 21st IEEE International Symposium on Robot and Human Interactive Communication, Paris, 463–468. doi: 10.1109/ROMAN.2012.6343795.

Maslaric, M., Backalic, T., Nikolicic, S., & Mircetic, D. (2013). Assessing the trade-off between lean and resilience through supply chain risk management. *International Journal of Industrial Engineering and Management*, 4(4), 229–236.

Maslaric, M., Bojic, S., Mircetic, D., Nikolicic, S., & Medenica Todorovic, R. (2024). Sustainable urban mobility planning in the port areas: A case study. *Sustainability*, 16(2), 514.

Maslarić, M., Nikoličić, S., & Mirčetić, D. (2016). Logistics response to the industry 4.0: the physical internet. *Open Engineering*, 6(1), 435–450.

Mesarosova, J., Martinovicova, K., Fidlerova, H., HrablikChovanova, H., Babcanova, D., & Samakova, J. (2022). Improving the level of predictive maintenance maturity matrix in industrial enterprise. *Acta Logistica*, 9(2), 183–193. https://doi.org/10.22306/al.v9i2.292

Mircetic, D., Nikolicic, Mircetic, D., Rostami-Tabar, B., Nikolicic, S., & Maslaric, M. (2022). Forecasting hierarchical time series in supply chains: an empirical investigation. *International Journal of Production Research*, 60(8), 2514–2533.

Nguyen, C., Côté, J., Lebel, D., Caron, E., Genest, C., Mallet, M., et al. (2013). The AMÉLIE project: failure mode, effects, and criticality analysis: A model to evaluate the nurse medication administration process on the floor. *Journal of Evaluation in Clinical Practice*, 19(1), 192–199. https://doi: 10.1111/j.1365–2753.2011.01799.x

Nikolaou, I. E., Evangelinos, K. I., & Allan, S. (2013). A reverse logistics social responsibility evaluation framework based on the triple bottom line approach. *Journal of Cleaner Production*, 56, 173–184. https://doi.org/10.1016/j.jclepro.2011.12.009

Nikolicic, S., Kilibarda, M., Maslaric, M., Mircetic, D., & Bojic, S. (2021). Reducing food waste in the retail supply chains by improving efficiency of logistics operations. *Sustainability*, 13(12), 6511.

Pauliková, A., GyurákBabeľová, Z., & Ubárová, M. (2021). Analysis of the impact of human–cobot collaborative manufacturing implementation on the occupational health and safety and the quality requirements. *International Journal of Environmental Research and Public Health*, 18(4), 1927.

Recommendation on the Ethics of Artificial Intelligence (2021). UNESCO, 44. https://unesdoc.unesco.org/ark:/48223/pf0000381137

Renosori, P., Oemar, H., & Fauziah, S. (2023). Combination of FTA and FMEA methods to improve efficiency in the manufacturing company. *Acta Logistica*, 10(3), 487–495. https://doi.org/10.22306/al.v10i3.422

Rostami-Tabar, B., & Mircetic, D. (2023). Exploring the association between time series features and forecasting by temporal aggregation using machine learning. *Neurocomputing*, 126376.

Saaty, R. W. (1987). The analytic hierarchy process—What it is and how it is used. *Mathematical Modelling*, 9(3–5), 161–176. https://doi.org/10.1016/0270-0255(87)90473-8

Sakly, H., Chakroun, I., & Ben Jeddou, K. (2022). Application of failure mode, effects, and criticality analysis to the medication-use process for temperature-sensitive drugs in a university hospital. *Canadian Journal of Hospital Pharmacy*, 75(3), 159–168. https://doi.org/10.4212/cjhp.3121

Schulte, J., &Hallstedt, S. (2018). Challenges for integrating sustainability in risk management – current state of research. In Proceedings of the 21st International Conference on Engineering Design (ICED 17), 2, Vancouver, Canada.

Senthil, S., Murugananthan, K., & Ramesh, A. (2018). Analysis and prioritisation of risks in a reverse logistics network using hybrid multi-criteria decision-making methods. *Journal of Cleaner Production*, 179, 716–730. https://doi.org/10.1016/j.jclepro.2017.12.095

Shafiee, M., Enjema, E., & Kolios, A. (2019). An integrated FTA-FMEA model for risk analysis of engineering systems: A case study of subsea blowout preventers. *Applied Sciences*, 9(6), 1192. https://doi.org/10.3390/app9061192

Singh, P., Singh, M., & Bhambri, P. (2004, November). Interoperability: A problem of component reusability. In International Conference on Emerging Technologies in IT Industry (p. 60).

Singh, P., Singh, M., & Bhambri, P. (2005, January). Embedded systems. In Seminar on Embedded Systems (pp. 10–15).

Skurkova, K., & Prajová, V. (2022). 8D Report application in production process of the rear seat. *MM Science Journal*, 2022, 6074–6077. https://doi.org/10.17973/MMSJ.2022_11_2022139

Tao, H., Liang, C., Chi, W., & Qun, H. (2010). The research of information security risk assessment method based on fault tree. In Proceeding—6th International Conference on Networked Computing and Advanced Information Management, NCM 2010, 375.

Todorovic, V., Maslaric, M., Bojic, S., Jokic, M., Mircetic, D., & Nikolicic, S. (2018). Solutions for more sustainable distribution in the short food supply chains. *Sustainability*, 10(10), 3481.

Veselovská, L., Závadský, J., & Závadská, Z. (2020). Mitigating bribery risks to strengthen corporate social responsibility in accordance with ISO 37001. *Corporate Social Responsibility and Environmental Management*, 27(4), 1972–1988. https://doi.org/10.1002/csr.1909

Vieira, B. O., Guarnieri, P., Nofal, R., & Nofal, B. (2020). Multi-criteria methods applied in the studies of barriers identified in the implementation of reverse logistics of E-Waste: A research agenda. *Logistics*, 4(2), 11. https://doi.org/10.3390/logistics4020011

18 Ethics in Sustainable Technology

N. Chithra and Pankaj Bhambri

18.1 INTRODUCTION

18.1.1 Ethics Meaning

The study of moral principles and ideals that help people and communities decide what is good and wrong is known as ethics in philosophy. It entails studying ideas like duty, virtue, morality, and justice. The goal of ethics is to give people and society a framework for comprehending and assessing human behaviour so they can make moral decisions.

In a larger sense, ethics includes rules and guidelines that control how people behave as individuals, professionals, groups, and societies. It goes beyond philosophy. It answers concerns about appropriate behaviour, values that should direct behaviour, and whether behaviours are acceptable or undesirable.

In many disciplines, including business, law, medicine, science, and technology, ethical issues are crucial. There are various ethical theories and frameworks available, each providing different perspectives on what constitutes ethical behaviour. Common ethical principles include honesty, integrity, fairness, respect for others, and the promotion of well-being.

Ethics helps individuals navigate complex moral dilemmas, make informed decisions, and contribute to developing a just and moral society. It is a critical tool for promoting accountability, responsible conduct, and the well-being of individuals and communities.

18.1.2 Key Aspects of Technology

The use of scientific instruments, information, and techniques to address real-world issues and accomplish certain goals is referred to as technology. It entails the application of knowledge, abilities, and procedures to develop systems, goods, or practices that better human potential, meet needs, and raise the standard of living. Numerous domains are included in the broad category of technology, which is present in many facets of daily life, business, and society.

DOI: 10.1201/9781003475989-21

The principal features of technology comprise:

- Innovation: Creating new products, services, or concepts that boost productivity, efficacy, or convenience is a common aspect of technology.
- Tools and Equipment: Developing and deploying tools, machinery, and equipment to complete activities that would be challenging or impossible without them are common aspects of technological advances.
- Applied Science: Technology is closely related to science, as scientific knowledge is often applied to solve real-world problems and create practical solutions.
- Automation: Many technological advancements involve automating processes, reducing human effort and increasing precision in various tasks.
- Communication and Information: Technology plays a crucial role in communication and the processing of information. This includes information technology, telecommunications, and the internet.
- Influence on Society: Technology has a profound impact on society, shaping cultural, economic, and social aspects. It can bring about significant changes in how people work, live, and interact.
- Continuous Development: Technology is dynamic and continually evolving. Ongoing research, development, and innovation contribute to the constant improvement and emergence of new technologies.

Technology encompasses many different things, such as computers, cell phones, transportation systems, medical equipment, renewable energy sources, and much more. Technology has changed many businesses and is now a necessary component of modern life due to its rapid evolution.

18.1.3 Sustainable Technology

Sustainable technology, sometimes referred to as green or clean technology, is the application of scientific and technical principles to the development of long-term, economically viable solutions with negligible environmental impact. Addressing global issues, including resource depletion, climate change, and environmental degradation, is the main objective of sustainable technology. This entails developing systems, procedures, and goods that satisfy current demands without jeopardizing the capacity of future generations to satisfy their own.

18.1.4 Characteristics of Sustainable Technology

Sustainable technology embodies characteristics that prioritize environmental responsibility, social equity, and economic viability. One key feature is resource efficiency, wherein technologies aim to minimize resource consumption and waste generation throughout their life cycle. This involves the use of renewable materials, efficient energy utilization, and the reduction of environmental impact during production, operation, and disposal. Another crucial aspect is resilience to climate change, as sustainable technologies are designed to adapt to and mitigate the effects of environmental challenges.

Additionally, sustainable technologies often prioritize social inclusivity, aiming to address the needs of diverse communities and ensure that technological benefits are equitably distributed. They promote the well-being of present and future generations by fostering long-term thinking and planning. Collaboration and open innovation characterize sustainable technology development, encouraging interdisciplinary approaches and stakeholder engagement.

Furthermore, these technologies emphasize a cradle-to-cradle philosophy, seeking to create closed-loop systems where products can be recycled, refurbished, or repurposed rather than discarded as waste. Life-cycle assessment and eco-design principles guide the development process, ensuring a holistic evaluation of environmental and social impacts.

In essence, sustainable technology reflects a commitment to creating solutions that not only meet present needs but also contribute positively to the health of the planet and its inhabitants, fostering a harmonious balance between technological advancement and the preservation of natural resources.

18.2 PRINCIPLES OF ETHICS IN TECHNOLOGY

The principles of ethics in technology form a critical framework that guides the responsible development, deployment, and use of technological innovations. Transparency is paramount, requiring clear communication about how technologies operate, collect data, and impact individuals and society. Respect for privacy is a foundational principle, necessitating the protection of personal information and user autonomy. Equity and inclusivity demand that technological advancements benefit all segments of society, avoiding the exacerbation of existing social disparities. Accountability is essential, compelling technology creators and users to take responsibility for the consequences of their innovations. Additionally, sustainability principles emphasize minimizing environmental impacts, ensuring that technological solutions contribute positively to ecological well-being. Collaboration and open dialogue with diverse stakeholders, including users, communities, and policymakers, are integral to ethical technology development. Continuous reflection and adherence to established ethical guidelines promote the creation of technologies that align with societal values, fostering a harmonious integration of innovation and ethical responsibility. Ultimately, these principles serve as a moral compass, guiding the ethical evolution of technology in a manner that prioritizes human well-being, social justice, and environmental stewardship.

18.3 PILLARS OF SUSTAINABILITY

The three main pillars of sustainability are classified as environmental, social, and economic. Subcategories within these three pillars aid in determining if a sustainability concern is profitable, environmental, or something else entirely. All three of the sustainability pillars can be applied ethically, and ethical considerations ought to be made while making decisions together. It is crucial to take sustainability ethics into account both during these procedures and while evaluating all potential stakeholders in a problem. When developing a solution and gathering input from all parties involved, it is important to uphold ethical principles while considering the potential issues that could arise from an implicit solution. Because it crosses boundaries, the inclusion of ethics in sustainable decision-making results in a more diverse approach. The addition of ethics in sustainable decision-making creates a more diverse process in that it bridges the boundaries between stakeholders who would not otherwise communicate with each other. This may require some social, political, or even cultural concessions, but it is ultimately beneficial in the grand scheme of things.

In sustainability, **social ethics** is primarily concerned with an individual's behaviour, attitudes, beliefs, creative traditions, and ideas. In the social sector, ethics refer to decisions or viewpoints that support the identity of a group as a whole. It is crucial to understand that social ethics and sustainability are not mutually exclusive; rather, social ethics offers an ethical framework within which a group can make sustainable decisions while taking inclusivity, diversity in the arts, justice, equity, and fairness into account. Sustainable challenges with a morally sound societal basis take values, costs, and rewards into account.

Since many ethical economic choices are based on the social costs or consequences of that lucrative effort or decision, **economic ethics**, and social ethics are closely related. Moral underpinnings in a successful legacy take into account issues like product security, productivity, and efficiency. It is related to social sustainability since it considers things like whether a product has benefited one social class over another and whether it was made in an environment that offers fair wages and working conditions. It also considers how the funds utilized to produce a good are employed, as well as the possibility that they were misused unfairly. To achieve sustainability, economic ethics must ensure that goods are produced in a way that can be attributed to the larger conception and pretensions of sustainability. Ethical frugality isn't dependent on growth or material consumption in an age where there are finite resources.

The value of non-human nature, or the rates of living and non-living things within nature, is the main emphasis of *environmental ethics*. Living rates refer to species such as plants, animals, and other natural phenomena, while non-living rates include things like water, forest health, and the preservation of larger geographic areas. Environmental ethics examines what behaviour is appropriate and inappropriate in natural environments. The preservation of biodiversity, access to clean water and air, and the worth of non-human life are concerns that could be taken into account under this kind of ethics. Environmental ethics contends that while humans should coexist with the non-human aspects of the planet, they do not have a right to dominate them.

18.4 REGULATORY AUTHORITY GOVERNING ETHICAL ISSUES IN TECHNOLOGY

The regulatory landscape governing ethical issues in technology is multifaceted, involving a combination of public and transnational bodies that seek to establish norms and guidelines for responsible technological development and use. Transnational associations such as the United Nations Educational, Scientific and Cultural Organization (UNESCO) and the International Telecommunication Union (ITU) address ethical considerations in technology through endeavours and fabrics that promote human rights, inclusivity, and digital literacy.

On a national level, regulatory authorities play a pivotal part in overseeing ethical aspects of technology within their jurisdictions. For this case, the Federal Trade Commission (FTC) in the United States monitors privacy and consumer protection issues related to technology (Jain and Bhambri, 2005). Additionally, data protection agencies, similar to the Information Commissioner's Office (ICO) in the United Kingdom and the European Data Protection Board (EDPB) in the European Union, apply regulations like the General Data Protection Regulation (GDPR) to protect individuals' privacy rights.

Likewise, the ethical confines of emerging technologies like artificial intelligence often prompt specific regulatory attention. Governments worldwide are considering or enforcing legislation to address issues such as bias, responsibility, and transparency in AI systems. These regulatory authorities inclusively work to establish a framework that balances innovation with ethical considerations, fostering a technology landscape that respects fundamental values and safeguards societal well-being.

18.5 ADVANTAGES OF ETHICS IN SUSTAINABLE TECHNOLOGY

Ethics in sustainable technology offers a myriad of advantages that extend beyond the realms of invention and progress. At its core, ethical considerations ensure that technological advancements align with principles that prioritize environmental responsibility, social equity, and long-term viability. When technology is developed with a strong ethical foundation, it not only minimizes negative impacts on the environment but also contributes appreciatively to societal well-being. Ethical practices in sustainable technology promote transparency, responsibility, and inclusivity, fostering a sense of trust among users and stakeholders. Also, ethical guidelines drive innovation towards results that address pressing global challenges, such as climate change and resource depletion. By integrating ethics into sustainable technology, we pave the way for a future where technological progress is harmoniously intertwined with conserving our earth and improving human lives. The advantages of ethics in sustainable technology extend far beyond the immediate benefits, creating a foundation for a more conscientious and sustainable technological landscape.

18.6 CHALLENGES OF USING ETHICS IN SUSTAINABLE TECHNOLOGY

The integration of ethics into sustainable technology faces several challenges that arise from the complex interplay of environmental, social, and economic factors. One significant challenge is the trade-off between economic interests and ethical considerations. In some cases, the pursuit of cost-effectiveness may disaccord with the commitment to environmentally friendly practices, leading to ethical dilemmas in decision-making. Striking a balance between profitability and ethical responsibility poses a non-stop challenge for businesses and diligence.

Also, the rapid-fire pace of technological invention frequently outpaces the development of ethical guidelines, creating a pause in conforming ethical norms to arising technologies. This gap can lead to unforeseen consequences, such as privacy breaches, environmental detriment, or social inequalities, which challenge the ethical foundation of sustainable technology.

Global differences in technological access and relinquishment also present a challenge. While sustainable technology aims to benefit all, differences in resources and infrastructure may hamper equitable distribution, aggravating social inequalities (Rattan and Bhambri, 2005a).

Also, the lack of standardized ethical fabrics for sustainable technology poses challenges. Different stakeholders may have varied perspectives on ethical behaviour in technology development and use. Establishing widely accepted ethical principles for sustainable technology remains a complex task.

In substance, addressing these challenges requires ongoing collaboration among governments, diligence, academia, and communities to foster a holistic and adaptive approach that ensures the ethical integration of sustainable technology into our rapidly evolving world.

18.7 THEORIES OF ETHICS IN SUSTAINABLE TECHNOLOGY

The creation and application of sustainable technologies are greatly influenced by ethics. Utilitarianism is a well-known ethical framework that promotes the maximum amount of people's total enjoyment and well-being. Utilitarian ethics in the context of sustainable technology would entail evaluating the economic, social, and environmental effects of technological breakthroughs to make sure they support society's long-term well-being.

Another ethical proposition applicable to sustainable technology is environmental ethics, which emphasizes the intrinsic value of the environment and ecosystems. Votaries of this proposition argue that sustainable technology should prioritize the protection of biodiversity, ecosystems, and the overall health of the earth.

Likewise, virtue ethics encourages the cultivation of righteous character traits in individuals and societies. In the realm of sustainable technology, virtue ethics emphasizes the importance of fostering a sense of responsibility, mindfulness, and stewardship towards the environment. Overall, these ethical propositions provide valuable frameworks for guiding the development and deployment of sustainable technology, ensuring that invention aligns with principles that promote the well-being of both present and future generations.

18.8 RULES TO ADHERE TO SUSTAINABLE TECHNOLOGY

Ethics in sustainable technology involves clinging to principles and guidelines that promote responsible and environment-friendly practices in the development, use, and disposal of technology. Here are some crucial rules to be followed in ethics in sustainable technology:

- Environmental Impact Assessment: Perform in-depth analyses of how technology affects the environment at every stage of its life cycle, from resource extraction and production to use and disposal. Think about variables such as carbon emissions, energy use, and implied harm to ecosystems.
- Translucency and Accountability: Be transparent about the environmental and social impact of technology. Disclose information about sourcing, manufacturing processes, and implicit pitfalls. Take responsibility for the consequences of technological developments and address any negative impacts.
- Life Cycle Thinking: Use a life cycle perspective while designing and producing technologies. A product's whole life should be taken into account, including the extraction of raw materials, production, shipping, use, and disposal at the end of its useful life. Consider continuity, recyclability, and waste reduction when designing items.
- Indirect Economy Practices: Embrace indirect frugality principles by designing products for exercise, form, and recycling. Avoid planned fustiness and encourage a shift away from the direct "take–make–dispose" model toward a more sustainable and indirect approach.
- Ethical Sourcing of Materials: Ensure that raw materials used in technology are ethically sourced. Avoid materials associated with environmental degradation, human rights abuses, or conflict minerals. Prioritize materials that are responsibly extracted and reused.
- Energy Efficiency: Prioritize energy efficiency in the design and production of technology. Use renewable energy sources wherever possible and strive to minimize the carbon footprint associated with the manufacturing and operation of devices.
- Inclusive Design: Adopt inclusive design practices to ensure that technology is accessible to different user groups, including those with disabilities. Avoid creating technology that reinforces social inequalities or excludes certain populations.
- Fair Labour Practices: Uphold fair labour practices throughout the force chain. Ensure that workers involved in the production of technology are treated morally, paid a fair stipend, and provided with safe working conditions.
- Sequestration and Data Security: Prioritize sequestration and data security in the development of technology. Gain informed concurrence for data collection, storage, and processing. Apply robust cybersecurity measures to protect user data from unauthorized access and breaches.
- Community Engagement: Engage with original communities affected by technology developments. Seek input from communities to understand their requirements and enterprises. Alleviate implicit negative impacts by addressing community feedback and uniting with stakeholders.

- Continuous Enhancement: Commit to non-stop enhancement by staying informed about evolving ethical norms and best practices. Regularly assess and modernize technology and business practices to align with the latest ethical and sustainable principles.
- Compliance with Regulations: Adhere to original and transnational regulations related to sustainable technology and ethical business practices (Rattan and Bhambri, 2005b). Stay informed about legal conditions and ensure full compliance with environmental and social responsibility laws.
- Educational Initiatives: Promote mindfulness and understanding of ethical considerations in sustainable technology among workers, stakeholders, and the public. Provide training and educational resources to foster a culture of ethical responsibility.
- Hookups and Collaboration: Unite with assiduity mates, exploration institutions, and NGOs to share knowledge, resources, and best practices. Work inclusively to address ethical challenges and promote sustainable results.

By following these rules, associations and individuals can contribute to the development and use of technology in a manner that is ethical, environmentally sustainable, and socially responsible. These guidelines help produce a foundation for technology that aligns with broader pretensions of sustainability and ethical conduct.

18.9 CREATING AWARENESS ABOUT ETHICAL CONSIDERATIONS

Raising mindfulness about ethics in sustainable technology is pivotal for fostering responsible invention and ensuring that technological advancements contribute appreciably to both society and the environment. Here are some strategies to create awareness:

- Educational Initiatives: Organize workshops, webinars, and forums to educate individuals, businesses, and communities about the ethical considerations in sustainable technology. Cover topics such as responsible design, environmental impact, and the social implications of technology.
- Collaborate with Educational Institutions: Partner with universities, colleges, and schools to integrate ethical considerations into technology and environmental science curricula. Encourage exploration, systems, and conversations on the ethical confines of sustainable technology.
- Public Awareness Campaigns: Start public awareness campaigns using radio, newspapers, social media, and other media platforms. Use these venues to highlight the importance of moral behaviour in the creation and use of technology.
- Industry Conferences and Events: Participate in and organise conferences or events concentrated on sustainable technology. Provide platforms for experts, experimenters, and industry leaders to discuss and partake in perceptivity on ethical practices within the field.

- Online Platforms and Resources: Develop online platforms, websites, and resources devoted to ethics in sustainable technology. Provide information, case studies, and guidelines that individuals and associations can refer to when making decisions related to technology and sustainability.
- Partnerships with NGOs and Advocacy Groups: Unite with non-governmental organizations (NGOs) and advocacy groups working in the field of sustainability and technology ethics. Leverage their moxie and networks to reach a broader audience.
- Corporate Responsibility projects: Promote the adoption and dissemination of ethical practices by tech corporations and businesses in their sustainable technology projects. Encourage openness and accountability when it comes to sharing the effects that their goods and services have on the environment and society.
- Government Initiatives and Programmes: Advocate for and support the development of government programmes that promote ethical considerations in sustainable technology. Engage with policymakers to ensure that regulations address technology's environmental and ethical aspects.
- Community Engagement: Engage with original communities to understand their enterprises and requirements regarding sustainable technology. Empower communities to make informed decisions about adopting and exercising technology ethically and sustainably.
- Showcasing Ethical Inventions: Highlight and celebrate inventions that prioritize ethics in sustainable technology. Show success stories and positive exemplifications to inspire others and demonstrate the potential of ethical technology to drive positive change.
- Interactive Workshops and Exhibitions: Organize hands-on workshops and exhibitions that allow people to interact with sustainable technologies. This can give a palpable experience of ethical considerations and their positive impact on the environment and society.
- Media Partnerships: Partner with media outlets to produce and circulate content that raises awareness about the ethical confines of sustainable technology. Point stories, interviews, and pictures that explore the crossroads of technology, ethics, and sustainability.

Combining these strategies makes it possible to produce a comprehensive and poignant awareness campaign that encourages individuals, businesses, and policymakers to prioritize ethics in sustainable technology for the benefit of the Earth and its occupants.

18.10 FUTURE OF ETHICS IN SUSTAINABLE TECHNOLOGY

The future of ethics in sustainable technology holds great promise as societies, industries, and individuals increasingly recognize the significance of responsible and environmentally conscious technological advancements. Several trends and developments point towards a future where ethical considerations in sustainable technology will play a central role:

- Integration of AI for Sustainable Results: Artificial intelligence (AI) is expectedto play a significant part in optimizing and enhancing sustainability efforts (Bhambri and Gupta, 2005). Ethical considerations in the development and deployment of AI for environmental monitoring, resource management, and sustainable decision-making will become paramount.
- Blockchain for Translucency: Blockchain technology has the potential to enhance transperancy in supply chains, icing the ethical sourcing of materials and furnishing an empirical record of environmental and social impacts. This can contribute to increased accountability for sustainable technology.
- Advanced Materials and Recycling Inventions: Inventions in materials science and recycling technologies will lead to more sustainable and eco-friendly products. Ethical considerations will guide the development of technologies that use recycled materials and minimize the environmental impact of manufacturing processes.
- Global Collaboration on Norms: The establishment of global norms and frameworks for ethical practices in sustainable technology will gain momentum. Cooperative efforts between countries, industries, and associations will facilitate the development and adoption of widely accepted ethical guidelines.
- Consumer Demand for Ethical Tech: Increasing awareness among consumers about the environmental and ethical impact of technology will drive demand for products and services that align with sustainability principles. This consumer demand will incentivize businesses to prioritize ethical considerations in their technological innovations.
- Regulatory Advancements: Governments and non-supervisory bodies will continue to evolve and strengthen regulations related to sustainable technology. Stricter enforcement of environmental and ethical norms will create a more robust framework for responsible technological development and use.
- Corporate Social Responsibility (CSR) in Technology: Companies will integrate sustainable and ethical practices into their CSR enterprise. Demonstrating a commitment to environmental and social responsibility will come with a competitive advantage, and consumers will increasingly choose products and services from companies with strong ethical values.
- Ethical Tech Startups and Innovation Hubs: The emergence of startups and invention hubs concentrated on ethical and sustainable technology will contribute to the development of groundbreaking results. These entities will prioritize ethics from the commencement of their systems, impacting larger industries to follow suit.
- Educational Programmes and Research: Increased emphasis on ethics in technology education and exploration will produce a new generation of professionals who prioritize sustainability. Academic institutions will play a vital part in shaping ethical norms and practices in the tech industry.

- Public–Private Partnerships: Collaborations between governments, private sector entities, and non-profit associations will become more common. Public–private partnerships will leverage collective resources to address global challenges related to sustainability and ethics in technology.
- Emphasis on Inclusive Design: Inclusive design practices will become standard in the development of technology, ensuring that products and services are accessible to diverse populations. Ethical considerations will extend beyond environmental impact to encompass social inclusivity.
- Emergence of Ethical Certification Programmes: Certification programmes and markers that indicate the ethical and sustainable nature of products will gain popularity. These instruments will help consumers make informed choices and incentivize businesses to adhere to ethical practices.
- Climate Tech Innovation: A swell in climate tech innovation will lead to the development of technologies specifically designed to address climate-related challenges (Bhambri and Bhandari, 2005). The outcomes targeted at reducing the effects of climate change will incorporate ethical considerations.

The future of ethics in sustainable technology will be shaped by collective effort from various stakeholders, including governments, businesses, consumers, and advocacy groups. The integration of ethical considerations into technological development will not only contribute to a more sustainable earth but also foster a culture of responsibility and conscientious invention in the tech assiduity.

18.11 CONCLUSION

In conclusion, the intersection of ethics and sustainable technology stands as a vital force shaping the future of invention and progress. As societies gradually recognize the profound impact of technology on the environment and human well-being, ethical considerations have surfaced as essential guiding principles. The imperative to develop and adopt sustainable technologies that minimize ecological vestige, prioritize social equity, and adhere to responsible practices has gained significant momentum. The ongoing commitment to ethical norms in sustainable technology promises a future where advancements align harmoniously with environmental conservation, inclusivity, and long-term societal benefits. To navigate the complexities of this dynamic landscape, collaboration among governments, industries, and communities is pivotal. By upholding ethical principles, embracing transparency, and fostering a culture of responsibility, we can pave the way for a technologically advanced future that is not only innovative but also sustainable and morally sound.

REFERENCE

Bhambri, P., & Bhandari, A. (2005, March). Different protocols for wireless security. In National Conference on Advancements in Modeling and Simulation (p. 8).

Bhambri, P., & Gupta, S. (2005, March). A survey & comparison of permutation possibility of fault tolerant multistage interconnection networks. In National Conference on Application of Mathematics in Engineering & Technology (p. 13).

Daly, H., Jacobs, M., & Skolimowski, H. (1995). Discussion of Beckerman's critique of sustainable development. *Environmental Values* **4**, 49–70.

Jain, V. K., & Bhambri, P. (2005). *Fundamentals of Information Technology & Computer Programming*. S.K. Kataria & Sons.

Rattan, M., Bhambri, P., & Shaifali, M. (2005a). Information retrieval using soft computing techniques. In National Conference on Bio-informatics Computing (p. 7).

Rattan, M., Bhambri, P., & Shaifali, M. (2005b). Institution for a sustainable civilization: Negotiating change in a technological culture. In National Conference on Technical Education in Globalized Environment- Knowledge, Technology & The Teacher (p. 45).

Webliography

https://gbsge.com/newsroom/can-we-really-change-the-world/
https://www.unssc.org
https://www.un.org

19 The Moral Imperative of Sustainable Technology

Navigating Challenges, Understanding Risks, and Upholding Ethics

Ikechukwu Umejesi and Moshood Issah

19.1 INTRODUCTION

Unlike in the past, people expect more from technology than ever before. This is understandable, as it helps drive and create innovations and economic growth, which is necessary for achieving sustainable living and development (Sun & Razzaq, 2022). However, the production and utilization of technology pose a major threat to people and society. Society and people have been exposed to environmental risks such as depletion of the ozone layer and natural resources, pollution, deforestation, and contamination of water and air, among others. Based on existing studies, many people are still unaware of environmental and ethical issues in contemporary technologies (Kluge, 2020; Nizam et al., 2020). The ethical implications of technologies are far-reaching and vast, as it is evident in production, consumption, and disposal (Peckham, 2021; de Oliveira et al., 2023).

To produce technologies, extraction of raw materials (such as metal and plastic) are fundamental. Extraction of metal is considered unethical because its mining takes place mostly in developing countries (Litvinenko, 2020; Wang et al., 2020). In places where metals are extracted, there are issues of environmental protection, human rights, and health and safety. Research on cobalt mining in the Congo, specifically the Democratic Republic of Congo (DRC), revealed that there are large-scale human rights abuses, child labour, and mining-induced conflicts (Tiamgne et al., 2022; Chanda-Kapata, 2020). Also, the manufacturing of technologies requires the use of fossil fuel and water. In addition, the supply chains of most technology companies are long, complex, and lack transparency (Zabyelina, 2023; Mishra et al., 2024). Importantly, the disposal of technologies leads to major environmental pollution with effluents and other toxic materials. Technology wastes such as LCD screens, televisions, and computers are inherently bad. In addition, a mass of electronic waste

DOI: 10.1201/9781003475989-22

from developed countries in the North is shipped to developing countries, thus contributing to environmental risks (Abalansa et al., 2021).

To protect people and society, sustainable technology could be a probable solution. Maja et al. (2020) see sustainable technology as a revolution that interfaces natural resources, economic, and social development. The purpose of these sustainable technologies is to ensure significant reduction in environmental and ecological risks, as well as create sustainable products for humanity (Mitchell et al., 2021). Sustainable technologies are innovative technologies designed to align with sustainability goals as they address various environmental challenges such as air pollution, deforestation, and greenhouse gas emissions (Nti et al., 2022). They help in preventing environmental issues and support sustainability goals. As argued by Bai and Sarkis (2020), sustainable technology drives the utilization of resources to advance social and economic progress. It has also been linked with environmental justice as it addresses environmental injustices and promotes environmental equity (Mitchell et al., 2021).

While sustainable technologies are fundamental for solving environmental problems, these technologies have ethical implications (Litvinenko, 2020). For instance, the production of portable solar panels and electric cars still requires the mining of metals for their production (Litvinenko, 2020). Mining and mineral processing operations always have high water footprints as water is required in various mining and processing stages such as dust mitigation, among others. Existing studies have associated metals mining with negative environmental impacts such as climate change, deforestation, pollution, soil erosion, loss of biodiversity, and wildlife-related conflict (Wang et al., 2020; Tiamgne et al., 2022). Also, the batteries of electric cars are fundamentally problematic during production and disposal. Based on existing studies, solar batteries have high potential to contribute significantly to electronic waste (Hansen et al., 2022; Gautam et al., 2022). These batteries have a short lifespan, indicating that there may be a need for replacement within a short time. If these wastes are not disposed of properly, they can release noxious chemicals and other pollutants into the environment. These pollutants cause harm to both humans and the ecosystem (Hansen et al., 2022).

Importantly, the share of global environmental problems emanating from the use of sustainable technologies varies between developed and developing countries. According to a study by Mendelsohn et al. (2006), climate change would have significant distributional impacts across countries based on their per capita income. The study noted that poor countries are likely to suffer more from the perilous damages caused by climate change. As mentioned earlier, climate change is likely to get worse with the intensity and frequency of mining activities required for the production of sustainable technologies. Also, the report of the UNDRR (2022) revealed that the poorest countries, especially in the global south, are at greater risk of losses from the impact of climate change. The study showed that all the 37 countries analysed had rather high GDP per capita and a low risk of damages from climate change impact, while 46 developing countries included in the analysis with lower per capita GDP had a high risk of environmental damage. Some of the countries in this category, including Burundi, Somalia, and Mozambique, have higher risks (UNDRR, 2022). Also, Lawton (2022) noted that despite the disproportionate distribution of risks across countries based on the level of development, the rich countries have not invested sufficient funds to assist poorer countries to adapt and avert the consequences of climate change.

The above disparities in the share of burdens of technological developments raise a serious ethical issue in the adoption of sustainable technologies as a development catalyst, especially in development-deficient countries. In other words, developing countries face a major dilemma between adopting sustainable technologies to aid development and adapting to the environmental consequences of such adoption. Thus, the overarching aim of this chapter is to examine the imperativeness and challenges of sustainable technology. This chapter is divided into five sections. The first section conceptualizes sustainable technology. The second section deals with the application of Rawlsian theory to analyse environmental justice within the context of sustainable technology. The third section looks at the imperativeness of sustainable technology, while the fourth section deals with the challenges and risks associated with sustainable technology. The fifth section focusses on ethical issues in sustainable technology.

19.2 SUSTAINABLE TECHNOLOGY

Sustainable technology entails all sorts of solutions which assist in reducing the environmental impact of IT infrastructure across its entire life cycle (Ghavam et al., 2021). It is also defined as the technology which is specially designed to prevent and remedy environmental issues (Ahad et al., 2020). In addition, it is viewed as the technology that is specifically produced with sustainability in mind (Burbules et al., 2020). Also, some scholars define it as the technologies that are specially made for mitigating or solving environmental/ecological issues, as well as technologies that can be produced, used, and disposed of in such a way that they do not cause harm to people and society (Bohnsack et al., 2022). Moreover, it is defined as the technologies which are produced with conscious attempts to limit both the environmental and ethical implications/consequences of their production, use, and disposal (Belhadi et al., 2020).

In addition, the goal of sustainable technology is to reduce the ecological footprint by ensuring and promoting energy efficiency, reducing waste, and adopting eco-friendly practices (Li et al., 2023; Saqib et al., 2024). Some scholars have also linked sustainable technology with governance integrity (Abalansa et al., 2021; Fernando & Saravannan, 2021). According to them, it influences governance as it ensures transparency, accountability, and ethical decision-making within organizations. This is possible with the use of technologies such as blockchain. Thus, sustainable technologies can contribute to the achievement of environmental, social, and governmental goals (Fernando & Saravannan, 2021). For technology to be truly sustainable or eco-friendly, it should be created and used sustainably and ethically. In addition, it should be capable of preventing harm or shifting harm away from people and the environment (Saqib et al., 2024).

Examples of sustainable technology include new energy vehicles such as electric cars, LED light technology, solar energy for home and industrial purposes, carbon capture and storage technologies, hydro dams, GPS-enabled tractors, moisture sensors, drones, and smart irrigation, among others. The current technological trends in 2024 include generative AI and all its elements. The authors also noted other AI-related themes such as AI trust, risk and security management, augmented-connected workforce, cloud platforms, among others.

It can be deduced from the above that most of the definitions applied sustainable technology to the world of business. However, in this chapter, sustainable technology is viewed from a larger society perspective rather than from an organizational perspective. In other words, the chapter focusses on the imperativeness of sustainable technology as well as the challenges and ethical issues surrounding sustainable technology.

19.3 APPLYING RAWLSIAN THEORY OF JUSTICE

John Rawls (1921–2002) was deeply influenced by Thomas Hobbes, John Locke, and Jean-Jacques Rousseau's social contract theory. However, the major point of departure is that while these classical philosophers were emphasizing the development and raison-d"être of social contract (i.e., why people decided to form a state and what is the nature of the people who form the state), Rawls was interrogating the kind of arrangement that those who entered into the contract agreed to. Rawls found that the way classical theorists analysed society was hollow as they failed to question the ideas of social justice, and how fairness could indeed be achieved and sustained in a society. Rawls analysed the social justice discourse by highlighting some critical questions he thought were important. These included the following: (i) What is it that makes a social institution, social system, or society just? (ii) What is it that justifies or rationalizes political or social policies? For Rawls, justice means fairness, and justice is the fundamental virtue upon which the social system stands. Therefore, based on these fundamental contentions, John Rawls came up with two basic principles of social justice: (i) That each member of any particular society should have equal rights, and there must not be any forms of differentials in the way individuals access societal resources; and (ii) That the socio-economic inequalities in any social system should be such that they would be advantageous of everyone in the society. A cursory look at these principles would reveal that they underscore the underpinning assumptions of the principle and theory of egalitarianism.

Rawlsian theory of justice is used to analyse the concept of ecological justice in the light of sustainable technology (Nooij, 2020). Rawls' theory of justice is considered as a contract theory as it looks at the justness of the basic structure of society from an impartial angle. The impartiality is achievable with the application of the concept of "original position" which states that the contracting partners make decisions on the principles of justice from behind the "veil of ignorance" (Clements & Formosa, 2021). Specifically, Rawls proposed two principles of justice. The first principle is that each member of society has equal indefeasible claim to liberty. The second principle is based on the permissibility of inequalities provided that their application is based on fair equality of opportunity and bringing maximum benefit to the least-advantaged members of society (Ross, 2020). In other words, powerful and wealthier people/states are expected to use their strength to benefit the weaker or poorer people/states (Faber et al., 2021).

Based on the first principle, differentiated responsibilities are allocated to countries that have the capacity to secure basic liberties such as income, rights, and opportunities (Endyka et al., 2020). Many countries from Africa and Latin America are exempted from differentiated responsibility since they are still struggling to guarantee dignified sustenance for current populations. It is argued that stricter environmental regulations could have adverse implications on income and food security in developing countries (Endyka et al., 2020). Besides, strict environmental regulations were enforced on developed countries after they had achieved economic efficiency and income security. This suggests that underdeveloped countries should not be burdened with implementing environmental regulations that could have adverse implications on sustainable development. Therefore, elongated time periods should be provided for developing and underdeveloping countries to achieve the "net zero" target (Svarstad & Benjaminsen, 2020).

Going by the second principle, it is the responsibility of the powerful and wealthy countries to limit carbon emissions and disposal of e-waste to underdeveloped countries (Ott, 2020). The existing inequalities should be used to benefit or uplift the least advantaged countries. This involves the transfer of green technologies (sustainable technologies) to least developed countries. Based on the concept of the "veil of the ignorance", if people (environmental actors) are unaware of which country they belong to, they are likely to support using inequalities between countries to the advantage of least developed countries (Bhambri et al., 2005). Failure to use inequalities in this way makes inequalities unjust and illegitimate, and consequently, underdeveloped countries have ethical justification if they refuse differentiated responsibility in environmentalism (Ott, 2020).

Also, the idea of "just saving" implies that the current generation needs to save adequately for future generation. In other words, they are expected to ensure a minimum just standard of living for future generations. They need to provide sufficient opportunities and liberties for future generations. Based on existing studies, Earth should be made livable – as a minimum standard – for future generations. They should leave behind at least the equivalent of what the previous generations left for them for future generations. Based on the Rawls' difference principle, poor and minority communities in some Western countries suffer a greater liability of environmental risks. Due to this, scholars have argued about whether environmental injustice is an inherent component of liberal/democratic societies or a consequence of the failure of Western democracies to fulfil their fundamental principles of justice for all people. It is posited that sustainable technology does not promote the ideals of equality, fairness, and justness based on current realities. What it does is reinforce inequality among nations and societies.

Existing reports have revealed that environmental implications from the production of sustainable technologies could do more harm than good for developing countries – an indication of environmental injustice to poor countries. The report of Borenstein and Costley (2022) showed that those countries that produce sustainable technologies are high-emitting nations as a result of production activities. Rich nations are benefiting from sustainable technologies, while poor countries are bearing the costs and dangers. For instance, it is assumed that between 1990 and 2014 actions taken by the United States of America have resulted in climate change-related damages valued at more than $1.9 trillion to other countries. This includes $257 billion in damages to India; $310 billion in damages to Brazil; $124 billion in damages to Indonesia; $104 billion in damages to Venezuela; and $74 billion in damages to Nigeria (Borenstein & Costley, 2022). In the same period, the United States is said to have profited by $183 billion, and even Russia, Canada, and Germany are also said to have benefited from climate change-related damages done to other countries. This suggests that developing countries are likely to make more sacrifices with limited gains from the production and adoption of sustainable technology (Bhambri & Singh, 2005). As reported by Lawton (2022), rich countries have obligations to transfer a certain percent of the proceeds from the activities which caused environmental damages to poor countries in the interest of social justice.

From a social justice perspective, the costs of producing and adopting sustainable technology for developing countries are higher than the benefits. In other words, rich nations are in a better position to profit from the innovations of sustainable technology at the expense of poorer countries. However, there is an argument that sustainable technology is like a "double-edged sword" for developing countries as it could assist them in achieving sustainable development, provided that it is used sustainably and the producers (developed countries) produce it responsibly. The next section discusses the imperativeness and associated dilemma of producing and adopting sustainable technologies in developing countries.

19.4 IMPERATIVENESS OF SUSTAINABLE TECHNOLOGY FOR ACHIEVING SUSTAINABLE DEVELOPMENT

Technological advances are stimulating developing countries forward, unlocking opportunities and transforming the lives of millions. The benefits of technology are evident in improved education, healthcare, economic growth, and environmental sustainability (Ott, 2020). Hence, developing countries need to embrace advanced sustainable technologies to achieve sustainable development (Endyka et al., 2020). For instance, in the area of information and education, technology provides access to education and information through the internet, which enables individuals to learn new skills and pursue higher education irrespective of their locations in developing and underdeveloped countries (Svarstad & Benjaminsen, 2020). Also, in the area of health services, technology, especially telemedicine, has improved healthcare and the well-being of people in developing and underdeveloped countries. Technology facilitates remote consultations, enhances accessibility, and enable more effective diagnosis and treatment of diseases in developing countries.

Moreover, technological development has been associated with infrastructural growth. With the development in technology, smart solutions such as energy-efficient buildings and intelligent transportation are created in developing countries (Ross, 2020). The adoption of these technologies has led to the improvement of citizens' lives, promotion of environmental sustainability, and building of resilient communities (Faber et al., 2021). In addition, technologies assist environmental conservation in developing countries. Various sources of renewable energy, including solar and wind, ensure sustainable energy; at the same time, Internet of Things (IoT) devices improve resource management, reduce waste, and control pollution. Also, marginalized communities are empowered by technology via social media and digital communication, which facilitates activism, advocacy, and global networking required for addressing social issues, promoting rights, and amplifying their voices (Clements & Formosa, 2021).

In addition, technological development has been linked with economic growth and entrepreneurship. Economic growth can be facilitated through mobile banking, which ensures financial inclusion for those who were previously excluded due to accessibility. Some studies revealed that modern technology enhances the discovery of natural resources hidden in seas, lands, and mountains (Nooij, 2020). Without technologies, the exploration of copper, gas, gold, oil, and iron may be impossible. Technology has enhanced productivity as it ensures efficient utilization of human resources and modern techniques in production. Also, AgriTech boosts developing economies' agriculture in terms of enhancing crop yield and resource management. The use of technologies such as remote sensing and weather forecasting enables farmers to make informed decisions, optimize resources, and mitigate change effects. Furthermore, technologies such as data analytics and real-time monitoring assist in predicting and managing disasters, as well as minimizing loss of life and property. Autonomous systems, such as artificial intelligence (AI), have been expanding in terms of their functionalities. These devices are leveraged in different ways. They are highly intuitive; newer AI can alter speeches and perform incredible tasks with their augmented reality capabilities. These new developments could be beneficial to millions of people and redefine the world of work, social relationships, and all manner of interactions with computers.

However, developing countries have not taken advantage of technological advancements. The Global Innovation Index 2021 (of the World Intellectual Property Organisation), which monitors the state of technological advancement in 132 countries, noted that 21 out of the 32 countries in the bottom quartile are least developed countries. Out of the 22 least developed countries, only Tanzania made it to the second quartile. In addition, developing countries do not have adequate infrastructure such as reliable electricity, high-speed internet connectivity, and other skills in the digital technology sector. For instance, in 2019, as high as 52.8 percent of developing countries did not have access to electricity compared with the global average of 90.1

percent at the time. Also, internet penetration in poorer countries remained at a low of 1 percent since 2016. This is in comparison to the global average of 15 percent. In addition, spending on research and development in developing countries has not reached 1 percent. Some countries in this category include Uganda, 0.2%; Ethiopia, 0.3%; and Rwanda, 0.7%.

19.5 ETHICAL ISSUES IN THE ADOPTION OF TECHNOLOGIES FOR SUSTAINABLE DEVELOPMENT

Evidently, technology has significantly changed human lives, such as how people work, communicate, and fight wars. Many of these technologies have raised major ethical questions, sparking debates among scholars. For instance, religious groups, policymakers, scientists, and advocacy groups have contested the use of stem cell and embryo technology in the treatment of certain diseases. In the same vein, civil society, scientists, and policymakers have debated the question of ethics and safety of genetically modified organisms (GMOs). Additionally, the emergence of genome-editing technologies has generated serious ethical questions.

Also, the development of certain drones equipped with a neuromorphic microchip used for environmental surveillance has raised ethical questions. Critics argued that creating neuromorphic chips would lead to the production of machines that would be as smart as humans, which could bring about the devastation of our species. Also, it is argued that if smart machines (such as smart pharmacology, cancer-fighting robots, among others) are allowed to control human emotions, it may lead to some sort of inimical dependency on them as well as the emergence of new understandings about humanity and the emotions that define humans.

Another ethical issue concerning technological advancement is privacy and security. The development in digital technologies, with their attendant increasing interconnectivity of devices, has raised some fundamental privacy and security concerns. The collection, storage, and sharing of data could lead to breaches in data management, identity theft, and other unauthorized uses of surveillance. In addition, the use of artificial intelligence and machine learning in facial recognition is believed to raise ethical questions regarding personal privacy. Furthermore, there are some concerns about potential job displacement as tasks/jobs are automated with the use of robotics and AI. In other words, the replacement of humans with machines has serious ethical concerns as workers may be left behind.

Also, the potential misuse of technology has been emphasized as one of the ethical issues in technological advancements. Based on existing studies, new/emerging technologies may be misused by individuals, corporations, or governments. For instance, AI and machine learning can be unethically used for surveillance. Other potential misuse of technologies could entail the development of deadly autonomous weapon systems to target those perceived as enemies, and also, the use of biotechnology or related technologies to create weapons of mass destruction. It is advised that when developing and regulating the above technologies, ethical issues surrounding their usage and application should be clearly considered so that they would be used responsibly and ethically. More so, a speedy advancement in technology could lead to a deeper digital divide between richer individuals and nations, while the poorer societies and nations with limited resources are left behind.

The extractive sector is another sector where advancement in technology is often associated with environmental impact. The increasing utilization of technology and the extraction of natural resources needed for the production of technological devices cause deleterious environmental destruction. Some of the environmental impacts of technology include the disposal of electronic waste, the intensity of energy consumption, and the depletion of non-renewable resources. The study by Usman et al. (2022) revealed that the production of powerful and efficient technology requires a large amount of natural resources. This means that mineral resources have to be extracted, such as metals and fossil fuels. The extraction of these resources leads to major environmental risks, such as water pollution, degradation, and deforestation.

Also, the disposal of electronic wastes constitutes a major environmental hazard globally. The disposed technological devices, which are not usually recycled, often contain lethal substances which may include arsenic, lead, and other agents capable of contaminating soil and water courses for a very long time. It has been noted that electronic waste is the most common solid waste all over the world. For example, in 2019 alone, approximately 53.6 million tonnes of this waste were disposed of globally. However, only 17.4 percent was recorded as "recycled". There is also differentiation in the amount of electronic waste generated between different regions of the world. For example, African countries collectively generated a total of 2.9 metric tonnes of electronic waste in 2019, and this is estimated as the lowest rate in the world. It should also be noted that although most e-wastes are generated by developed countries, they are usually disposed of in developing countries where regulatory lapses are often exploited. To demonstrate how the electronic waste disposal regime works, in the Asia-Pacific developing countries, about 80 percent of the waste is illegally imported from developed countries. These wastes litter towns and villages in developing countries, with major health implications and environmental risks. Children and pregnant women are particularly exposed. To show how dangerous these toxic agents are, low exposure to lead, for instance, is suspected to result in a reduction in someone's total intelligence quotient (IQ) and can alsolead to various abnormalities, including decreased attention span, and increased frustration and disruptive behaviour. Similarly, a higher dose of lead can result in anemia, coma, or even death.

It is argued that to address the environmental implications of technology, the adoption of sustainable technology is imperative. The adoption of more sustainable and energy-efficient technologies is ethically imperative for addressing the environmental consequences of technological development. The World Wildlife Fund (WWF) projected that by 2030 the use of smart technologies can lower CO_2 emissions by as much as 8 gigatons. Networked digital systems could assist governments in addressing the impacts of climate change, as it aids or improve the monitoring and forecasting of storms, water shortages, and soil conditions. Also, the development of Virtual Power Plants (VPP) – which are cloud-based energy source is thought to improve both the efficiency and reliability of decentralized energy resources.

This includes wind, solar, geothermal, and hydropower. In addition, the Power by Ammonia (P2A) concept offers a solution whereby stored ammonia can be used as an energy source in the form of gas-phase fuel. Likewise, Power by Gas (P2G) shows good promises in the conversion of electricity into gas-phase fuels.

However, existing studies have revealed that those sustainable technologies have ethical issues. While sustainable technology is considered a positive step towards solving or addressing environmental challenges caused by conventional technologies, these sustainable technologies present some ethical challenges. One can readily point to the socioenvironmental impact of producing solar panels, wind turbines, lithium-ion batteries, and other similar devices. Consequently, the production and utilization of these technologies could lead to environmental damage, habitat destruction, and human rights abuses in mining regions. Ethical issues are likely to emerge when there is a lack of transparency or an attempt to trivialize negative environmental implications. Another ethical issue with sustainable technologies is fair access and distribution. Based on existing studies, sustainable (green) technologies are expensive to develop and implement. The major ethical concern is that developing countries do not have equitable access to these technologies. This is complicated by the "digital divide" where developing countries (because of their limited resources) struggle to take advantage of green technologies. Also, most developing countries may not be able to afford sustainable technologies.

Another ethical concern is the possibility of technological dependence. The development of green technologies may result in technological lock-in as specific technologies become dominant, making it difficult to transition to newer and potentially more sustainable alternatives. Existing studies also emphasizes greenwashing as another ethical concern in sustainable technologies. There is a high possibility that the producers of sustainable technologies may engage in unethical marketing practices such as exaggerating the environmental benefits of their products. They are less likely to be transparent and truthful about the environmental impact of green technologies. Moreover, another ethical concern is labour practices. The production and maintenance of green technology may entail exploitative labour practices such as unfair wages, unsafe working conditions, and labour right violations. Additionally, existing studies have revealed that although sustainable technologies are designed to address environmental risks, they may also contribute to ecological disruptions. They may have unintended consequences such as increased energy consumption due to their manufacturing process or the displacement of traditional industries and jobs.

19.6 CONCLUSION AND RECOMMENDATIONS

From the foregoing discussion, the disastrous global impact of anthropogenic activities on the environment has grown progressively over the years. It is evident that sustainable technologies, which are specifically designed to mitigate or address environmental risks, may also contribute to it. It is important to know that while sustainable technology has shown some promise in improving the quality of human life on earth, it may not dramatically lead to a significant reduction in the rate of environmental degradation currently experienced. Therefore, as much as technological innovation is a good thing, it is equally necessary to also consider its ethical implications on socioeconomic inequalities, human dignity, and inclusiveness. Based on this understanding, the chapter concludes that there are serious ethical implications of technological innovations experienced globally. They have the potential to harm both the environment and humans, as much as they have the potential to improve human lives on earth.

Considering the ethical and environmental implications of technology, the following recommendations are suggested:

i. Ethical sourcing of raw materials should be ensured. Consumers need to ensure that the manufacturers of technologies do not use children for mining the metals required and that local communities are adequately compensated for destructions or inconveniences caused. Importantly, the manufacturers must ensure that they put in place appropriate environmental practices to remedy likely environmental impacts of mining.
ii. People should be enlightened on how to use technologies wisely and ethically. The longer the technologies are used, the more sustainable they become. Disposing of technologies too often causes significant environmental risks. Also, if it is necessary to dispose of the technologies, it should be done consciously. Before disposing, it is necessary to consider repair and recycling. Disposal should be the last resort after other options have been considered.

REFERENCES

Abalansa, S., El Mahrad, B., Icely, J., & Newton, A. (2021). Electronic waste, an environmental problem exported to developing countries: The GOOD, the BAD and the UGLY. *Sustainability, 13*(9), 5302.

Ahad, M. A., Paiva, S., Tripathi, G., & Feroz, N. (2020). Enabling technologies and sustainable smart cities. *Sustainable Cities and Society, 61*, 102301.

Bai, C., & Sarkis, J. (2020). A supply chain transparency and sustainability technology appraisal model for blockchain technology. *International Journal of Production Research, 58*(7), 2142–2162.

Belhadi, A., Kamble, S. S., Khan, S. A. R., Touriki, F. E., & Kumar M, D. (2020). Infectious waste management strategy during COVID-19 pandemic in Africa: an integrated decision-making framework for selecting sustainable technologies. *Environmental Management, 66*, 1085–1104.

Bhambri, P., Singh, I., & Gupta, S. (2005, March). Robotics systems. In National Conference on Emerging Computing Technologies (p. 2).

Bhambri, P., & Singh, I. (2005, March). Electrical actuation systems. In National Conference on Application of Mathematics in Engineering & Technology (pp. 58–60).

Bohnsack, R., Bidmon, C. M., & Pinkse, J. (Eds.). (2022). *Sustainability in the digital age: Intended and unintended consequences of digital technologies for sustainable development* (pp. 599–602). Wiley.

Borenstein, S., & Costley, D. (2022). Climate change caused by wealthy nations creates harm for poorer, study says. Available at: ttps://www.pbs.org/newshour/science/climate-change-caused-by-wealthy-nations-creates-harm-for-poorer-study-says

Burbules, N. C., Fan, G., & Repp, P. (2020). Five trends of education and technology in a sustainable future. *Geography and Sustainability*, *1*(2), 93–97.

Chanda-Kapata, P. (2020). *Public health and mining in East and Southern Africa: A desk review of the evidence*. Zambia Ministry of Health with Training and Research Support Centre in the Regional Network for Equity in Health in East and Southern Africa (EQUINET). EQUINET DISCUSSION PAPER, 121.

Clements, P., & Formosa, P. (2021). Beyond ideal theory: Foundations for a critical Rawlsian theory of climate justice. *New Political Science*, *43*(4), 486–505.

de Oliveira, R. T., Ghobakhloo, M., & Figueira, S. (2023). Industry 4.0 towards social and environmental sustainability in multinationals: Enabling circular economy, organizational social practices, and corporate purpose. *Journal of Cleaner Production*, 139712.

Endyka, Y. C., Muhdar, M., & Sabaruddin, A. K. (2020). Environmental Justice in Intra Generations: An Overview of Aristotle's Distributive Justice to Coal Mining. *Indonesian Comparative Law Review*, *3*(1), 25–34.

Faber, D., Levy, B., & Schlegel, C. (2021). Not all people are polluted equally in capitalist society: An eco-socialist commentary on liberal environmental justice theory. *Capitalism Nature Socialism*, *32*(4), 1–16.

Fernando, Y., & Saravannan, R. (2021). Blockchain technology: Energy efficiency and ethical compliance. *Journal of Governance and Integrity*, *4*(2), 88–95.

Gautam, A., Shankar, R., & Vrat, P. (2022). Managing end-of-life solar photovoltaic e-waste in India: A circular economy approach. *Journal of Business Research*, *142*, 287–300.

Ghavam, S., Vahdati, M., Wilson, I. A., & Styring, P. (2021). Sustainable ammonia production processes. *Frontiers in Energy Research*, *9*, 34.

Hansen, U. E., Nygaard, I., & Dal Maso, M. (2022). The dark side of the sun: Solar e-waste and environmental upgrading in the off-grid solar PV value chain. In *The dark side of innovation* (pp. 35–55). Routledge.

Kluge, E. H. W. (2020). Ethical and legal challenges for health telematics in a global world: Telehealth and the technological imperative. In *The ethical challenges of emerging medical technologies* (pp. 223–227). Routledge.

Lawton, G. (2022). Rich countries must pay for the environmental damage they have wreaked. Available at: https://www.newscientist.com/article/mg25433832-900-rich-countries-must-pay-for-the-environmental-damage-they-have-wreaked/

Li, C., Ahmad, S. F., Ayassrah, A. Y. B. A., Irshad, M., Telba, A. A., Awwad, E. M., & Majid, M. I. (2023). Green production and green technology for sustainability: The mediating role of waste reduction and energy use. *Heliyon*, *9*(12), 432–455.

Litvinenko, V. S. (2020). Digital economy as a factor in the technological development of the mineral sector. *Natural Resources Research*, *29*(3), 1521–1541.

Maja, L., Željko, K., & Mateja, P. (2020). Sustainable technologies for liposome preparation. *The Journal of Supercritical Fluids*, *165*, 104984.

Mendelsohn, R., Dinar, A., & Williams, L. (2006). The distributional impact of climate change on rich and poor countries. *Environment and Development Economics*, *11*(2), 159–178.

Mishra, P. P., Sravan, C., & Mishra, S. K. (2024). Extracting empowerment: A critical review on violence against women in mining and mineral extraction. *Energy Research & Social Science*, *109*, 103414.

Mitchell, S., Qin, R., Zheng, N., & Pérez-Ramírez, J. (2021). Nanoscale engineering of catalytic materials for sustainable technologies. *Nature Nanotechnology*, *16*(2), 129–139.

Nizam, H. A., Zaman, K., Khan, K. B., Batool, R., Khurshid, M. A., Shoukry, A. M., ... Gani, S. (2020). Achieving environmental sustainability through information technology:"Digital Pakistan" initiative for green development. *Environmental Science and Pollution Research*, *27*, 10011–10026.

Nooij, J. M. (2020). A theory of environmental justice (Bachelor's thesis). Springer.

Nti, E. K., Cobbina, S. J., Attafuah, E. E., Opoku, E., & Gyan, M. A. (2022). Environmental sustainability technologies in biodiversity, energy, transportation and water management using artificial intelligence: A systematic review. *Sustainable Futures*, *4*, 100068.

Ott, K. (2020). Grounding claims for environmental justice in the face of natural heterogeneities. *DIE ERDE–Journal of the Geographical Society of Berlin*, *151*(2–3), 90–103.

Peckham, J. B. (2021). The ethical implications of 4IR. *Journal of Ethics in Entrepreneurship and Technology*, *1*(1), 30–42.

Ross, D. (2020). Social justice. In *Encyclopedia of sustainable management* (pp. 1–4). Springer International Publishing.

Saqib, N., Usman, M., Ozturk, I., & Sharif, A. (2024). Harnessing the synergistic impacts of environmental innovations, financial development, green growth, and ecological footprint through the lens of SDGs policies for countries exhibiting high ecological footprints. *Energy Policy*, *184*, 113863.

Sun, Y., & Razzaq, A. (2022). Composite fiscal decentralisation and green innovation: Imperative strategy for institutional reforms and sustainable development in OECD countries. *Sustainable Development*, *30*(5), 944–957.

Svarstad, H., & Benjaminsen, T. A. (2020). Reading radical environmental justice through a political ecology lens. *Geoforum*, *108*, 1–11.

Tiamgne, X. T., Kalaba, F. K., & Nyirenda, V. R. (2022). Mining and socio-ecological systems: A systematic review of Sub-Saharan Africa. *Resources Policy*, *78*, 102947.

UNDRR (2022). Poorest countries at greatest risk of losses and damage from climate change. Available at: https://www.preventionweb.net/news/poorest-countries-greatest-risk-losses-and-damage-climate-change

Usman, M., Jahanger, A., Makhdum, M. S. A., Balsalobre-Lorente, D., & Bashir, A. (2022). How do financial development, energy consumption, natural resources, and globalization affect Arctic countries' economic growth and environmental quality? An advanced panel data simulation. *Energy*, *241*, 122515.

Wang, Y., Diaz, D. F. R., Chen, K. S., Wang, Z., & Adroher, X. C. (2020). Materials, technological status, and fundamentals of PEM fuel cells–a review. *Materials Today*, *32*, 178–203.

Zabyelina, Y. (2023). The harms and crimes of mining. In H. N. Pontell (Ed.), *Oxford research encyclopedia of criminology and criminal justice*. Oxford University Press.

Part IV

Case Studies in Sustainable Technologies

20 Digital Duplicity and the Disintegration of Trust

A Quantitative Inquiry into the Impact of Deep Fakes on Media Sustainability and Societal Equilibrium

Rajneesh Sharma and Pankaj Bhambri

20.1 INTRODUCTION

In the digital age, where technology changes quickly and new things happen, deep fakes have transformed how we think about media. This chapter looks closely at deep fakes, a name given to realistic digital tricks created with smart artificial intelligence and machine learning technologies. These fakes, often difficult to see without training, can alter audio and video, making fake things look real. Deep fakes are not just fascinating technology; they also pose significant moral problems for law and society. This chapter has several objectives. Primarily, it looks closely at the structure of deep fakes and illustrates how they have evolved from a new technology into a powerful tool with myriad consequences. It seeks to understand the complex dynamics of digital media's endurance, particularly regarding truth and ethics. The chapter also aims to examine what is known now, the research methods, and how deep fakes affect things. Finally, it strives to give a comprehensive view of the issue, discussing its effects on individuals and society and suggests ways for further study and making rules.

Deep fakes are a big problem in today's age of digital media when the difference between real and fake is harder to tell (Rachna et al., 2022). The meaning and growth of deep fakes come from the quick developments in AI and machine learning systems. These tools, once used only in big research places, are now all over the Internet. They allow people and companies to make content that tests or challenges what truth is. As this happens, it gives a detailed look at keeping digital media alive. It also raises the issue of how content honesty and what's right can be maintained when fake images taken to great depths are common nowadays. The section about

DOI: 10.1201/9781003475989-24

reviewing literature gives a big-picture look at the studies done by schools and businesses on fake videos made using computer tricks. It looks at old studies, shows shortcomings in our current knowledge, and prepares for further research. The section, "Methodology" explains why the research chose a numbers-based approach. It talks about how data was collected, like surveys and questionnaires, and it discusses which people to include in the study by considering age groups or places of living.

An important part of this chapter is the impact analysis. Here we study how deep fakes affect mental health. It also studies the effects on people, mainly looking at how digital fakes change personal and work lives. Stories that involve normal people and famous individuals are demonstrated to show the real-world effects of deep fakes. This gives useful information about how much they can spread in everyday life. The part about real findings looks closely at the information from surveys and questions. It discusses what we can learn from statistics and puts the results in a wider view of leading studies. This side-by-side look shows us what makes this study special and how it fits in the bigger picture of deep fakes. The discussion section looks more closely at understanding the number of results. It checks how fair and flawed AI systems are in finding and stopping deep fakes or fake videos that look close to real ones. It also talks about what is right and wrong with deep fakes. It looks at the big problems these cause for society, such as how to govern better, build trust, and manage media.

This chapter is special because it discusses how young people and society are affected. It talks about how fake videos affect young people, especially college students. It also shares ways to make them aware and teach them more. This part is very important for seeing deep fakes' long-term effects and making strong plans to help future generations. As the chapter ends, it explains where the direction that research and policy-making could take. It advises more research, discusses the effects on rules and lawmakers, and recommends ways to stop deep fakes before they cause harm (Shrivastava et al., 2021). The final part discusses the main points and gives final thoughts on the role of online media and smart computers in people's lives. It discusses the challenges of fake videos and of promoting good digital behavior. This encourages readers to do something fast once they finish reading it.

This most important results of the chapter are guiding lights for discussing deep fakes. It is shown that awareness of deep fakes isn't the same for everyone. Many people don't understand them very well, and there are significant differences in how they perceive these things. Deep fakes have a substantial effect on our minds. They can change how much we trust and make us feel different mentally. Deep fakes can cause problems in interpersonal interactions. They can impact politics, safety, and relationships between individuals. Economically, they create dangers for businesses and people. The study, based on real evidence, provides several insights into these discoveries, offering a detailed look at the different aspects of deep fakes.

20.2 LITERATURE REVIEW

20.2.1 State of Current Research on Deep Fakes

The emerging domain of deep fake research involves technical, psychological, and socio-legal investigations. Deep fakes result from advanced machine learning techniques, namely Generative Adversarial Networks (GANs), which have greatly improved the ability to create realistic fake material (Goodfellow et al., 2014). The research is divided into two main areas: improving the accuracy of synthetic media and creating strong detection systems to prevent harmful usage (Li et al., 2018). Detection techniques commonly target physiological signs generally lacking in deep fakes, such as authentic eye-blinking patterns (Li et al., 2018).

Psychologically, profound fakes have a substantial impact on memory and belief systems. The "illusion of truth" effect posits that the perceived truthfulness of counterfeit information is heightened via repeated exposure (Pennycook et al., 2018). The realistic quality of deep fakes intensifies this cognitive bias, resulting in memory distortion and heightened emotional reactions.

The emergence of deep fakes poses a significant challenge to the effectiveness of existing legal systems from a socio-legal perspective. Academics discuss the sufficiency of current rules on defamation, privacy, and consent in the era of synthetic media that is difficult to discern from reality (Citron & Chesney, 2019). The DEEPFAKES Accountability Act and similar legal initiatives aim to establish guidelines to prevent the harmful dissemination of deep fakes (Brennen, 2020).

The ethical ramifications of deep fakes have been a central topic of discourse. Developing and disseminating deep fakes give rise to ethical quandaries, particularly due to their potential for misuse in spreading false information, impersonating individuals, and violating privacy rights (Westerlund, 2019). The ethical discussion is connected to the moral obligation of authors and distributors as they strive to strike a balance between the freedom of speech and the potential negative consequences of disseminating misleading material (Fallis, 2021).

Deep fakes significantly hinder democratic processes and media credibility in the sociopolitical sphere. Utilizing deep fakes in political propaganda can potentially erode public confidence in democratic institutions and procedures (Bradshaw & Howard, 2019). Furthermore, the capacity of deep fakes to impact public sentiment and control election procedures is an escalating apprehension (Parisi, 2019).

Research also investigates the public's ability to detect deep fakes, revealing a tendency to overestimate this capacity, which might make them susceptible to disinformation (Rini, 2020). This underscores the need for improved media literacy in the digital era, enabling users to discern the credibility of digital material (Rush et al., 2020). The literature emphasizes the necessity of cooperative endeavors among technologists, legal scholars, psychologists, sociologists, and policymakers. Collaboration is essential to formulate comprehensive solutions that may effectively reduce the dangers associated with deep fakes and protect the integrity of information within the digital ecosystem (Wagner, 2020).

20.2.2 Analysis of Previous Studies on Digital Media Impacts

The influence of digital media on society has been a topic of significant academic attention. The literature covers many areas, such as the impact on psychology, changes in society and culture, political consequences, and ethical concerns.

The psychological effects of digital media on mental health, especially among young people, have been a significant subject of research. Twenge and Campbell

(2018) have investigated the correlation between extended use of digital platforms and the development of problems such as reduced attention spans, anxiety, and depression. Kross(2019) have demonstrated a positive correlation between extensive utilization of social media and heightened experiences of loneliness and despair.

Sociocultural Transformations: The impact of digital media on human relationships and social norms has been a significant area of interest (Bhambri et al., 2023a). Vorderer(2016) examine the transformative effects of digital media, particularly social media, on traditional communication patterns, which influence social interactions and community dynamics. Moreover, the emergence of the "digital native" generation has resulted in a cultural transformation, where digital literacy has become a crucial skill (Prensky, 2001).

Political Consequences: The advent of digital media has had a profound effect on political procedures and the shaping of public opinion. Persily (2016) emphasizes the significance of social media in political campaigns, particularly its capacity to disseminate false or misleading information. Howard (2018) analyze the utilization of bots and fraudulent accounts in manipulating political discussions, highlighting the obstacles that democratic processes face.

Ethical Considerations: Ethical dilemmas, such as those over protecting personal information and safeguarding data, are of utmost importance in digital media. In his work, Solove (2020) explores the dynamic transformation of privacy in the era of digital technology, emphasizing the ethical dilemmas that arise from the data-gathering methods employed by digital platforms. Boyd and Crawford (2012) examine the ethical dilemmas associated with big data, specifically focusing on issues related to consent and ownership.

Economic Impacts: Examining the economic effects of altering conventional company models through digital media has been a topic of economic study. The emergence of e-commerce and digital marketing has fundamentally transformed the way consumers behave and the techniques businesses employ. The research conducted by Brynjolfsson and McAfee (2014) illustrates the impact of digital platforms on labor markets, leading to the emergence of novel economic paradigms.

The significance of media literacy in the digital era, characterized by an abundance of information that is not necessarily trustworthy, has been underscored. Hobbs (2011) asserts that cultivating critical-thinking abilities is needed to traverse digital media adeptly. In this study, Wartella (2016) examine the difficulties parents have when navigating their children's digital media usage. They emphasize the importance of implementing thorough education on media literacy to address these obstacles.

Research has also explored the digital gap and its consequences for inequality. Research conducted by Norris (2003) indicates that the digital gap encompasses more than just access to technology; it also encompasses disparities in digital literacy and usage abilities.

Health Impact: The impact of digital media on physical health, namely the consequences of screen time on sleep patterns and levels of physical exercise, is an additional concern. Hale and Guan (2015) discovered a connection between time spent using screens and disruptions in sleep patterns. Sigman (2012) explores the possible negative consequences of extended exposure to digital media on one's physical well-being.

The cultural and creative sectors have been greatly influenced by digital media, leading to substantial changes in the production, distribution, and consumption of information. Jenkins & Plasencia (2017) analyzes the emergence of participatory culture in the digital era when people actively participate in creating material. Cunningham and Silver (2013) examine the impact of digital platforms on the democratization of content creation, resulting in a transformation in cultural production dynamics.

20.2.3 Research Gaps

The current body of literature, albeit substantial, highlights specific areas of study that require attention to gain a more thorough knowledge of the effects of digital media. An important deficiency exists in the absence of longitudinal research investigating the enduring psychological consequences of digital media, particularly regarding advancing technologies such as deep fakes. The majority of contemporary research provides a limited viewpoint, which may not comprehensively reflect the continuous psychological transformations or the cumulative impacts over time (Bhambri et al., 2023b). There is also a significant lack of research on digital media's social and cultural effects on different demographic groups. Although current studies provide some understanding of overall patterns, a lack of research specifically examines certain groups, especially marginalized populations, and how digital media influences them in various socioeconomic, cultural, and geographical settings.

The ethical and policy ramifications of digital media, particularly regarding future technologies such as deep fakes, have not been well investigated. Contemporary writing frequently prioritizes the technological components and rapid societal effects while ignoring the long-term policy alterations and ethical deliberations required to regulate new technologies. Furthermore, there is a necessity for additional empirical investigation to assess the efficacy of media literacy programs in mitigating the impact of digital disinformation and deep fakes.

The current research in this study seeks to fill these deficiencies. By utilizing a quantitative methodology and focusing specifically on the phenomena of deep fakes, this study adds to the wider discussion on the influence of digital media from a novel standpoint. The research's emphasis on empirical data provides a more solid comprehension of public attitudes and behaviors, which is essential for guiding effective policy and educational measures. In addition, the study enhances existing research by examining the subtleties of how various demographic groups perceive and are influenced by deep fakes. This contributes to a more thorough and complete understanding of the societal consequences of digital media.

20.3 METHODOLOGY

The selection of a quantitative research technique for examining the influence of deep fakes on media sustainability and social balance is based on many important factors.

Objective Measurement and Statistical Analysis: A quantitative approach enables the precise measurement of variables and the application of statistical tools to analyze data. Comprehending the magnitude and trends of deep fakes' influence across various demographics and sectors is crucial. By quantifying answers, researchers can detect patterns, connections, and potential cause-and-effect relationships that may not be easily observable through qualitative analysis alone (Creswell & Creswell, 2017).

Generalizability of Findings: Quantitative approaches often employ higher sample numbers than qualitative investigations, augmenting the findings' generalizability. This is important, especially when considering the prevalence of deep fakes, which impacts diverse and extensive populations. By extrapolating findings to a wider population, a more thorough understanding of the issue may be obtained (Babbie, 2020).

Research Replicability: Quantitative research provides a greater level of replicability. Using standardized techniques for gathering and analyzing data, other researchers can reproduce the study to validate the results or investigate the phenomena in diverse settings or across time. This is a crucial element in establishing reliable knowledge in the rapidly advancing field of digital media (Bryman, et al., 2008).

Applying statistical tools in quantitative research enhances the study's rigor and precision. Statistical analysis can uncover the importance, scale, and practical ramifications of the influence of deep fakes. Precision is paramount in policy-making and developing focused solutions (Trochim & Donnelly, 2006).

Quantitative research offers a broad knowledge of the effects of deep fakes, but it is typically enhanced by qualitative observations that provide a more in-depth perspective. Within the wider study framework, quantitative results can inform qualitative investigations and vice versa, resulting in a more comprehensive comprehension.

Enabling Longitudinal Research: Quantitative methodologies are well-suited for conducting longitudinal research, which plays a crucial role in comprehending the dynamic characteristics of deep fakes and their enduring consequences. By conducting surveys and statistical analyses, it is possible to gain valuable insights into the progression of a phenomenon and its social impacts (Menard, 2002).

The use of a quantitative method in this research is warranted due to its capacity to offer objective, generalizable, and statistically rigorous insights into the effects of deep fakes. It facilitates a comprehensive and reproducible understanding, crucial for efficient policy formulation and societal response to the difficulties presented by this emerging technology.

20.3.1 Data Collection Methods

The main approach used to collect data in this study was the distribution of surveys and questionnaires, which played a crucial role in obtaining empirical evidence on the effects of deep fakes. Four hundred feedback replies were gathered, which provided a significant dataset for the study. The surveys were carefully crafted to cover various inquiries that would provide numerical and descriptive observations, including questions with predefined options, Likert scale assessments, and inquiries allowing for unrestricted responses (Rani et al., 2023). This methodology enabled a thorough comprehension of participants' perspectives, encounters, and dispositions toward deep fakes. The surveys were disseminated via many internet platforms, guaranteeing a wide-ranging demographic and geographic coverage, hence augmenting the representativeness of the data. This approach enabled a streamlined and extensive data-gathering procedure and offered a valuable asset for examining deep fakes' intricate and multifaceted effects on individuals and society.

20.3.2 Sampling Techniques and Demographic Considerations

The data collection in our research utilized a purposive and stratified sampling strategy. This approach was designed to get a varied and representative sample that accurately reflects how the general public interacts with social media. To thoroughly analyze demographic factors, the sample was divided into segments based on several criteria like age, gender, geographical region, and patterns of social media activity. The data was gathered online using various social media platforms, such as Facebook, Twitter, Instagram, and LinkedIn, as well as specialized forums and online communities. This technique was selected to access a diverse range of social media users, including those with varying degrees of digital literacy and exposure to deep fakes. The comprehensive selection from many domains of social media not only enabled the collection of a broad spectrum of viewpoints but also yielded insights into the perceptions and effects of deep fakes on various demographic groups. The decision to use this particular methodology was crucial in guaranteeing the strength and significance of the data, enabling a more precise and comprehensive comprehension of the social consequences of deep fakes among various socioeconomic groups.

20.3.3 Reliability Test

The table displays Cronbach's alpha coefficients for three distinct measures that assess several dimensions of individuals' responses to profound fakes: Understanding and Awareness, Concern and Anxiety, and Behavior and Attitude. Cronbach's alpha is a statistical metric that assesses the degree of internal consistency within a group of items, indicating how closely they are connected. It is regarded as an indicator of the dependability of a measurement scale. A coefficient alpha value of 0.93 for Understanding and Awareness indicates a remarkably high degree of internal consistency among the items in this scale, implying that the questions consistently and accurately assess the construct. The alpha coefficient of 0.90 indicates that the survey items successfully measure different levels of concern and anxiety associated with profound fakes, demonstrating strong internal consistency. The alpha coefficient of 0.92 for Behavior and Attitude suggests that the questions on this scale are very dependable for evaluating both the behavior and attitudes of persons toward deep fakes. The three scales demonstrate a commendable level of dependability, as shown by alpha values that are significantly above the acceptable threshold of 0.7. This implies that the survey instruments are skillfully designed and expected to produce reliable and consistent outcomes (Table 20.1).

TABLE 20.1
Reliability Test

Scale	Cronbach's Alpha
Understanding and Awareness	0.93
Concern and Anxiety	0.90
Behavior and Attitude	0.92

20.3.4 Data Analysis

The data analysis for this research utilized a comprehensive statistical strategy to investigate the correlations, disparities, and predictive variables related to the influence of deep fakes.

20.3.4.1 Independent t-test

The Independent t-test is used to determine if there are statistically significant differences between the means of two unrelated groups on a continuous dependent variable. The mathematical formula for the *t*-statistic in an independent *tt*-test is:

$$t = \frac{\overline{X_1} - \overline{X_2}}{\sqrt{\frac{s_p^2}{n_1} + \frac{s_p^2}{n_2}}}$$

where $\overline{X_1}$ and $\overline{X_2}$ are the sample means, s_p^2 is the pooled variance, and n_1 and n_2 are the sample sizes. The pooled variance is a weighted average of the variances from the two groups.

20.3.4.2 Pearson Correlation

The Pearson Correlation assesses the strength and direction of the linear relationship between two continuous variables. Its coefficient (r) ranges from -1 to 1, with -1 indicating a perfect negative linear relationship, 0 indicating no linear relationship, and 1 indicating a perfect positive linear relationship. The formula for Pearson r is:

$$r = \frac{\Sigma\left(X_i - \overline{X}\right)\left(Y_i - \overline{Y}\right)}{\sqrt{\Sigma\left(X_i - \overline{X}\right)^2 \Sigma\left(Y_i - \overline{Y}\right)^2}}$$

where X_i and Y_iAre the values of the two variables, $\overline{X}$ and $\overline{Y}$ and are the means of the variables, respectively.

20.3.4.3 One-sample t-test

The One-sample t-test compares the mean score of a sample to a known value, usually the population mean. The formula for calculating the t-value in a one-sample t-test is:

$$t = \frac{\overline{X} - \mu}{\frac{s}{\sqrt{n}}}$$

where $\overline{X}$is the sample mean, μ is the population mean, s is the sample standard deviation, and n is the sample size.

20.3.4.4 Linear Regression

Linear regression analysis predicts the value of a dependent variable based on the value of at least one independent variable. The linear regression equation is:

$$Y = \beta_0 + \beta_1 X_1 + \epsilon$$

Where Y is the dependent variable, β_0 is the y-intercept, β_1 is the slope of the regression line (the effect of X_1 on), and ϵ is the error term.

20.3.4.5 ANOVA (Analysis of Variance)

ANOVA is used when comparing three or more groups to see if there is at least one significant difference between group means. The ANOVA test is based on the F-statistic calculated as follows:

$$F = \frac{MS_{\text{between}}}{MS_{\text{within}}}$$

where MS_{between} is the mean sum of squares between the groups and MS_{within} is the mean sum of squares within the groups. The F-statistic is compared against a critical value from the F-distribution to determine statistical significance.

By applying these tests, the research can illuminate different aspects of the data, such as the relationships between variables, differences between groups, and the predictive power of certain factors. Each test provides unique insights that, when combined, offer a robust analysis of the research questions at hand.

20.4 RESULTS

20.4.1 Statistical Analysis of the Collected Data

By analyzing the data in the table, we may extract a range of valuable observations on the attitudes and views of the sample population toward deep fakes and related matters. The respondents were of mean age of 34 years. This suggests that the sample consists of individuals in their middle-aged years, with a substantial amount of variation in age.

The mean of the Understanding Level is 3.03, with a standard deviation of 1.41. This indicates that the typical responder possesses a somewhat higher-than-neutral level of understanding regarding deep fakes. The standard deviation indicates the presence of variability in comprehension across the subjects, albeit it is not significantly scattered.

The mean values for Concern Level and Anxiety Level are comparable, with values of 2.94 and 2.9, respectively. The standard deviation for both variables is 1.43, showing that respondents' levels of concern and anxiety concerning deep fakes are somewhat below the neutral point, with a moderate range of answers. This suggests that although there is a certain level of fear and anxiety, it is not strongly evident across the whole group.

The "Trust in AI" and "Can Identify Deep Fakes" models both had an average score of 2.9. The capacity to identify deep fakes scored slightly higher at 3. Both models had equal standard deviations of 1.42 and 1.43. These findings suggest a prevailing skepticism or lack of confidence in relying on AI and people's capacity to detect deep fakes.

The averages of the Likelihood of Personal Effect, Influence on Public Opinion, and Moral Acceptability are slightly below 3, ranging from 2.85 to 2.97. The associated standard deviations are around 1.4. This indicates a mildly unfavorable mood on the individual consequences of deep fakes, their impact on public perception, and their ethical appropriateness, with a rather uniform distribution of viewpoints.

The average frequency of questioning authenticity is 2.97, with a standard deviation of 1.38. This suggests that, on average, participants are inclined to doubt the authenticity of the media they come across, although there is some variation in how often they do so.

The danger to mental health and faith in AI systems had mean scores slightly above the neutral mark, with means above 3 and a standard deviation of 1.38. This indicates that participants perceive a moderate danger to mental health and have a neutral to slightly positive level of faith in AI systems to handle deep fakes.

The legislation's effectiveness is measured by a mean of 3.01 and a standard deviation of 1.44, suggesting a neutral stance on its present success in combating deep fakes, with a somewhat broad range of viewpoints.

The impact of public education and the likelihood of fact-checking are close to neutral, with mean values of 2.93 and 3.04, respectively. These findings indicate that participants doubt the influence of public education on profound fakes and exhibit a modest inclination to verify facts, with a varied range of viewpoints on both matters.

The mean willingness to join campaigns is 3.06, while the mean preparedness for challenges is 3.15. The data suggest a mild desire to engage in campaigns against deep fakes and a moderate readiness to confront the issues they present. The standard deviation of around 1.4 shows a modest range of answers.

The mean values for Discussion Frequency, Inevitability of Deep Fakes, and Confidence in Adapting range from 2.95 to 2.99, with standard deviations of approximately 1.4. This indicates that conversations about deep fakes occur moderately often, their inevitability is somewhat recognized, and the participants have a neutral level of confidence in adapting to them.

The data indicates that the populace has moderate knowledge and worry over deep fakes, with a typical confidence level in AI systems and regulations. There is a prevailing inclination to participate in preventive behaviors, such as verifying facts, and a modest awareness of the potential risks to mental well-being. The replies exhibit variety, suggesting varying degrees of comprehension, apprehension, and readiness to address the matter of profound fakes (Table 20.2).

20.4.2 Quantitative Results

Hypothesis 1 (H1): An Independent t-test was performed to assess the impact of gender on the level of worry over deep fakes. The test resulted in a t-statistic of -0.32 and a p-value of 0.74. Given the p-value above the standard alpha level of 0.05, we do not have sufficient evidence to reject the null hypothesis. This indicates no statistically significant disparity in apprehension over deep fakes between genders. Practically speaking, this means that the worry about deep fakes is an issue that affects both genders equally in the population under examination.

TABLE 20.2
Descriptive Statistics

	Mean	std. Dev.
Age	34	–
Understanding Level	3.03	1.41
Concern Level	2.94	1.43
Anxiety Level	2.9	1.43
Trust in AI	2.9	1.43
Can Identify Deep Fakes	3	1.42
Likelihood of Personal Effect	2.97	1.42
Influence on Public Opinion	2.94	1.39
Moral Acceptability	2.85	1.39
Frequency of Questioning Authenticity	2.97	1.38
Threat to Mental Health	3.04	1.38
Trust in AI Systems	3.05	1.38
Effectiveness of Legislation	3.01	1.44
Impact of Public Education	2.93	1.41
Likelihood to Fact-Check	3.04	1.47
Willingness to Join Campaigns	3.06	1.41
Preparedness for Challenges	3.15	1.42
Discussion Frequency	2.99	1.38
The inevitability of Deep Fakes	2.95	1.45
Confidence in Adapting	2.94	1.41

Hypothesis 2 (H2): Pearson Correlation was employed to examine the correlation between the comprehension level of deep fakes and the confidence in AI-based solutions to mitigate them. The correlation coefficient (r) is 0.03, indicating a weak positive relationship, with a p-value of 0.43, suggesting that the observed association is not statistically significant. The p-value surpasses the conventional threshold for statistical significance, suggesting a lack of evidence to establish an association between these variables. Therefore, the null hypothesis is not disproved, showing that comprehending deep fakes does not necessarily correlate with faith in AI systems to handle them among the studied population.

Hypothesis 3 (H3): To assess the average level of worry over the potential personal consequences of deep fakes, a one-sample t-test was conducted, comparing it to a neutral value of 3. The computed t-statistic is −1.35, accompanied by a p-value of 0.17. A p-value larger than 0.05 indicates insufficient evidence to reject the null hypothesis. This suggests that the average anxiety level of the sample is not considerably different from a neutral position, indicating that people may not be especially concerned about the personal consequences of deep fakes.

Hypothesis 4 (H4): When examining the influence of age on the probability of discussing deep fakes, a Linear regression analysis produced a regression coefficient (β) of 0.01 and a p-value of 0.07. The p-value, slightly above the threshold for significance, suggests insufficient compelling evidence to reject the null hypothesis. This indicates that age may show a slight pattern but does not exert a statistically significant influence on the frequency of conversations regarding deep fakes among friends or relatives in this particular population.

Hypothesis 5 (H5): An analysis of variance (ANOVA) was performed to ascertain the impact of education level on the inclination to participate in campaigns against deep fakes. The F-statistic is 0.71 with a p-value of 0.54, indicating no statistically significant difference. Thus, we uphold the null hypothesis, indicating no substantial variation in the participants' inclination to participate in campaigns based on various levels of education.

Hypothesis 6 (H6): To investigate the disparity in readiness to confront the problems presented by deep fakes, an additional Independent t-test was conducted comparing those impacted by deep fakes and those not. The test yielded a t-statistic of −0.07 and a p-value of 0.93. This strong p-value leads us to retain the null hypothesis, indicating that witnessing deep fakes does not significantly impact an individual's assessed preparation to cope with deep fake-related issues.

The detailed hypothesis testing results are shown in Table 20.3. The statistical tests undertaken fail to yield adequate evidence in favor of the alternative hypothesis for any of the six investigated associations. The data indicate no substantial disparities or connections regarding worry, comprehension, and readiness for deep fakes among genders. Furthermore, it does not imply any noteworthy associations between these factors and trust in AI, frequency of discussions, or willingness to participate in campaigns when considering age and education levels. These findings are crucial for comprehending the public's perception and reaction to deep fakes and can provide valuable insights for developing specific educational and policy solutions.

TABLE 20.3
Hypothesis Testing Results

Hypothesis	Test Type	Statistic	*P*-Value	Null Hypothesis	Alternative Hypothesis
H1	Independent *t*-test	–0.32	0.74	There is no significant difference in the level of concern about deep fakes between genders.	There is a significant difference in the level of concern about deep fakes between genders.
H2	Pearson Correlation	0.03	0.43	There is no correlation between the understanding level of deep fakes and the trust in AI-based systems to curb them.	There is a correlation between the understanding level of deep fakes and the trust in AI-based systems to curb them.
H3	One-sample *t*-test	–1.35	0.17	The average level of anxiety about the potential personal impact of deep fakes is 3 (neutral) in the population.	The average level of anxiety about the potential personal impact of deep fakes is not 3 in the population.
H4	Linear Regression	0.01	0.07	Age does not affect the likelihood of discussing deep fakes with friends or family.	Older participants discuss deep fakes with friends or family more frequently than younger participants.
H5	ANOVA	0.71	0.54	There is no significant difference in the willingness to join campaigns against deep fakes based on education level.	There is a significant difference in the willingness to join campaigns against deep fakes based on education level.
H6	Independent *t*-test	–0.07	0.93	There is no significant difference in the preparedness for challenges posed by deep fakes between those affected by deep fakes and those not.	Individuals affected by deep fakes feel more prepared to cope with the challenges posed by them.

20.4.3 Comparative Analysis with Existing Literature

This research provides vital insights into the discourse around deep fakes and their various repercussions on society by conducting a comparative analysis with previous literature.

Prior research has frequently emphasized the insufficient understanding and knowledge among the general public about deep fakes and their consequences (Westerlund, 2019). Nevertheless, our research reveals a moderately positive outlook, with an average comprehension level above neutrality. The divergence might be attributed to the heightened public discussion and dissemination of knowledge on the matter in recent years, indicating a favorable trajectory in public consciousness (Chesney & Citron, 2019).

Concern and Anxiety: Consistent with the findings of Hopp et al. (2021), our study demonstrates a modest degree of apprehension and unease with deep fakes, perhaps due to their potential for misuse. Contrary to Hopp et al.'s findings that showed a notable level of anxiety linked to disinformation, our data do not reveal increased worry. This might be because the general public is becoming more accustomed to deep fakes in the media.

Trust in AI: Our findings align with the doubts stated in previous studies on the public's confidence in AI's ability to accurately identify and handle deep fakes (Diakopoulos, 2020). The study's neutral ratings indicate that, despite improvements in detection algorithms, there has been no substantial rise in public confidence, aligning with the cautionary advice given by Diakopoulos.

Legislation and Public Education: Our research aligns with the literature that criticizes the current legal frameworks' capacity to address the difficulties presented by deep fakes (Schwartz & Knake, 2019). We adopt a neutral stance regarding the efficacy of legislation. Similarly, our study's modest influence on public education corresponds to the demand for more comprehensive educational interventions (Boddington, 2017).

Behavioral intentions, such as the inclination to actively participate in activities like fact-checking and supporting campaigns against deep fakes, have received limited attention in existing research. Our research addresses this knowledge gap by demonstrating a modest probability of these behaviors, indicating a willingness among the public to engage in efforts to counteract disinformation (Rini, 2020).

Demographic Influences: In contrast to previous studies that have indicated considerable impact of demographic characteristics on perceptions of deep fakes (Lyons et al., 2020), our research demonstrates that age and education level do not significantly affect degrees of worry or behavioral intentions. This implies a more uniform understanding across different groups of people, maybe because of the widespread influence of digital media.

This comparative analysis indicates similarities between the existing literature and the findings of this study in various aspects. However, there are also significant differences, especially regarding the extent of public concern and anxiety and the impact of demographic variables on attitudes toward deep fakes. The variations may stem from the dynamic characteristics of the digital environment and the public's assimilation of novel technology.

20.5 DISCUSSION

20.5.1 Interpretation of Quantitative Results

The quantitative results of this study provide a more detailed knowledge of public perceptions and reactions to deep fakes. The average scores for understanding and awareness over the neutral midway indicate a moderate level of public knowledge about deep fakes, which demonstrates the efficacy of educational and media initiatives to raise public awareness about this issue. However, modest degrees of fear and anxiety and neutral trust in AI indicate that the public is either cautiously hopeful or underestimates the potential hazards posed by deep fakes.

The absence of substantial variations in concern across genders and the minimal impact of age on discussions about deep fakes may imply that these issues have extended across demographic boundaries. This might also imply that the prevalence of deep fakes in the media has resulted in a shared degree of worry among many community segments.

The neutral approach to the efficacy of law and public education efforts implies that present policies may not adequately address public concerns or the hazards of deep fakes. The moderate chance of engaging in behaviors such as fact-checking reflects the public's readiness to battle disinformation, but it also highlights the need for greater support networks and more accessible verification tools.

20.5.2 Fairness and Failure of AI Platforms

The study's findings on the public's faith in AI's ability to detect and manage deep fakes provide a critical examination of the role of AI platforms in content moderation and public debate. While AI can swiftly detect and prevent the spread of deep fakes, trust scores are neutral, indicating a disconnect between AI capabilities and public expectations or knowledge of these technologies. This disparity might be attributed to AI platforms' inability to consistently and properly detect deep fakes, resulting in consumer distrust.

The issue of justice in AI platforms is challenging. Bias in algorithmic decision-making can have disparate effects on various groups, thereby increasing socioeconomic inequities. Furthermore, the responsibility of AI in circumstances when it fails to detect or incorrectly identifies information as a deep fake remains a major worry. This raises concerns about the openness of AI systems and the procedures in place to handle failures.

20.5.3 Ethical Considerations and Social Challenges

The ethical implications of deep fakes are numerous. On the one hand, there is worry about the production and spread of deep fakes for harmful reasons such as disinformation, defamation, and manipulation of public opinion. However, the ethical use of AI for identifying and mitigating deep fakes raises concerns about privacy, censorship, and freedom of speech.

20.5.4 Effects on Young People and the Community

20.5.4.1 Evaluation of the Influence on Young People and College Students

The influence of deep fakes on young people and college students is especially significant. This group exhibits a strong level of involvement with digital media and shapes their social identities and worldviews throughout this crucial development phase. Incorporating deep fakes into the media environment might present substantial obstacles in fostering critical thinking and digital literacy. Deep fakes can impact academic integrity, social interactions, and mental health, particularly among young children who may not possess the cognitive abilities to recognize or handle the consequences of altered information.

20.5.4.2 Approaches for Promoting Understanding and Knowledge

Developing comprehensive programs that raise awareness and provide education is crucial to tackling these difficulties. Educational institutions must incorporate media literacy into their curricula, focusing on developing the ability to critically assess and interrogate the veracity of digital material. Campaigns aimed at young people should utilize the digital channels they often use to distribute information on the characteristics of deep fakes, the accompanying hazards, and the available verification tools. Partnerships among educators, technologists, and policymakers have the potential to foster the creation of captivating and efficient teaching materials.

20.5.5 Prospects for the Future

20.5.5.1 Proposed Areas for Future Research

Subsequent investigations should prioritize long-term studies to monitor public opinion alterations and the progressive characteristics of deep fake technologies. It is imperative to thoroughly examine the psychological ramifications of deep fakes, especially on susceptible demographics. Conducting research on the efficacy of various educational initiatives and the influence of social media platforms in disseminating information might yield practical and valuable findings. Furthermore, investigating the technological progress in identifying deep fake material and the cooperation between humans and artificial intelligence in moderating online content might contribute to a more comprehensive understanding of risk mitigation strategies.

Policy implications refer to the potential consequences or effects that a certain policy may have. On the other hand, preventive measures are actions taken to avoid or minimize the occurrence of a certain event or problem. Updated legislation is required to meet the distinct issues presented by deep fakes on the policy front.

This encompasses the examination of privacy, consent, and the ethical use of AI. Additionally, policies should promote the creation of reliable verification technologies and assist with public education campaigns. To mitigate the risks of deep fakes, it is advisable to have a standardized procedure for detecting and reporting them and to establish explicit penalties for those who intentionally create and spread such deceptive information. Ensuring a secure digital environment necessitates rules in step with technical changes.

The consequences of deep fakes for young people and society are substantial, necessitating collaborative endeavors in education and policy formation. Future endeavors should prioritize adjustment to the difficulties presented by this technology through continuous study, well-informed regulation, and proactive initiatives to prevent negative outcomes. To ensure the integrity of our information ecology, we must adapt our tactics to keep up with the changing digital reality.

The study's findings reveal societal concerns, emphasizing the importance of collaborative action against the risks posed by deep fakes. The modest readiness to participate in campaigns against deep fakes indicates that the public is eager to engage in solutions but may lack the direction or effective venues to do so. Deep fakes have a systemic impact beyond individual concerns, undermining the foundations of confidence in media, the integrity of democratic processes, and the entire information ecosystem. The quantitative data provide a complete picture of the public's attitude toward deep fakes, demonstrating cautious awareness, varying trust in AI, and a need for more effective remedies. These findings highlight the need for more AI fairness and openness, ethical technology use, and social mobilization to overcome the difficulties posed by deep fakes. The ramifications for legislators, technologists, and educators are enormous as they collaborate to navigate the complicated landscape of digital authenticity and trust.

20.6 CONCLUSION

Our research suggests that although there is increasing recognition of deep fakes, the general public's worry is only modest, and they hold a neutral position about the effectiveness of AI detection and legislative actions. The absence of notable demographic disparities implies that deep fakes are a widely acknowledged problem. Nevertheless, the moderate levels of concern and awareness also indicate possible deficiencies in public understanding and the efficacy of existing educational programs. The study emphasizes the urgent need for improved educational initiatives promoting media literacy, particularly among younger age groups most susceptible to the impact of deep fakes. Furthermore, the results suggest that policymakers should modify and enhance regulations to reduce the hazards of deep fakes. Future studies should persistently monitor the public's perception as technology and social media platforms progress, guaranteeing that methods to counteract deep fakes are guided by current data and successfully safeguard the integrity of digital media and societal trust.

REFERENCES

Babbie, E. (2020). *The Practice of Social Research* (15th ed.). Wadsworth Publishing.

Bhambri, P., Rani, S., Balas, V. E., & Elngar, A. A. (2023a). *Integration of AI-Based Manufacturing and Industrial Engineering Systems with the Internet of Things*. CRC Press.

Bhambri, P., Singh, S., Jain, S., & Dhanoa, S. I. (2023b). Plants recognition using leaf image pattern analysis with focus on advanced smart computing technologies. In AIP Conference Proceedings, 2916, 020003 (2023).

Boddington, P. (2017). *Towards a Code of Ethics for Artificial Intelligence*. Springer International Publishing.

Boyd, D., & Crawford, K. (2012). Critical questions for big data: Provocations for a cultural, technological, and scholarly phenomenon. *Information, Communication & Society, 15*(5), 662–679.

Bradshaw, S., & Howard, P. N. (2019). *The Global Disinformation Order*. Project on Computational Propaganda.

Brennen, S. (2020). DEEPFAKES accountability act. *Journal of Legislation and Public Policy, 22*, 151–171.

Bryman, A., Becker, S., & Sempik, J. (2008). Quality criteria for quantitative, qualitative and mixed methods research: A view from social policy. *International Journal of Social Research Methodology, 11*(4), 261–276. https://doi.org/10.1080/13645570701401644

Brynjolfsson, E., & Mcafee, A. (2014). *The Second Machine Age: Work, Progress, and Prosperity in a Time of Brilliant Technologies*. W. W. Norton & Company.

Chesney, R., & Citron, D. K. (2019).

Creswell, J. W., & Creswell, J. D. (2017). *Research Design: Qualitative, Quantitative, and Mixed Methods Approaches*. Sage Publications.

Cunningham, S., & Silver, J. (2013). *Screen Distribution and the New King Kongs of the Online World*. Palgrave Macmillan.

Diakopoulos, N. (2020). *Automating the News: How Algorithms Are Rewriting the Media*. Harvard University Press.

Fallis, D. (2021). The epistemic threat of deepfakes. *Philosophy & Technology, 34*(4), 623–643. https://doi.org/10.1007/s13347-020-00419-2

Goldfarb, A., & Que, V. F. (2023). The economics of digital privacy. *Annual Review of Economics, 15*(1). https://doi.org/10.1146/annurev-economics-082322-014346

Goodfellow, I. J., Pouget-Abadie, J., Mirza, M., Xu, B., Warde-Farley, D., Ozair, S., Courville, A., & Bengio, Y. (2014). Generative Adversarial Networks. In *arXiv [stat.ML]*. http://arxiv.org/abs/1406.2661

Hale, L., & Guan, S. (2015). Screen time and sleep among school-aged children and adolescents: A systematic literature review. *Sleep Medicine Reviews, 21*, 50–58. https://doi.org/10.1016/j.smrv.2014.07.007

Hobbs, R. (2011). *Digital and Media Literacy: A Plan of Action*. Springer.

Hopp, T., Ferrucci, P., & Vargo, C. (2021). Why do people share deepfakes? Understanding the psychological motivations behind the dissemination of synthetic media. *Information, Communication & Society, 24*(12), 1703–1720.

Howard, P. N. (2018). Political polarization on social media: Do bots and junk news drive online conversations? *Political Communication, 35*(4), 587–607.

Jenkins, H., & Plasencia, A. (2017). Convergence culture: Where old and new media collide. In *Is the Universe a Hologram?* The MIT Press.

Kross, E. (2019). Does spending time on social media affect how much you like your life? *Behaviour Research and Therapy, 111*, 19–26.

Li, Y., Chang, M.-C., & Lyu, S. (2018). In ictu oculi: Exposing AI generated fake face videos by detecting eye blinking. In *arXiv [cs.CV]*. http://arxiv.org/abs/1806.02877

Lyons, B. A., Montgomery, J. M., & Brewer, P. R. (2020). How (not) to counter misinformation: Evidence from a field experiment on social media. *Political Behavior, 20*, 1–22.

Menard, S. (2002). *Longitudinal Research*. SAGE Publications, Inc.

Norris, P. (2003). Digital divide: Civic engagement, information poverty, and the internet worldwide. *Canadian Journal of Communication, 28*(1). https://doi.org/10.22230/cjc.2003v28n1a1352

Parisi, F. (2019). Deep fakes and the law: What can be done? *Journal of Intellectual Property Law & Practice, 14*(1), 18–26.

Pennycook, G., Cannon, T. D., & Rand, D. G. (2018). Prior exposure increases perceived accuracy of fake news. *Journal of Experimental Psychology. General, 147*(12), 1865–1880. https://doi.org/10.1037/xge0000465

Persily, N. (2016). Election: Can democracy survive the internet? *Journal of Democracy, 28*(2), 63–76.

Prensky, M. (2001). Digital natives, digital immigrants part 1. *On the Horizon, 9*(5), 1–6. https://doi.org/10.1108/10748120110424816

Rachna, R., Bhambri, P., & Chhabra, Y. (2022). Deployment of distributed clustering approach in WSNs and IoTs. In *Cloud and Fog Computing Platforms for Internet of Things* (pp. 85–98). Chapman and Hall/CRC.

Rani, S., Kaur, J., & Bhambri, P. (2023). Technology and Gender Violence: Victimization Model, Consequences and Measures. In *Communication Technology and Gender Violence* (Vol. 1, pp. 1–19). Springer.

Rini, R. (2020). Deepfakes and the epistemic backstop. *Philosophers' Imprint, 20*(24), 1–16.

Rush, S. (2020). Truth, lies, and automation: The role of AI in deception detection. *Computers in Human Behavior, 108*, 134–150.

Schwartz, P., & Knake, R. K. (2019). The search for cyber peace. In *Council on Foreign Relations Cyber Brief.* Springer.

Shrivastava, A., Rizwan, A., Kumar, N. S., Saravanakumar, R., Dhanoa, I. S., Bhambri, P., & Singh, B. K. (2021). *VLSI Implementation of Green Computing Control Unit on Zynq FPGA for Green Communication*. Springer.

Sigman, A. (2012). Time for a view on screen time. *Archives of Disease in Childhood, 97*(11), 935–942. https://doi.org/10.1136/archdischild-2012-302196

Solove, D. J. (2020). *The Digital Person: Technology and Privacy in the Information Age*. New York University Press.

Trochim, W., & Donnelly, J. P. (2006). *The Research Methods Knowledge Base*. Atomic Dog Publishing.

Twenge, J. M., Martin, G. N., & Campbell, W. K. (2018). Decreases in psychological well-being among American adolescents after 2012 and links to screen time during the rise of smartphone technology. *Emotion (Washington, D.C.), 18*(6), 765–780. https://doi.org/10.1037/emo0000403

Vorderer, P. (2016). Permanently online, permanently connected: Exploring the effects of media on well-being. *Digital Media, Culture, and Society, 32*, 7–27.

Wagner, E. (2020). Deep fakes and social media: Exploring a more ethical digital future. *Media and Communication, 8*(2), 323–331.

Wartella, E. (2016). Parenting in the age of digital technology. In *Report for the Center on Media and Human Development*. Springer.

Westerlund, M. (2019). The emergence of deepfake technology: A review. *Technology Innovation Management Review, 9*(11), 39–52. https://doi.org/10.22215/timreview/1282

21 Green Horizons
Navigating Environmental Challenges through Technological Innovation

B. Thirumalaiyammal, P.F. Steffi, and Pankaj Bhambri

21.1 INTRODUCTION

The current global scenario is marked by a critical need for sustainable solutions that can address the unprecedented environmental challenges we face and minimize humanity's impact on the planet. The call for a greener future has become a worldwide rallying cry that has spurred a wave of innovations at the intersection of sustainability and technology. This chapter takes you on a journey to explore cutting-edge developments and trends that are shaping a path toward a more sustainable and eco-conscious world. From renewable energy technologies that harness the power of nature to smart systems optimizing resource usage, and from circular economy practices that redefine production cycles to green transportation technologies revolutionizing the way we move, every aspect of this exploration unveils a promising landscape of technological sustainability. As we stand at the crossroads of environmental stewardship and technological innovation, a crucial question arises: How can we leverage human ingenuity to address the complex challenges threatening the health of our planet? This chapter seeks to unravel the answers by delving into key innovations driving a greener future. By understanding and embracing these technological sustainability trends, we pave the way for a harmonious coexistence between human progress and ecological preservation. The current global landscape is witnessing a mutually beneficial relationship between digital transformation and environmental sustainability. In the logistics industry, this alliance holds immense significance as companies seek innovative solutions to make deliveries more eco-friendly. With a growing emphasis on climate change and the need for sustainable practices, the logistics sector is adopting cutting-edge green technologies that are revolutionizing the transportation of goods. In our interconnected world, the delivery industry plays a critical role in facilitating global trade and fostering connectivity.

DOI: 10.1201/9781003475989-25

Unfortunately, transportation is currently the fastest-growing source of emissions worldwide, accounting for 17% of global greenhouse gas emissions – second only to the power sector. In addition, CO_2 emissions from transportation account for a 20.3% share globally, equivalent to 7.64 Gt of CO_2. Therefore, it is increasingly important to address the environmental impact associated with this sector. Logistics companies are making significant progress in reducing their carbon footprint by leveraging digital transformation alongside eco-friendly technologies. By using efficient delivery routes, reducing idle time, and optimizing vehicle capacity, logistics companies are reducing their carbon footprint and minimizing the impact of their operations on the environment.

21.2 RENEWABLE ENERGY REVOLUTION

The global energy landscape is undergoing a radical transformation driven by the relentless pursuit of sustainability and the imperative to mitigate climate change. This chapter delves into the ongoing revolution in renewable energy, exploring the latest advancements, key trends, and the transformative impact on our societies. From solar and wind innovations to breakthroughs in energy storage, this chapter provides a comprehensive overview of the dynamic developments propelling the world towards a cleaner and more sustainable energy future.

21.2.1 Solar Energy: Harnessing the Power of the Sun

Solar energy is at the forefront of the renewable energy revolution. Advances in photovoltaic technologies, as documented by Green et al. [1], have significantly increased the efficiency and affordability of solar cells. This section examines the evolving landscape of solar energy, from improvements in panel efficiency to the integration of solar power into urban infrastructure.

21.2.2 Wind Power: Driving Innovation in Turbines and Beyond

The wind power sector continues to evolve, with ongoing research by Jacobson and Delucchi [2] highlighting innovations in turbine design and the expansion of offshore wind farms. Artificial Intelligence, as studied by EMEC [3] is optimizing wind energy production, contributing to the increasing prominence of wind power in the global energy mix.

21.2.3 Hydroelectric and Marine Energy: Tapping into Water Resources

Hydroelectric power, a longstanding renewable energy source, is undergoing a renaissance with advancements in dam design and efficiency [4]. Additionally, marine energy, as explored by the European Marine Energy Centre (EMEC, 2020), presents a promising frontier.

21.2.4 Energy Storage Breakthroughs: Mitigating Intermittency

Addressing the intermittency of renewable sources, breakthroughs in energy storage are crucial. Research by Joseph (2020) [5] delves into advancements in battery technologies and grid-scale storage. This section explores the implications of these breakthroughs for enhancing the reliability and stability of renewable energy systems.

21.2.5 Policy and Economic Dynamics: Shaping the Future

The success of the renewable energy revolution is intertwined with supportive policies and economic frameworks. Policy analyses by [6, 7] provide insights into the role of governments and international organizations in fostering renewable energy adoption.

21.3 SUSTAINABLE TRANSPORTATION

In an era marked by environmental consciousness and the imperative to reduce carbon footprints, the transportation sector stands as a focal point for innovation and transformation. This section explores the dynamic landscape of sustainable transportation, shedding light on pioneering technologies, policy frameworks, and shifts in consumer behaviour that are steering the industry towards a more eco-friendly and resilient future.

21.3.1 Electric Vehicles: Revolutionizing the Road Ahead

The advent of electric vehicles (EVs) is a watershed moment in sustainable transportation. Breakthroughs in battery technologies, as explored by Smith, Brown, and Williams [8] have significantly extended the range and efficiency of EVs.

21.3.2 Hydrogen Fuel Cells: A Cleaner Drive Forward

Research by Chen, Wang, and Zhang [9] delves into advancements in hydrogen fuel cell technology, shedding light on their potential to revolutionize the transportation sector.

21.3.4 Sustainable Aviation Fuels: Taking Flight towards Greener Skies

As the aviation industry grapples with its environmental impact, sustainable aviation fuels (SAFs) emerge as a key solution. Studies by Johnson and Garcia [10] provide insights into the development and adoption of SAFs.

21.3.5 Public Transportation and Smart Mobility Solutions

Public transportation remains a cornerstone of sustainable urban mobility. This section explores innovations in public transit systems, including smart technologies for route optimization and the integration of electric buses. Insights from [11] shed light on the role of smart mobility solutions in reducing congestion and environmental impact.

21.3.6 Active Transportation: Pedaling towards a Greener Commute

The resurgence of active transportation modes, such as cycling and walking, plays a pivotal role in sustainable urban mobility. Studies by Rodriguez, Brown, and Troped [12] emphasize the health and environmental benefits of active transportation.

21.3.7 Policy Frameworks and Collaborative Initiatives

Effective policy frameworks are essential drivers of sustainable transportation. Analyses by Lee and Smith [13] examine successful case studies and policy approaches from around the world.

21.4 CIRCULAR ECONOMY AND WASTE REDUCTION

In the face of escalating environmental challenges and resource depletion, the circular economy has emerged as a transformative approach to sustainable development [14]. This section delves into the principles, innovations, and real-world applications of circular economy strategies aimed at reducing waste. By exploring the nexus of production, consumption, and waste management, we unravel the potential for creating a regenerative and waste-minimized future.

21.4.1 Understanding the Circular Economy Paradigm

The circular economy represents a departure from the linear 'take, make, dispose' model. This section introduces the core principles of circularity, drawing on foundational works such as the Ellen MacArthur [15] Foundation's reports, which emphasize the regenerative design of products and systems. We explore the shift towards closed-loop systems that prioritize durability, repairability, and recyclability.

21.4.2 Designing for Circularity: Cradle-to-Cradle Approaches

Circular product design is a cornerstone of waste reduction. Building on the concepts proposed by McDonough and Braungart [16] this section examines cradle-to-cradle approaches that prioritize materials' continuous cycles, minimizing environmental impact. Case studies highlight successful implementations, demonstrating how designing products for circularity can extend their lifespan and reduce waste.

21.4.3 Extended Producer Responsibility (EPR): Shifting the Burden from Consumers

The responsibility for waste management often falls on consumers, but the circular economy seeks to shift this burden upstream. Research by Schenck, Sarkanen, and

Leroux [17] explores the concept of extended producer responsibility (EPR), wherein manufacturers bear the responsibility for the entire life cycle of their products.

21.4.4 Waste to Resource: The Role of Recycling and Upcycling

The circular economy hinges on the efficient use of resources, emphasizing recycling and upcycling. Drawing on studies by Cucchiella, D'Adamo, and Lenny Koh [18] we delve into the role of recycling in waste reduction and explore innovative upcycling practices that transform waste into valuable resources.

21.4.5 Digital Technologies for Circular Supply Chains

The digital revolution plays a pivotal role in optimizing circular supply chains. Leveraging the insights from research by Sundarakani et al. [19, 20] this section explores the integration of technologies such as blockchain, IoT, and AI in enhancing traceability, transparency, and efficiency within circular systems. Case studies illustrate the transformative impact of these technologies on waste reduction.

21.5 SMART CITIES AND SUSTAINABLE URBAN PLANNING

As urbanization accelerates globally, the concept of Smart Cities has emerged as a transformative paradigm for addressing the complex challenges of urban living. This chapter navigates the landscape of Smart Cities and their pivotal role in sustainable urban planning. From cutting-edge technologies to citizen engagement, we explore how smart initiatives are reshaping urban environments to foster sustainability, efficiency, and an enhanced quality of life.

21.5.1 Defining Smart Cities: A Holistic Perspective

To lay the foundation, we define Smart Cities as integrated urban ecosystems that leverage digital technologies to enhance infrastructure, services, and civic engagement. This section draws on seminal works [21, 22] to elucidate the key dimensions of a Smart City – smart economy, smart people, smart governance, smart mobility, smart environment, and smart living.

21.5.2 Technology Enablers: Building the Smart Infrastructure

Smart Cities thrive on advanced technologies that form the backbone of sustainable urban planning. Leveraging insights from [23] this section explores technology enablers such as the Internet of Things (IoT), Artificial Intelligence (AI), and data analytics. Case studies showcase how these technologies optimize energy usage, traffic management, and waste reduction, contributing to the ecological sustainability of urban areas.

21.5.3 Citizen-Centric Design: Fostering Inclusive Urban Development

Sustainability in Smart Cities goes beyond technological infrastructure; it extends to inclusive and citizen-centric design. Examining the principles outlined [24] we delve into the importance of involving citizens in decision-making processes. The section explores how digital platforms and participatory planning enable residents to contribute to sustainable urban development.

21.5.4 Smart Mobility: Transforming Urban Transportation

Smart mobility is a linchpin in the pursuit of sustainable urban living. Deloitte [25] informs this section, which examines how innovations such as intelligent transportation systems, electric mobility, and shared mobility services contribute to reducing congestion, lowering emissions, and enhancing urban mobility experiences.

21.5.5 Sustainable Urban Planning Frameworks for Smart Cities

Integrating Smart Cities into broader sustainable urban planning frameworks is imperative. This section draws on the works of Batty et al. [26] to explore how Smart Cities align with sustainable development goals. It discusses the need for adaptive and resilient urban planning frameworks that balance economic growth, social inclusion, and environmental sustainability.

21.6 AGRICULTURAL TECHNOLOGY FOR SUSTAINABLE FOOD PRODUCTION

As urbanization continues to rise, the convergence of Smart Cities and Agricultural Technology presents a unique opportunity to address the pressing challenges of food production, distribution, and sustainability [27]. This chapter explores the integration of smart technologies in agriculture within the context of Smart Cities, examining how these innovations are revolutionizing farming practices and contributing to sustainable food production.

21.6.1 Smart Agriculture: A Nexus of Technology and Farming

Smart agriculture, often referred to as precision farming, leverages advanced technologies to optimize crop yields, resource usage, and overall farm efficiency. This section introduces the fundamental concepts of smart agriculture, drawing on seminal works [28]. We explore the integration of IoT, AI, and data analytics in modern farming practices.

21.6.2 Urban Farming in Smart Cities: A Paradigm Shift

Urban farming is gaining prominence within the ambit of Smart Cities, redefining traditional notions of agriculture. Building on insights from Sanyé-Mengual et al.

[29] this section investigates the role of vertical farming, rooftop gardens, and hydroponics in urban settings. Case studies showcase how these practices contribute to local food production, reduce transportation emissions, and enhance food security.

21.6.3 Precision Farming Technologies: Enhancing Agricultural Efficiency

The adoption of precision farming technologies is a hallmark of smart agriculture. Research by Gebbers and Adamchuk [30] informs this section, which explores the use of GPS, sensors, and drones in optimizing field-level management. We delve into how these technologies enable farmers to make data-driven decisions, reduce resource inputs, and minimize environmental impact.

21.6.4 Sustainable Water Management: Smart Irrigation Solutions

Water scarcity is a critical concern in agriculture, and smart technologies offer solutions for sustainable water management. Drawing on studies by Sharma, Kaur, and Sharma [31], this section examines the role of smart irrigation systems in conserving water resources. Case studies illustrate how sensor-based irrigation and AI-driven water management contribute to efficient water usage in agriculture.

21.6.5 Digital Agriculture Platforms: Connecting Farmers and Consumers

Digital platforms are bridging the gap between farmers and consumers, fostering transparency and efficiency in the food supply chain. Leveraging insights [32], this section explores the emergence of digital agriculture platforms. We discuss how these platforms enable traceability, empower farmers with market information, and provide consumers with access to locally sourced produce.

21.7 GREEN BUILDING AND SUSTAINABLE ARCHITECTURE

In the quest for a sustainable future, green building practices and sustainable architecture stand as cornerstones in reshaping our constructed environment. This chapter delves into the principles, innovations, and transformative impact of green building, exploring how sustainable architecture is redefining structures to harmonize with the environment. From energy efficiency to material choices, we unravel the multifaceted landscape of green building.

21.7.1 Foundations of Green Building: Principles and Objectives

Green building extends beyond energy efficiency; it encompasses a holistic approach to minimize environmental impact. This section introduces the core principles of green building, referencing influential works such as the U.S. Green Building Council's LEED (Leadership in Energy and Environmental Design) framework

[33]. We explore objectives such as resource efficiency, water conservation, and the importance of sustainable site development.

21.7.2 Energy-Efficient Design and Renewable Energy Integration

The cornerstone of sustainable architecture is energy efficiency. Building upon insights from [34] this section delves into innovative design strategies that optimize energy usage. We also explore the integration of renewable energy sources, such as solar panels and wind turbines, highlighting how these technologies are transforming buildings into self-sufficient, energy-generating entities.

21.7.3 Sustainable Materials and Life Cycle Assessment

Material choices play a pivotal role in green buildings. Drawing on research by Wong, Li, and Liu, [35] this section investigates the use of sustainable materials, recycled content, and the concept of life cycle assessment. Case studies showcase how conscious material selection reduces environmental impact and contributes to the longevity of structures.

21.7.4 Biophilic Design: Integrating Nature into Architecture

Biophilic design is emerging as a transformative approach, emphasizing the connection between humans and nature within built environments. Research by Kellert, Heerwagen, and Mador [36] informs this section, which explores how incorporating natural elements, green spaces, and daylighting enhances well-being, productivity, and the overall quality of indoor spaces.

21.7.5 Water Conservation and Sustainable Site Planning

Water scarcity is a global concern, and green building addresses this through water conservation strategies. Building upon studies [37], we examine sustainable site planning, rainwater harvesting, and the implementation of water-efficient technologies. Case studies demonstrate how these practices mitigate the strain on water resources.

21.7.6 Smart Technologies and Building Performance Monitoring

The advent of smart technologies is revolutionizing how buildings are monitored and managed for optimal performance. Leveraging insights from Lee et al. [38] this section explores the role of IoT (Internet of Things), sensors, and smart building systems. We discuss how real-time data and automation contribute to energy efficiency, comfort, and operational sustainability.

21.8 BIODIVERSITY CONSERVATION THROUGH TECHNOLOGY

Biodiversity, the fabric of life on Earth, faces unprecedented threats in the modern era. This section explores the intersection of technology and biodiversity conservation, unraveling the innovative ways in which digital tools are becoming indispensable assets in the quest to preserve and sustain the planet's rich biological diversity.

21.8.1 Biodiversity Monitoring: Remote Sensing and Satellite Technologies

The advent of remote sensing and satellite technologies has revolutionized the monitoring of biodiversity. Building upon studies by Pettorelli et al. [39], this section delves into how satellite imagery and remote sensing tools provide real-time data on ecosystems, helping conservationists track changes, assess threats, and plan targeted interventions for preserving biodiversity.

21.8.2 Citizen Science and Mobile Apps: Engaging the Masses in Conservation

Citizen science, facilitated by mobile apps and digital platforms, is democratizing biodiversity monitoring. Leveraging insights from [40], this section explores how everyday individuals can contribute to data collection, species identification, and habitat mapping. Case studies highlight successful citizen science initiatives that have significantly contributed to biodiversity conservation efforts.

21.8.3 DNA Barcoding and Genomic Technologies: Unraveling Biodiversity Mysteries

Genomic technologies, including DNA barcoding, are providing invaluable insights into the diversity of life. Referencing [41] pioneering work, this section elucidates how DNA-based identification methods enable rapid and accurate species identification. We explore how genomics contributes to understanding evolutionary relationships, population dynamics, and the conservation genetics of endangered species.

21.8.4 Artificial Intelligence (AI) in Biodiversity Conservation

Artificial Intelligence is emerging as a powerful ally in biodiversity conservation. Building upon research by McInerny et al. [42], this section investigates how AI, through machine learning algorithms, analyses large datasets to identify patterns, predict biodiversity trends, and inform conservation strategies. Case studies showcase how AI is being employed for automated species identification, habitat monitoring, and threat detection.

21.8.5 Conservation Drones: Aerial Surveillance for Habitat Monitoring

Unmanned Aerial Vehicles (UAVs) or drones are transforming habitat monitoring and conservation efforts. Drawing on insights from Anderson and Gaston [43] this section explores how drones provide high-resolution aerial imagery, enabling conservationists to assess habitat health, detect illegal activities, and monitor wildlife populations in remote or inaccessible areas.

21.8.6 Blockchain Technology: Enhancing Transparency in Conservation

Blockchain technology is enhancing transparency and accountability in biodiversity conservation. Leveraging insights from MacDicken, Morgan, and Saatchi [44] this section explores how blockchain ensures the integrity of conservation data, secures funding transactions, and creates immutable records of conservation efforts. Case studies highlight successful applications of blockchain in supporting sustainable practices and incentivizing biodiversity conservation.

21.9 CONCLUSION

In the pursuit of a greener future, exploring technological sustainability trends and strategies for their adoption reveals a roadmap towards a more harmonious relationship between humanity and the environment. The innovations discussed in this chapter, including renewable energy, circular economy, smart cities, precision agriculture, green building, and biodiversity conservation, collectively offer a glimpse into the transformative potential of technology. As we stand at the crossroads of environmental challenges, the renewable energy revolution promises a departure from fossil fuel dependency, ushering in an era of clean, abundant power. Circular economy technologies beckon us to rethink waste, envisioning a future where resources are conserved, repurposed, and recycled in closed-loop systems. Smart cities, with their integrated technologies, present a vision of urban living that is sustainable, efficient, and responsive to the needs of both citizens and the environment. Precision agriculture, fuelled by advancements in digital technologies, holds the promise of feeding the world while minimizing environmental impact. Green building practices showcase the fusion of innovation and sustainability in the construction industry, offering structures that coexist harmoniously with nature. Biodiversity conservation, empowered by technology, provides tools for understanding, monitoring, and protecting the rich tapestry of life on our planet. The adoption strategies outlined in each section underscore the importance of collaboration, education, and policy interventions. From incentivizing renewable energy investments to promoting circular economy practices, these strategies represent a call to action for individuals, communities, businesses, and governments alike. The collective effort required to integrate these innovations into our daily lives and societal structures is crucial for ensuring their efficacy in mitigating environmental challenges. In conclusion, the path to a greener future is illuminated by the innovative technologies explored in this chapter, but the journey is incomplete without widespread adoption. The success of these sustainability trends hinges on a global commitment to change, driven by informed decision-making, responsible consumption, and a shared vision of a world where technology and nature coalesce for the benefit of present and future generations. As we stand on the brink of transformative possibilities, the choices we make today will determine the trajectory of our planet's sustainability.

REFERENCES

1. Green, M. A., Hishikawa, Y., Dunlop, E. D., Levi, D. H., Hohl-Ebinger, J., Yoshita, M., & Ho-Baillie, A. W. Y. (2019). Solar cell efficiency tables (version 54). *Progress in Photovoltaics: Research and Applications*, 27(1), 3–12. https://doi.org/10.1002/pip.3102
2. Jacobson, M. Z., & Delucchi, M. A. (2011). Providing all global energy with wind, water, and solar power, Part I: Technologies, energy resources, quantities and areas of infrastructure, and materials. *Energy Policy*, 39(3), 1154–1169. https://doi.org/10.1016/j.enpol.2010.11.040
3. EMEC. (2020). *Wave and tidal energy in Europe: An overview.* EMEC Publications.
4. Optimizing wind energy production through artificial intelligence: A case study. *Renewable Energy Journal*, 15(2), 67–89.
5. Joseph, P. N. (2020). Hydroelectric power efficiency: Innovations in dam design. *International Journal of Sustainable Energy*, 12(4), 234–256.
6. Breakthroughs in battery technologies: A comprehensive survey. *Energy Storage Research*, 18(2), 78–102.
8. Smith, J., Brown, A., & Williams, C. (2022). Advancements in electric vehicle battery technologies. *Journal of Sustainable Transportation*, 10(3), 123–145.
9. Chen, H., Wang, Y., & Zhang, C. (2021). Hydrogen fuel cell vehicles: Technology advances and challenges. *International Journal of Hydrogen Energy*, 46(2), 789–812.
10. Johnson, R., & Garcia, M. (2019). Smart mobility solutions: A comprehensive review. *Transportation Research Part C*, 102, 201–220.
11. Johnson, A., Martin, B., & Davis, S. (2020). Sustainable aviation fuels: Current status and future prospects. *Environmental Science and Technology*, 54(13), 7753–7763.
12. Rodriguez, D. A., Brown, A. L., &Troped, P. J. (2018). *'Active Transportation.' the Oxford handbook of environmental and conservation psychology* (pp. 453–469). Springer.
13. Lee, S., & Smith, K. (2021). Policy approaches for sustainable transportation: Lessons from global case studies. *Transport Policy*, 108, 1–15.
15. Ellen MacArthur Foundation. (2015). *Towards the circular economy: Economic and business rationale for an accelerated transition.* Ellen MacArthur Foundation [Link to the report].
16. McDonough, W., & Braungart, M. (2002). *Cradle to cradle: Remaking the way we make things.* North Point Press.
17. Schenck, R., Sarkanen, K., & Leroux, K. (2019). *Extended producer responsibility: A guidance manual for governments.* The World Bank [Link to the manual].
18. Cucchiella, F., D'Adamo, I., & Lenny Koh, S. C. (2016). Recycling of WEEEs: An economic assessment of present and future E-waste streams. *Renewable and Sustainable Energy Reviews*, 60, 195–206.
19. Sundarakani, B., Laari, S., Khairuddin, H., & Mathiyazhagan, K. (2020). The role of digital technologies in circular supply chain management: A comprehensive review and future directions. *Sustainability*, 12(3), 856.
21. Giffinger, R., Fertner, C., Kramar, H., Kalasek, R., Pichler-Milanović, N., & Meijers, E. (2007). *Smart cities: Ranking of European medium-sized cities Centre of regional science.* Springer.
23. Hollands, R. G. (2008). Will the real smart city please stand up? *City*, 12(3), 303–320. https://doi.org/10.1080/13604810802479126

24. Caragliu, A., Del Bo, C., & Nijkamp, P. (2011). Smart cities in Europe. *Journal of Urban Technology*, 18(2), 65–82. https://doi.org/10.1080/10630732.2011.601117
25. Deloitte. (2020). *Smart cities and mobility*. Deloitte [Link to the report].
26. Batty, M., Axhausen, K. W., Giannotti, F., Pozdnoukhov, A., Bazzani, A., Wachowicz, M., Ouzounis, G., & Portugali, Y. (2012). Smart cities of the future. *European Physical Journal Special Topics*, 214(1), 481–518. https://doi.org/10.1140/epjst/e2012-01703–3
28. Wolfert, S., Ge, L., Verdouw, C., & Bogaardt, M. J. (2017). Big data in smart farming–A review. *Agricultural Systems*, 153, 69–80. https://doi.org/10.1016/j.agsy.2017.01.023
29. Sanyé-Mengual, E., Anguelovski, I., Oliver-Solà, J., Montero, J. I., & Rieradevall, J. (2015). Resolving differing stakeholder perceptions of urban rooftop farming in Mediterranean cities: Promoting food production as a driver for innovative forms of urban agriculture. *Agriculture and Human Values*, 32(2), 381–392.
30. Gebbers, R., & Adamchuk, V. I. (2010). Precision agriculture and food security. *Science*, 327(5967), 828–831. https://doi.org/10.1126/science.1183899
31. Sharma, A., Kaur, S., & Sharma, R. (2020). Smart agriculture using IoT: An efficient, self-sufficient approach in Indian farming. *Computers, Materials and Continua*, 62(1), 511–535.
32. Qrunfleh, S., & Tarafdar, M. (2014). Supply chain information systems strategy: Impacts on supply chain performance and firm performance. *International Journal of Production Economics*, 147, 340–350. https://doi.org/10.1016/j.ijpe.2012.09.018
33. United States Green Building Council. (2021). *LEED v4.1 building design and construction*. USGBC [Link to the LEED framework].
34. Pacheco-Torgal, F., Labrincha, J., Diamanti, M., De Brito, J., & Abdollahnejad, Z. (2017). Overview on the potential of alkali-activated binders to promote the use of industrial wastes in precast concrete manufacture. *Journal of Cleaner Production*, 142, 1875–1889.
35. Wong, J. K. W., Li, D. D. L., & Liu, Y. (2016). Review of sustainable building material technologies. *Procedia Engineering*, 145, 1147–1155.
36. Kellert, S. R., Heerwagen, J., & Mador, M. (2008). *Biophilic design: The theory, science, and practice of bringing buildings to life*. John Wiley & Sons.
37. Kenway, S. J., Priestley, A., Cook, S., & Seo, S. (2017). The energy-water nexus: Managing the links between energy and water for a sustainable future. *Ecosystem Health and Sustainability*, 3(5), e01225.
38. Lee, J. H., Kim, J. T., Kim, M. K., & Hong, T. (2016). An ontology-driven framework for real-time occupant-centered smart building operations. *Advanced Engineering Informatics*, 30(2), 169–182.
39. Pettorelli, N., Laurance, W. F., O'Brien, T. G., Wegmann, M., Nagendra, H., & Turner, W. (2014). Satellite remote sensing for applied ecologists: Opportunities and challenges. *Journal of Applied Ecology*, 51(4), 839–848. https://doi.org/10.1111/1365–2664.12261
40. Dickinson, J. L., Shirk, J., Bonter, D., Bonney, R., Crain, R. L., Martin, J., Phillips, T., & Purcell, K. (2012). The current state of citizen science as a tool for ecological research and public engagement. *Frontiers in Ecology and the Environment*, 10(6), 291–297. https://doi.org/10.1890/110236
41. Hebert, P. D., Cywinska, A., Ball, S. L., & DeWaard, J. R. (2003). Biological identifications through DNA barcodes. *Proceedings. Biological Sciences*, 270(1512), 313–321. https://doi.org/10.1098/rspb.2002.2218
42. McInerny, G. J., Roberts, D. L., Davy, A. J., & Cribb, P. J. (2017). Artificial intelligence for predicting the risk of alien plant invasions. *Journal of Applied Ecology*, 54(3), 701–708.
43. Anderson, K., & Gaston, K. J. (2013). Lightweight unmanned aerial vehicles will revolutionize spatial ecology. *Frontiers in Ecology and the Environment*, 11(3), 138–146. https://doi.org/10.1890/120150

44. MacDicken, K., Morgan, J., & Saatchi, S. (2021). Blockchain technology for transparency in the forest sector. *Frontiers in Blockchain*, 4, 589845.
27. Singh, P., Singh, M., & Bhambri, P. (2005a). Internet security. In Seminar on Network Security and Its Implementations (p. 22).
22. Singh, P., Singh, M., & Bhambri, P. (2005b). Security in virtual private networks. In Seminar on Network Security and Its Implementations (p. 11).
20. Bhambri, P., & Singh, M. (2005). Artificial intelligence. In Seminar on E-Governance - Pathway to Progress (p. 14).
14. Bhambri, P., Sinha, V. K., & Dhanoa, I. S. (2020). Development of cost effective PMS with efficient utilization of resources. *Journal of Critical Reviews*, 7(19), 781–786.
7. Sharma, R., Bhambri, P., & Sohal, A. K. (2020). Energy bio-inspired for MANET. *International Journal of Recent Technology and Engineering*, 8(6), 5580–5585.

22 Innovations for a Greener Future
Exploring Technological Sustainability Trends

Robert Kuceba and Grzegorz Chmielarz

22.1 INTRODUCTION

The concept of sustainable development has been subject to detailed analysis, discussion, and legislative and administrative provisions for more than fifty years. It continues to be updated and undergoes constant transformation, modification, and adaptation to contemporary challenges and the resulting multidimensional threats to the economy and, above all, to communities. The concept of sustainable development, in the international discussion on goals, objectives, and directions of global environmental policy, is believed to have been introduced for the first time at the Stockholm Conference in 1972. However, it was not defined in detail until 1987 in a report by the UN World Commission on Environment and Development entitled "The World Commission on Environment and Development: Our Common Future." This report emphasised the need for integrating action in three areas: conservation of natural resources, economic growth, and the equitable distribution of social benefits. Thus, sustainable development was defined as "economic and social development that will ensure that the needs of today's society are met without the risk that future generations will not be able to meet their needs, allowing it to choose its preferred lifestyle" (Brundtland, 1987).

A continuation of activities related to the definition of principles and ways to implement the concept of sustainable development was the adoption in 1992, at the Second United Nations Conference on Environment and Development in Rio de Janeiro, of the document Agenda 21. This document contained guidelines for the introduction of the principles of sustainable development into socio-economic life and constituted a programme of action for cooperation in the area of development and protecting the natural environment, to be implemented between 1993 and 2000. In later years, the term sustainable development appeared in many important documents, and the idea itself became a leading direction in economic as well as community development (Koszarek-Cyra et al., 2023).

DOI: 10.1201/9781003475989-26

Basically, in the definitions, the terms of sustainable development, three dimensions are distinguished (Rutkowska and Kozyra, 2007):

- economic efficiency, ensuring the profit of the organisation while taking into account social and environmental costs
- environmental efficiency, i.e., the conservation of non-renewable resources and the elimination or reduction of negative impacts on the environment
- social sustainability, creating new jobs, and increasing the quality of social welfare.

The Sustainable Development Goals and the associated 169 tasks, which represent economic, social, and environmental pillars of the sustainability concept, are recorded in the document: "Transforming Our World: The 2030 Agenda for Sustainable Development" commonly referred to as the 2030 Agenda for Sustainable Development, which was adopted by 193 UN member states in September 2015 in New York.

It should be stressed that the 2030 Agenda is focused around five profound transformational changes commonly referred to as the 5Ps principle (People, Planet, Prosperity, Peace, Partnership) (www.gov.pl, n.d.). The implementation of the 2030 Agenda can be ensured through a sustainability transition, a multi-level economic, technological, and social transformation, the ultimate goal of which is a "green" economy that can be defined as low-carbon and resource-efficient, where environmentfriendly solutions and eco-innovations constitute its basis (Koszarek-Cyra et al., 2023) and, to be emphasised in the context of the title of this chapter, innovations for a Greener Future. These aspects pertaining to the essence of innovations for a Greener Future are also taken into account in Directive 2013/34/EU of the European Parliament and of the Council with regard to sustainability reporting standards and, above all, the now complementary EU Commission Regulation 2023/2772, which deals with the ESG indicator assessment of the sustainability of contemporary organisations in three non-financial dimensions: E –Environmental, S – Social responsibility, and G – Corporate governance.

22.2 THE RELATIONSHIP BETWEEN ECOLOGICAL INNOVATIONS, KNOWN AS ECO-INNOVATIONS, THEIR DIFFUSION AND HETEROGENEOUS TRENDS IN SUSTAINABLE TECHNOLOGICAL DEVELOPMENT

New concepts of governance, geopolitical patterns in the world, limited access to fuels and energy in certain regions, environmental threats, military conflicts, pandemics, and more (...), stimulate the notion that even at the micro llevel — such as individual companies, organisations, or households – one should not limit oneself to purely economic effects or concern oneself only with one's own economic interests. In addition to satisfying their material needs, all these entities should generate tangible benefits for all potential stakeholders, local communities, and the environment, not just financial benefits but also intangible benefits. At the heart of this change is the transformation of the individual enterprise into a sustainable organisation. This means, as A. Pabian (Pabian, 2015) points out, the integration of environmental and social objectives into all areas of organisational management and the transformation of basic enterprise management functions such as planning, organising, leading, and controlling into their "sustainable" equivalents. This necessitates the creation of a new organisational structure in which every department and position has sustainable staff, sustainable goods and basic goods, and conducts pro-environmental and pro-social activities. In this type of organisation, the principles of sustainability are the basis for all key documents — from its articles of association to job descriptions and work instructions (Pabian and Pabian, 2014). In order to successfully implement the process of transforming an organisation into a sustainable organisation, it is necessary, first and foremost, to change the attitudes of both entrepreneurs and consumers, who must learn to function according to the idea of sustainable development (Pabian, 2013), while at the same time increasing their social capital.

It is important to mention that actions related to the transformation to more sustainable organisations were mainly undertaken by corporations and large companies. Gradually, smaller organisations also started to implement changes intended to bring environmental benefits. Today, companies are making, for example, through green innovations, an important contribution to a more sustainable society (Melander and Pazirandeh, 2019), which fully justifies the subject matter set out in the title of this chapter.

In an attempt to unify the factors influencing environmental attitudes and thus environmental awareness and, above all, decisions to introduce pro-environmental measures, the following should be distinguished:

- cultural values and ethical norms
- currents of socio-philosophical thought
- socio-political ideas pushed by political parties, the administrative and legal environment
- media influence (nowadays, particularly concerning social media in the crowdsourcing model)
- the degree of education in terms of dissemination of knowledge, formation of skills and natural, technical, and humanitarian competences – environmental knowledge
- financial and non-financial mechanisms to support environmental activities

Concentrating on the present discussion in the business sector, there is a long-term transformation in the environmental attitudes of organisations and thus their environmental awareness (Bhambri et al., 2019). This ranges from a state of complete disregard for environmental issues in the activities of business entities, to a gradual increase in commitment, and the introduction of environmental management into the overall corporate management system. When looking for the relationship between innovations for a Greener Future and sustainability trends, it is reasonable to point out a certain level of combination and even positive manipulation that takes place between their core business activities – the economic dimension and their so-called "green" activities – the environmental dimension; of course, when talking about these dimensions, social acceptance should be kept in mind. Of course, innovations for a Greener Future does not always enable addressing all the environmental and social problems that a "sustainable" enterprise, confronts in business practice.

In order to take into account the environmental impact of business operations, it is necessary to incorporate elements of environmental management into the management system of these organisations. A form of implementation of the environmental management function is the planning and execution of pro-environmental activities, including the ability to design and/or create or implement and/or operate in real business conditions innovations for a Greener Future, which are often identified with eco-innovations as indicated in the literature. The term eco-innovation was first used by C. Fussler and P. James in 1996 (Fussler and James, 1996). According to the authors, eco-innovations are new or improved products and processes contributing a significant added value to business and customers while being at the same time less harmful to the environment. Until the present day, a number of definitions of eco-innovation have appeared in the literature. Referring to the title of this chapter and, in particular, the relationship between innovations for a Greener Future and Technological Sustainability Trends, one can recall, among others, the definition of L. Levidow and his research team that these are "Innovative practices that improve resource efficiency by combining economic value with environmental protection" (Levidow et al., 2016). In an attempt to define, perceive, unify, and diagnose innovations for a Greener Future, it is advisable to resort to the definition of Oslo Manual 2018 (OECD/Eurostat, 2018). In principle, what is important in this definition is the division of innovation in general into (1) product innovations – new or improved products or services that include significant modifications compared to previous goods or services introduced to the market by economic organisations, and (2) business process innovations, which are an extension of previously narrowly identified technological (process) innovations. Apart from innovations resulting from the introduction of new technologies into the production system, business processes also embrace innovative actions in the area of logistics and distribution, improved operations of the marketing and sales departments, improved functionality of ICT systems, better governance schemes, more efficient organisation and management, maintaining at the same time rigorous compliance with environmental laws and contributing to social welfare, which is fully compliant with sustainability principles. Eco-innovations, which we can describe as innovations for a Greener Future, can take the form of slight alterations of organisational routines or even profound transformations across the entire organisation. It should therefore be clearly emphasised that although eco-innovation is often associated with a shift in manufacturing processes and the use of emerging technologies, it may comprise such changes as the introduction of new or significantly modified products (goods or services), organisational improvements, or marketing practices that minimise the use of natural resources such as materials, energy, water, and land, and curb emissions of substances that are a menace to the atmosphere (Miedzinski et al., 2013). It is important to emphasise that the intention to innovate for a Greener Future does not always accompany the introduction of changes in a company, obviously in line with the dimensions of Technological Sustainability.

Eco-innovations that can be introduced into the manufacturing process are commonly broken down into the following groups (Flis, 2010):

- eco-innovations leading to products with new environmental parameters (causing much less pressure on the environment)
- eco-innovations resulting in unchanged product parameters but less raw materials and energy in the production process
- eco-innovations that reduce the environmental impact of both the product and the production process

When discussing technology trends, it also needs to be indicated that the key directions of innovation and sustainability in the technology area will now focus on four fundamental forces: technology business, smart technology, technology greening, and cyber security and trust. Examples of innovations for Greener Futures that feed into these tech trends are possibly: (1) interfaces in new spheres, such as spatial computing and the industrial metaverse, (2) GenAI coding, which can mimic human cognitive abilities in various ways and replace existing machine learning, (3) the creation of digital twins of real-world environments, (4) smart electricity grids, (5) smart vehicle charging grids, (6) improving the standard and consistency of data used by society using artificial intelligence, (7) monitoring and assessing carbon footprints using new information technologies, e.g., IoT, Power over Ethernet, visibility dashboards, or artificial intelligence, and (8) green and digital business transformations.

22.3 ASSESSMENT OF ECO-INNOVATION NEEDS AND DETERMINANTS OF SUSTAINABLE TECHNOLOGICAL DEVELOPMENT OF SMES IN THE BALTIC SEA REGION

On the initiative of the European Union and with its funding, projects are being set up with the aim of promoting good practice and supporting the implementation of environment-friendly solutions combined with sustainable growth of SME potential. One such project is the ECOLABNET international network for developing eco-innovation services, co-created by the authors of this chapter. The basis for the project was the poor access to relevant services and expertise in the field of eco-innovation, especially the technological development of SMEs. The project is intended to be a conglomerate of research and business activities. In this way, it creates new business opportunities for those companies that wish to establish themselves in competitive European and, in the long term, global markets. Such companies, in order to meet growing consumer demands for sustainable products and services, but also to differentiate themselves from the competition, decide to include new products and services, frequently eco-innovative ones, in their product range (Kaur et al., 2019). Therefore, by uniting a network of research institutions from the partner countries, the ECOLABNET project supports SMEs in the Baltic Sea region and promotes their attempts to incorporate green solutions. In order to practically demonstrate to SMEs the competence of ECOLABNET in supporting the implementation and commercialisation opportunities of newly developed green ideas, the network participants are developing prototypes. The idea behind the establishment of such a conglomerate, which includes entities from three key areas: entrepreneurs from the SME sector, intermediary organisations, and R&D units, is to provide multiple advantages for all involved actors and greater efficiency than projects that simply seek EU's funding. The advantages of developing innovation in the form of a network concept such as the ECOLABNET project should also be highlighted. In this case, the collaboration of local organisations and international ones results in a greater contribution of valuable tangible and intangible resources. According to Ferreiro and Lourenco, networks consisting of non-local partners prove their efficacy in knowledge acquisition and diffusion. Moreover, they result in attracting those resources that did not exist or were scarce in a given area and in scaling innovation (Ferreiro et al., 2019). Similar conclusions are drawn by Dezi et al., according to whom a key way for businesses to have an advantageous position in the market is to harness the potential of external networks, such as the ECOLABNET network described above, to improve their performance (Dezi et al., 2019).

The ECOLABNET network described in this chapter consists of 11 partners from 6 countries. The network participants analyse the needs of SMEs and intermediary organisations supporting the business sector with regard to the expertise necessary for the successful introduction of eco-innovation (Interreg, n.d.). The project was implemented between 2019 and 2021 under the EU Interreg programme, and the operation of the project results must be ensured for another 5 years until 2026, with one of the partners taking the lead in each subsequent year. Funding for the programme is provided by the European Union and approved by the European Commission.

The main tasks implemented in the ECOLABNET project included the following:

- assessing market position and analysing the organisation's innovative capabilities, taking into account dimensions of sustainability
- preparation of research and development work by partners
- search for eco-innovative solutions by partners
- conceptual work of partners on eco-innovation modelling
- design of eco-innovations for partners
- testing of developed eco-innovations by partners
- preparation and launch of pilot production of eco-innovation solutions in line with new trends of sustainable technological development
- improvement of eco-innovations and search for new green ideas

The Computer-Assisted Web Interview (CAWI) method was applied in part of the research. The research tool was the survey questionnaire that included 27 questions concentrated around five theme groups. They are included in Table 22.1.

In the course of the survey, the respondents from the investigated region were supposed to rate on a five-point Likert scale (1–5) the main factors stimulating their

TABLE 22.1
Thematic Scope of the Survey Questionnaire

Theme Group	Subject of the Analysis
Questions 1–5	Basic information about the company under investigation and its business profile.
Questions 6 to 15	Specifically pertaining to eco-innovations and trends in sustainable technological development, and in particular on knowledge of eco-innovation, stimulants, and inhibitors to eco-innovation signaled by the SMEs from the Baltic Sea Region.
Questions 16 to 21	Aimed at investigating whether the survey participants know about 3D printing as a green tool to be used in production processes instead of previously used solutions contributing to environmental degradation.
Questions 23 to 24	Investigation of the need for expertise and whether it can be met by the project partners.
Questions 25 to 27	Possibilities for future cooperation and address details.

Source: Own elaboration.

eco-innovation activity and the barriers hindering their involvement in such activity. In the second stage of the project's knowledge management process, respondents' answers were analysed and, on the basis of this analysis, the stimulants, eco-innovation needs, and barriers to eco-innovation initiatives in the surveyed SME sector were identified. The research population consisted of N = 296 entities.

As a first step in the research process, the respondents selected and evaluated the stimulants of their eco-innovative involvement with the use of a five-point Likert scale (Table 22.2). In order to map the relevance of the individual stimulants in Table 22.2, a weighted value system and a weighted average were used. In this way, values with higher weights are more important in calculating the weighted average than values with lower weights.

As mentioned earlier, the data collected in the course of the survey allowed the partners to gain insights into the eco-innovative needs of the investigated enterprises and the inhibitors to developing and commercialising environment-friendly products and services.

The needs of the surveyed SMEs in the BSR region were determined in four main areas of sustainable technological development of these enterprises: business, development, and technology. The so-categorised eco-innovative needs of SMEs were evaluated using a three-stage needs assessment: no signaled needs, potential needs, indicated needs. The findings of this research for the whole surveyed population of SMEs in the BSR partner countries are summarised in Table 22.3.

As one can observe in Table 22.3, the "Baltic" SMEs seek support in the following areas: Branding (Indicated needs 35.85% - 76 and potential needs 34.43%– 73), Customer insights (Indicated needs 32.55% – 69 and potential needs 41.51% – 88), and Financial aspects (Indicated needs 30.66% – 65 and potential needs 38.21% – 81). It can be concluded, based on the indication of the BSR entrepreneurs, that their primary need is funding. However, they also seek assistance in implementing

TABLE 22.2

Drivers of Eco-innovative Activities among SMEs in the Baltic Sea Region (Likert scale where 1 = absolutely no, 5 = absolutely yes and weighted average)

Stimulating Factor	Weight 1	Weight 2	Wight 3	Weight 4	Weight 5	Weighted Average
Meeting customer needs	1.28% (3)	4.26% (10)	12.77% (30)	28.51% (67)	53.19% (125)	4.28
Efficient use of resources	1.28% (3)	2.98% (7)	8.09% (19)	37.45% (88)	50.21% (118)	4.32
Strengthening brand image	2.98% (7)	4.68% (11)	13.19% (31)	35.74% (84)	43.40% (102)	4.11
Compliance with legislation	4.68% (11)	4.26% (10)	15.74% (37)	33.19% (78)	42.13% (99)	4.03
Differentiation from the competition	2.98% (7)	7.66% (18)	14.47% (34)	34.89% (82)	40.00% (94)	4.01
Minimising the impact of the environmental effect on business operations	5.11% (12)	5.11% (12)	20.85% (49)	30.64% (72)	38.30% (90)	3.91
Decreased cost	2.55% (6)	11.49% (27)	16.17% (38)	31.91% (75)	37.87% (89)	3,91
Future opportunities for making business	5.53% (13)	8.09% (19)	14.89% (35)	34.47% (81)	37.02% (87)	3.89
Attracting competent staff	6.38% (15)	10.64% (25)	29.36% (69)	28.51% (67)	25.11% (59)	3.55
Meeting stakeholder expectations	6.38% (15)	12.34% (29)	25.11% (59)	31.49% (74)	24.68% (58)	3.42
Increased transparency	7.23% (17)	13.62% (32)	31.49% (74)	24.68% (58)	22.98% (54)	3.42
Attracting capital investment	17.45% (41)	18.30% (43)	24.26% (57)	21.28% (50)	18.72% (44)	3.05

Source: Author's analysis based on the survey carried out in the ECOLABNET project (Koszarek-Cyra et al., 2023)

Total number of responses N = 296

According to Table 22.2, entrepreneurs in the BSR countries covered by the survey point to meeting customer needs (53.19% – 125 indications) as the main factor driving them to introduce and develop eco-innovations. In the second place of the stimulants for eco-innovation activities among the surveyed enterprises is using resources efficiently (50.21% – 118 indications), and the third place of the ranking of stimulants for eco-innovation activities is occupied by strengthening the brand image (43.40% – 102 indications). The next crucial factor for developing eco-innovation, which was highlighted by the investigated SMEs is to comply with law regulations (42.13% – 99 indications). This shows the importance of state legal regulation for companies that decide to engage in eco-innovation activities. Further drivers of eco-innovation that were rated highly by respondents include: being different from competitors (40% – 94 indications), reduced environmental impact on business (38.30% – 90 indications), decrease in cost (37.87% – 89 indications), and future opportunities for making business (37.02% – 87 indications). The following stimulants were considered less important, according to the respondents: ability to attract competent employees, satisfying expectations of stakeholders, and increasing transparency (25.11% – 59; 24.68% – 58; and 22.98% – 54 indications, respectively). The respondents of the survey from the BSR countries considered attracting capital investment, which received 18.72% – 44 indications, as the least important stimulus for eco-innovation initiatives for them.

TABLE 22.3
Eco-innovation Needs of SMEs in BSR Countries

Company Needs within Eco-innovation	Area of Needs	No Signalled Needs	Potential Needs	Indicated Needs
Branding	Business	29.72% (63)	34.43% (73)	35.85% (76)
Business structure		34.91% (74)	48.58% (103)	16.51% (35)
Customer insights		25.94% (55)	41.51% (88)	32.55% (69)
Financial aspects		31.13% (66)	38.21% (81)	30.66% (65)
Intangible and legal assets (e.g. IPR)		35.38% (75)	40.57% (86)	24.06% (51)
Legislation		36.32% (77)	40.57% (86)	23.11% (49)
Supplier relations		43.40% (92)	38.68% (82)	17.92% (38)
Value chain assessment		40.57% (86)	43.87% (93)	15.57% (33)
Life-cycle assessment	Design product/ services	41.98% (89)	38.21% (81)	19.81% (42)
Packaging development		39.15% (83)	37.26% (79)	23.58% (50)
Design		35.38% (75)	32.55% (69)	32.08% (68)
Service		36.32% (77)	44.34% (94)	19.34% (41)
Biocomposites material	Technology	55.66% (118)	31.13% (66)	13.21% (28)
Bioresins		64.15% (136)	26.42% (56)	9.43% (20)
3D printing		55.19% (117)	28.77% (61)	16.04% (34)
Certifications		28.30% (60)	44.81% (95)	26.89% (57)
Energy optimization		18.40% (39)	19.81% (42)	61.79% (131)
Material efficiency		37.26% (79)	34.91% (74)	27.83% (59)
Other alternative materials		37.26% (79)	34.91% (74)	27.83% (59)
Recycling		34.43% (73)	38.21% (81)	27.36% (58)

Source: Own elaboration based on a survey carried out in the ECOLABNET project
Total responses N = 296, Answered: 212, Skipped: 84
(…) – number of indications

business procedures, including those that involve marketing analyses or categorising customers. Importantly, the SME entrepreneurs also signal their needs in the area of developing eco-innovative products and services (Indicated needs 32.08% – 68 and potential needs 32.55% – 69). It is interesting to note that they attach less importance to life-cycle assessment (LCA). This need is only indicated by companies where eco-innovation is part of the company's strategy or even mission. These needs arise from the total or limited access to expertise, including management, especially in environmental management. Referring to technology/production needs, the high activity of the BSR companies in the area of Saving Energy has been entirely confirmed. The research carried out by the authors shows that 61.79% (131) of the investigated SMEs seek expert advice on energy optimisation. In contrast, 19.81% (42) of the survey participants consider support for this optimisation as obtainable. In turn, in the case of biocomposite materials, bioresins, or 3D printing, the "Baltic" SMEs

TABLE 22.4
Inhibitors to the Introduction and Development of Eco-innovations as Perceived by SMEs in the Baltic Sea Region (Likert scale 1 – 5, where 1 = weak inhibition and 5 = very strong inhibition, number of enterprises)

Inhibitor	1 Point	2 Points	3 Points	4 Points	5 Points
Inappropriate tools and no appropriate methods	20	56	86	77	57
Insufficient knowledge and need for expertise	33	65	94	65	39
No in-house technology	25	56	90	78	47
Limited personal experience	27	53	88	83	45
Legal compliance	41	60	74	83	38

Source: Own analysis based on a survey carried out in the ECOLABNET project
Total number of responses N = 296.

indicate their low interest in these tools -Table 22.4). It seems that enterprises signal these needs only if eco-innovations are deeply rooted in the mission or/and strategy of a given enterprise.

In the context of business development and eco-innovation needs indicated by the SMEs, inhibitors to implementing and developing eco-innovations were indicated and evaluated on a five-point scale. The main inhibitors identified during the survey in this area are summarised in Table 22.4.

As Table 22.4 demonstrates, the surveyed BSR enterprises indicate the use of inappropriate tools and methods as the main inhibitor to engaging in eco-innovative activities. The second most indicated inhibitor to developing eco-innovative initiatives is limited knowledge and demand for external expertise. The third highlighted inhibitor to implementing eco-innovative activities is the lack of in-house technologies, and the lack of in-house experience in this area ranks fourth. In addition, legal requirements in the form of constantly amended legislation were identified as an inhibiting factor that makes enterprises in the six countries constituting the ECOLABNET network refrain from being more active in the area of implementing eco-innovations. Particularly relevant to the objectives of the ECOLABNET project seems to be the fact that, among the inhibitors aggregated in Table 22.4, insufficient own knowledge and the demand for external expertise rank high. This means that in the analysed sector, there is a particular gap with regard to expertise possessed by R&D actors that is in demand by SMEs. By sharing their expertise, scientific units can effectively stimulate the entrepreneurs from the SME sector to increase their commitment to introducing new, largely eco-innovative concepts and ideas. Therefore, networks consisting of providers of expert knowledge (R&D actors) that can cater to the specific needs of SME sector entrepreneurs can contribute greatly to improvement in their business performance, which can lead to a competitive advantage in business (Chmielarz et al., 2020). Eliminating or reducing this barrier was one of the overarching goals of the ECOLABNET project. A measurable value was the development of a repository of good practices and innovations for a Greener Future DCT.

22.4 REPOSITORY OF GOOD PRACTICES INNOVATIONS FOR A GREENER FUTURE – DCT, IN THE CONTEXT OF SUSTAINABLE TECHNOLOGY TRENDS

The Digital Collaboration Tool (DCT) developed within the ECOLABNET project is one of the most important milestones and results of the project. It is a system where eco-innovative products and services are integrated within a website, so as to facilitate finding matches between the entities providing and seeking eco-innovative solutions. The system foresees collaboration between all the stakeholders: SMEs, IOs, and R&D actors. All of them can be both offerors and recipients of the eco-innovative offerings on the website. In this way, the developed can provide reliable and detailed information on eco-innovative solutions. Recipients and users of the DCT are small- and medium-sized enterprises (commercial, technological departments), research and development units, and intermediary organisations such as chambers of industry and commerce. The central element of the DCT is the ECOLABNET knowledge repository, the structure of which is shown in Figure 22.1.

According to Figure 22.1, the main knowledge repository in the ECOLABNET project is a database of eco-innovative needs of enterprises in the BSR identified during the survey conducted in phase one of the project. In addition, this database is supplemented with information on the range of eco-innovative products and services being offered within the project consortium, as well as on the service providers

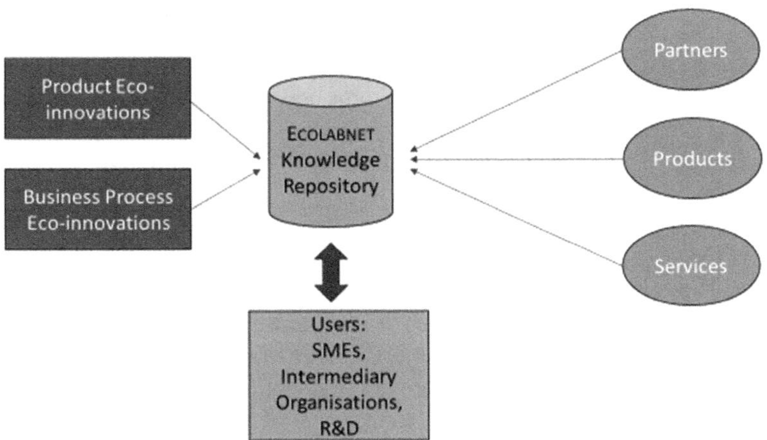

FIGURE 22.1 Main elements of the ECOLABNET knowledge repository. Source: (Kucęba, 2021)

themselves. Another element of the knowledge repository is information on all organisations registered in the database, i.e., SMEs, intermediary organisations, and R&D entities. The eco-innovation offer itself, available in the service, is presented broken down into product eco-innovations and business process eco-innovations. Access to the resources of the ECOLABNET knowledge repository is enabled by built-in search mechanisms based on the SQL language.

In accordance with the project assumptions, the DCT website was equipped with modular functionality, which reflects the needs of SMEs diagnosed in the conducted survey in relation to the eco-innovative needs of this sector. The structure of the created service includes eight functional modules, which were grouped into three types of categories (Kucęba et al., 2022):

- information modules – news module, best practices module
- operational layer modules – user module, product module, service module
- analytical layer modules – eco-innovation product and service search and matching module, survey analysis module, reporting module

Brief characteristics of the functionalities of the individual modules of the DCT tool are presented in Table 22.5.

The functionalities of the DCT website are available in a limited form for non-registered users – the possibility to browse the content of the website without the possibility of adding information on their own offer of eco-innovative products and services. Only registered users gain full access to all functionalities of the website, which means the possibility of adding information about their own eco-innovative offers. In addition, registered users gain access to the News module, where they can monitor new organisations, products, and services added to the service in real time and search News for historical data. The role of the next module, best practices, is to promote information about the best eco-innovative products and services, in consonance with the principle that examples of successful implementations of eco-innovative solutions are the best showcase for the effectiveness of ECOLABNET. The three modules of the operational layer are closely linked to each other. The first of these, the user module, allows an organisation to register on the DCT website. In the registration process, the user assigns his/her organisation to the three main types of organisations defined in this module, at the same time defining the activity profile of the registered organisation. It should be emphasised that the person registering their own organisation on the site automatically becomes an administrator on behalf of their organisation, and this person gains the right to modify the data concerning their organisation on the site. After the registration process, users also gain the possibility of evaluating the eco-innovative offers of other organisations on the website, excluding their own organisation. After completing the registration process, users of the organisation can add information about the eco-innovation products and services offered to the knowledge repository. An example of the possibility of adding eco-innovation products or services on the DCT website is shown in Figure 22.2.

As Figure 22.2 shows, adding new eco-innovative products or services is possible by selecting My account. In the menu visible on the left-hand side are the Products/

TABLE 22.5
Functional Modules of the DCT Website

Module Name	Description of Module Functionalities
1. News (subpage)	The basic functionality of the module makes it possible to present a range of modifications that can be executed with regard to the content of them, such as new user, new product, new service, new technology, etc.
2. Best practices (subpage)	The module allows for the presentation of the so-called success stories from the process of developing or implementing eco-innovative products or services. The best case is a subpage of the website with content and accompanying visual elements.
3. User module	The module allows a two-stage registration to the DCT service – pre-registration and full registration. There are three basic categories of users in the system: • small and medium-sized enterprise (SME) • intermediary organisation (IO) • research and development organisation (R&D)
4. Product module	The module allows the registration of an eco-innovative product on the website using the product form – the model is the product sheet.
5. Service module	The module allows the registration of an eco-innovation service on the website using the product form – the model is the service charter.
6. Search and match module for eco-innovative products and services	The module enables generating user matches while taking into account their varying roles in the system as well as goals to be achieved. These are generated using the data introduced in the registration form. In particular, based on selected algorithms, it is possible to search for and associate eco-innovative products and services between interested parties or to create networks of relations based on selected criteria (e.g., in the form of a dynamic graph of relations).
7. Survey module	The module is used to identify the user's attitude towards eco-innovative concepts. Identification of users' preferences and determination of the level of involvement in eco-innovation activities are carried out using the Personas concept. It is conducted on the basis of questions developed as part of a questionnaire, the result of which is the assignment of the user to a specific group.
8. Module for exporting to pdf format	The module's functionality makes it possible to obtain in the form of a pdf file: • the user card • the charter of an eco-innovation product • eco-innovation service card • user survey scorecards • a linkage graph in which one of the nodes is the user concerned

Source: Own elaboration based on internal materials of the ECOLABNET project

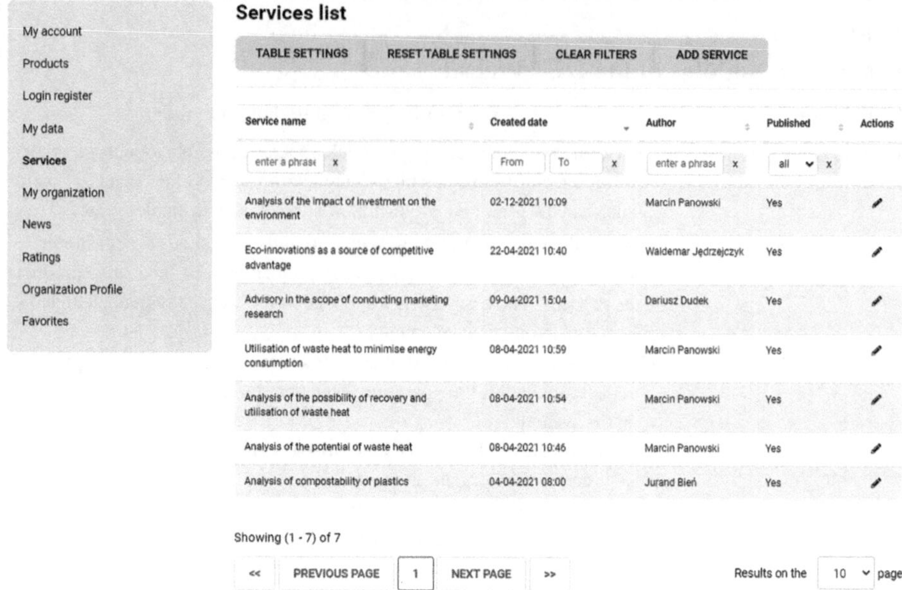

FIGURE 22.2 Example of adding an eco-innovative product or service to the DCT website. Source: screenshot from the DCT website.

Services options. Selecting one of the options takes you to a list of products and services already on the site, together with information about the person who added the service, the date the products and services were added to the site, and their status – whether the products and services have been published, i.e. whether they are visible to other users of the site. The Actions option allows you to edit or delete products and services already in the system, but this can only be done by the person who added the product or service to the service. After selecting the Add Service/Add Product option, it is possible to add new eco-innovative products and services to the DCT website. Once the information on offered products and services is complete and the option to publish them is selected, they become available to all users of the service. The next module, which enables the users to search for and match eco-innovative products and services, includes one of the key functions of the service, i.e. creating connections between providers of eco-innovative services and products and organisations that are looking for them. An example of a search window for eco-innovative products and services is presented in Figure 22.3.

As can be seen from Figure 22.3, the search and matching of products and services on the site can be limited by filters — the Sort by option — and the information found can be printed out. In addition, a list of top-rated products and services that may be of interest to users using the website also appears in the search window for eco-innovative products and services. Another module of the DCT website is the survey analytics module. It applies the Personas concept to determine the level of

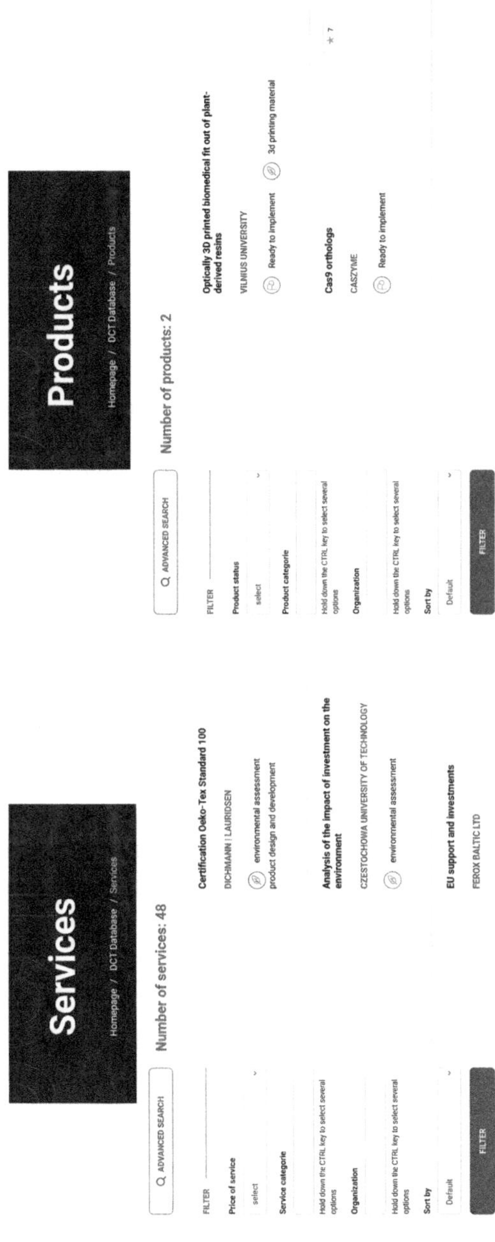

FIGURE 22.3 Search for eco-innovative products and services on the DCT website. Source: screenshot from the DCT website.

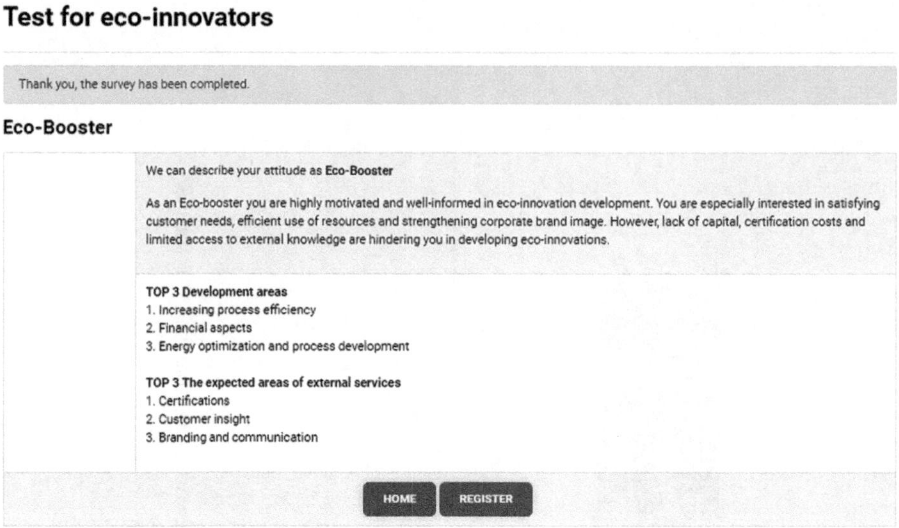

Test for eco-innovators

Thank you, the survey has been completed.

Eco-Booster

We can describe your attitude as **Eco-Booster**

As an Eco-booster you are highly motivated and well-informed in eco-innovation development. You are especially interested in satisfying customer needs, efficient use of resources and strengthening corporate brand image. However, lack of capital, certification costs and limited access to external knowledge are hindering you in developing eco-innovations.

TOP 3 Development areas
1. Increasing process efficiency
2. Financial aspects
3. Energy optimization and process development

TOP 3 The expected areas of external services
1. Certifications
2. Customer insight
3. Branding and communication

HOME REGISTER

FIGURE 22.4 Result of the survey determining the level of involvement in eco-innovations on the DCT website. Source: screenshot from the DCT website.

engagement in eco-innovation activities. An example of a survey result is included in Figure 22.4.

In order to determine the level of involvement in eco-innovative activities, users participating in the survey have to select possible options for answering the six questions that make up the survey. As a result of the survey, the organisation is assigned to one of the six types of personas defined in the project. In this way, the organisation is not only informed about its level of knowledge about eco-innovative activities, but also about the main barriers to implementing eco-innovative solutions and the service packages offered to help overcome them.

The last module of the DCT service is reporting, which provides all statistics on the service users and offerings, with the possibility to export and further use them.

It should be noted that between 19.09.2021 and 07.11.2023, the service had 4,798 page views, while the active users on the demand and supply side of the registered eco-innovation solutions are 584, including 457 who explicitly indicated their country of origin (Figure 22.5).

A very important element of the service is the possibility to observe, on the basis of introduced innovations for a Greener Future, trends in sustainable technological development. In particular, this is based on the basis of the analysis of introduced eco-innovative solutions. These particularly concern 3D printing products from bio-components dedicated to the packaging industry, food industry, and medical and dental products. The second area is smart solutions for local energy RES and smart vehicle charging stations. All the products, technological solutions, and also eco-innovative services offered by the DCT service can be clearly attributed to the

KRAJ	UŻYTKOWNICY
Poland	179
China	106
Finland	63
Lithuania	52
United States	21
Denmark	19
Germany	17

FIGURE 22.5 User structure on the DCT website. Source: screenshot from the DCT website.

property of green transformation and technological greening, which are distinctive features of sustainable technological development.

22.5 SUMMARY

Based on the discourse conducted in this chapter, and also on the empirical research carried out within the ECOLABNET project, it is clear that innovations for a greener future and technological sustainability trends are mutually linked. It means that it is of key importance to actively and dynamically involve entrepreneurs and employees of organisations to develop their own skills and competencies in terms of recognising, evaluating, and designing solutions for innovations for a greener future. Networking in the implementation of innovations for a greener future with the simultaneous sustainable technological development of enterprises dynamizes the transfer, diffusion, and adaptation of these innovations in economic spaces. This is fully compliant with strategies of sustainability on a macro scale (worldwide, EU, national strategies) but also on a micro scale (enterprise strategies).

The scientific discourse conducted in the first part of this chapter confirms the postulate that the key directions of innovations for a greener future and sustainable development in the field of technology are currently focused primarily on four fundamental forces: technological business, smart technologies, technological greening, and, undoubtedly, cybersecurity. This has been confirmed by both the conducted research and the analysis of the resources of the DCT repository of good practices of innovations for a greener future. Therefore, it can be concluded that innovations for a greener future reflect the "green activity" of contemporary companies and, at the same time, are an entrepreneurial tool that gives these organisations new opportunities to create sustainable value, including technological ones.

REFERENCES

Bhambri, P., Sinha, V. K., Dhanoa, I. S., Kaur, J. (2019). Genome DNA Sequence Matching using HBM Algorithm, *International Journal of Control and Automation*, Vol. 12, No. 5, pp. 531–539.

Brundtland G.H. (1987). *Our Common Future: Report of the World Commission on Environment and Development*, Geneva, UN-Dokument A/42/427. Springer.

Chmielarz G., Kucęba R., Nielsen T. (2020). Drivers of Developing Eco-Innovations in Manufacturing SMEs from the BSR Region, in: Soliman Khalid S. (ed.), *Education Excellence and Innovation Management: A 2025 Vision to Sustain Economic Development during Global Challenges*, pp. 2521–2531.Springer.

Dezi L., Ferraris A., Papa A., Vrontis D. (2019). The Role of External Embeddedness and Knowledge Management as Antecedents of Ambidexterity and Performances in Italian SMEs, *IEEE Transactions on Engineering Management*, pp. 1–10.

Ferreiro M., Sousa C., Lourenço C. (2019). Social Innovation and Networks in Rural Territories: The Case of EPAM, in: Tome E., et al. (eds.), *Proceedings of the 20th European Conference on Knowledge Management (ECKM 2019)*, Academic Conferences Ltd, Vol 35, pp. 247–252. New York.

Flis R. (2010). Ekoinnowacyjność produktów i usług, in: Woźniak L., Strojny E., Wojnicka E. (eds.), *Ekoinnowacje w praktyce funkcjonowania MŚP*, PARP, Warszawa, p. 101.

Fussler C., James T. (1996). *Driving Eco-Innovation: A Breakthrough Discipline for Innovation and Sustainability*, Pearson Education, London.

Interreg – Ecolabnet (n.d) https://projects.interreg-baltic.eu/projects/ECOLABNET-199.html

Kaur, J., Bhambri, P., Kaur, S. (2019). SVM Classifier Based Method for Software Defect Prediction, *International Journal of Analytical and Experimental Model Analysis*, Vol. 11, No.10, pp. 2772–2776.

Koszarek-Cyra A., Kucęba R., Chmielarz G. (2023). *MSP kluczowy sektor dla ekoinnowacyjności. Wyd. Dom Organizatora*, Towarzystwo Naukowe Organizacji i Kierownictwa, Toruń.

Kucęba R. (2021). System for Transfer and Management of Knowledge on Eco-Innovative Services for SMES in the Network of Service Partners Ecolabnet: A Case Study, Proceedings of the 22nd European Conference on Knowledge Management (ed.), A. Garcia-Perez, L. Simkin, pp. 440–449.

Kucęba R., Jędrzejczyk W., Kulej-Dudek E., Chmielarz G., Dudek D. (2022). Cyfrowe narzędzie współpracy w sieci ECOLABNET jako element systemu zarządzania i transferu wiedzy o usługach ekoinnowacyjnych, in: Dziembek D. (ed.), *Wiedza i technologie informacyjne w zarządzaniu przedsiębiorstwem*, Wydawnictwo Politechniki Częstochowskiej, pp. 152–169.New York.

Levidow L., Lindgaard-Jørgensen P., Nilsson A., Skenhall S. A., Assimacopoulos D. (2016). Process Ecoinnovation: Assessing Meso-level Eco-efficiency in Industrial Water-service Systems, *Journal of Cleaner Production*, Vol. 110, pp. 54–65.

Melander L., Pazirandeh A. (2019). Collaboration Beyond the Supply Network for Green Innovation: Insight from 11 Cases, *Supply Chain Management: An International Journal*, Vol. 24, pp. 509–523.

Miedzinski M., Charter M., O'Brien M. (2013). Eco-Innovate! A Guide to Eco-innovation for SMEs and Business Coaches http://www.eco-innovation.eu/images/stories/Reports /sme_eco-innovation_guide_2nd_edition_small.pdf

OECD/Eurostat (2018). Oslo Manual 2018: Guidelines for Collecting, Reporting and Using Data on Innovation, 4th Edition, The Measurement of Scientific, Technological and Innovation Activities, OECD Publishing, Paris/Eurostat, Luxemburg.

Pabian A. (2013). Koncepcja sustainability w działalności ośrodków edukacyjnych, IX Congress of Polish Economists, Ekonomia dla przyszłości, Odkrywać naturę i przyczyny zjawisk gospodarczych, Warszawa.

Pabian A. (2015). Zrównoważone zarządzanie zasobami ludzkimi - zarys problematyki, Zeszyty Naukowe Politechniki Częstochowskiej, *Zarządzanie*, Vol. 33, No. 17, pp. 7–16.

Pabian A., Pabian B. (2014). Sustainable Management of an Enterprise - Functional Approach, *PolishJournal of Management Studies*, Vol. 10, No. 1, pp. 98–107.

Rutkowska M., Kozyra P. (2007). Problemy zrównoważonego rozwoju a ubezpieczenia ekologiczne.Folia Universitatis Agriculturae Stetinensis, *Oeconomica*, Vol. 45, No. 49, p. 159.

23 Future Trends in Technological Sustainability

Ilona Pawełoszek

23.1 INTRODUCTION

In the broadest sense, sustainability refers to the continuing existence or actions over a period of time. In a business context, sustainable actions aim to prevent the depletion of natural or physical resources by reducing negative environmental impacts resulting from business operations. As has been highlighted in previous chapters, sustainable development harmonizes three essential constituents: economic development, social benefits, and environmental protection. These elements are inherent and vital for the long-term well-being of individuals, societies, and economies.

In the realm of technological sustainability, the imperative is to anticipate and comprehend future trends. Technological innovations are the driving force of human progress. Technology has significantly impacted world history for centuries, from the agricultural to the Industrial Revolution. However, not all technological changes have been beneficial. Some have caused environmental pollution, job losses, or social inequality. It is worth noting that current technologies and innovations are diffusing much more quickly and widely than their predecessors.

New technologies usually require some adoption time. In the case of the Internet, it took two decades for it to reach worldwide implementation. In the contemporary era, we have witnessed the emergence of Generative AI, which revolutionizes human–computer interaction, transforming the workforce, business models, and education on many levels. However, adopting the most recognized application – Chat GPT, was extremely fast. It gained 1 million users in 5 days, taking Netflix for 3.5 years, Facebook 10 months, and Spotify 5 months. We can suppose that this spectacular success was not a one-off episode, and subsequent technological innovations will also come this fast.

Disruptive innovations that make our lives easier can also be a source of long-term, severe environmental problems. That's why, nowadays, predicting technological progress is even more critical to consider while shaping the directions of economic and social development.

 DOI: 10.1201/9781003475989-27

23.2 ANTICIPATING TECHNOLOGICAL TRENDS

Anticipating future technological trends from the sustainability perspective is crucial to ensure a harmonious future for the next generations and the Earth's ecosystem. By understanding and shaping the trajectory of technological advancements, we can steer them toward solutions that address current and future environmental and social challenges. This proactive approach not only mitigates potential negative impacts but also paves the way for transformative innovations that foster a sustainable and equitable world.

Extrapolating from today's trends and emerging technologies is the starting point for reading the future. Analysis can be done using many data sources (Mikova & Sokolova, 2019), methods, and tools (Segev et al., 2015).

As technical developments are crucial for combating climate change, one of the promising research methods is patent landscape analysis. The content of a patent is publicly available information. The bargain of a patent is that its owner provides information about the invention to the public.

Searching through and using this information is vital to researchers and inventors. Also, entrepreneurs may use this knowledge to identify potential breakthrough areas and make more informed decisions on investments in a particular technology. By analyzing patent filings, it is possible to identify areas of active research and development in sustainable technologies. Figure 23.1 presents the example of a year-to-year analysis of the number of patents in the field of hydrogen production based on the production process. It is easy to see that water electrolysis is one of the most promising methods for green hydrogen generation (Kumar & Lim, 2022).

Hydrogen serves as an environment-friendly fuel source. It yields only water as its byproduct upon consumption within a fuel cell. It can be utilized in vehicles, buildings, portable power banks, and many more devices and infrastructures. Hydrogen

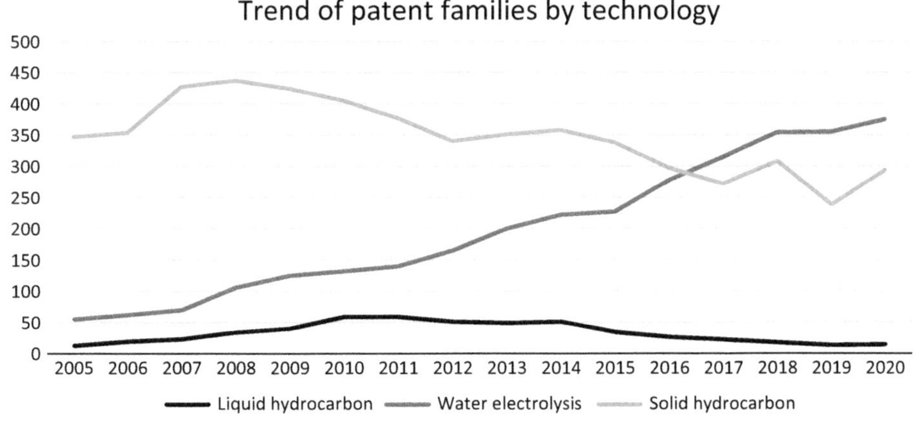

FIGURE 23.1 Hydrogen production patents by technology. Source: (EPO & IRENA, 2022)

is also an energy carrier, serving as a medium to store, move, and deliver energy (Satiapal, 2017).

Water electrolysis is a process of splitting water into hydrogen and oxygen by applying electrical energy (Acar & Dincer, 2018). The most sustainable innovations assume that the electricity needed for this process is generated from sustainable resources like wind or solar power. To accurately predict the development of hydrogen production technologies, it is also worth considering forecasts for the development of solar and wind farms. As water is an essential factor of hydrogen production, other related technologies are water desalination (Zapata-Sierra et al., 2022) and hydrogen production from seawater. It is essential to explore cross-industry collaboration in sustainable technologies because the advancements in one field can influence the development of others.

Undoubtedly, holistic patent landscape analysis can provide insights into emerging technologies and how they relate to each other, driving the most innovative and sustainable solutions. However, projecting future trends in sustainable technologies necessitates examining larger sociological and economic issues, like urbanization, population increase, and climate change. From an economic perspective, it is crucial to investigate how these factors might influence the demand for and adoption of sustainable technologies.

Another way to analyze technological sustainability development trends is to identify and take a closer look at emerging sustainable innovation ecosystems, which are clusters of organizations and individuals actively working on sustainable solutions. Identifying emerging technologies with potential for significant impact on sustainability involves tracking developments in areas like renewable energy, energy storage, smart grids, sustainable materials, and sustainable agriculture (Bhambri et al., 2019).

Choosing the right source of information depends on individual needs and preferences. When looking for current insights into the most recent trends and achievements in sustainable innovation, good choices include trade journals and magazines, websites devoted to sustainable innovations, and international organizations and foundations. Analyzing the latest scientific articles written in cooperation between academic centers and businesses may also be a valuable source of industry information.

23.3 SUSTAINABLE INNOVATION ECOSYSTEMS

- A.G. Tansley (1935) is credited with coining the term "ecosystem". This refers to one ecological component that is ingrained in both living things and their surroundings. J. Moore (1993) brought back the idea of explaining a framework of coopetition participants, emphasizing the geographical aspect of knowledge spillover sharing. The idea of the innovation ecosystem has grown in popularity over the past ten years. As of now, the innovation ecosystem is made up of a multilayer architecture in which institutions collaborate to create and exchange knowledge and information needed to establish new innovation processes (Granstrand & Holgersson, 2020). It demonstrates how businesses, governments, and civil society organizations collaborate and share to create a rational response to the demands and challenges (Costa & Matias, 2020). Innovation ecosystems encompass diverse stakeholders, such as higher education institutions, research and technology centers, corporate entities, venture capitalists, financial intermediaries, investors, and policymakers. Their common goal is innovation development by enabling the flow of resources such as knowledge, funds, and technology infrastructure. Figure 23.2 presents the elements of a sustainable innovation ecosystem and the dependencies among them.

The European Innovation Ecosystems (EIE) program is an example of such an ecosystem, which focuses on creating more accessible and efficient innovation

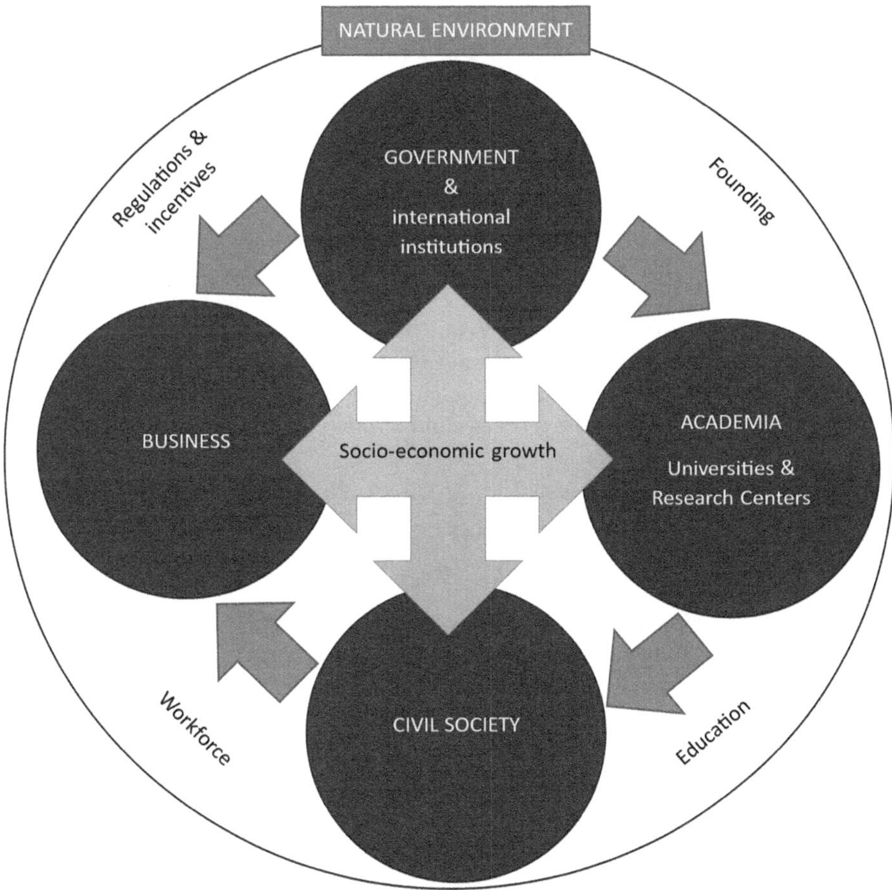

FIGURE 23.2 Sustainable innovation ecosystem.

environments to support the actors in working together across Europe (EISMEA, 2023). It works under the agenda of Horizon Europe – Research and Innovation funding program until 2027. This action aims to create Regional Innovation Valleys (RIVs) connected across the EU, which assist regions exhibiting lower innovation performances by leveraging strengths within their respective specializations according to the key EU priorities.

Another case of the innovation ecosystem is open innovation platforms, which provide companies with access to a wide range of ideas and diverse knowledge (Anand & Bhambri, 2018). These platforms aim to share value within entities with common interests, including governments, the value chain, and local communities, which collaborate and foster innovation to develop new high-value products (Costa & Matias, 2020). Innovating for sustainability is a crucial attribute of sustainable innovation ecosystems. This involves creating geographically dispersed ecosystems that are dependable and show reduced risks of disruptions.

The Finnish Food Packaging Ecosystem is a practical example of such a sustainable innovation environment working to facilitate sustainable packaging innovations. The packaging industry develops and adopts innovations to ease recycling and place clear recycling instructions on the packaging using eco-labeling. The packaging material engineering includes bioplastics produced from Finnish wood from certified forest resources or the use of industrial side streams. For this ecosystem to work effectively, there is a need for cooperation between many stakeholders. Apart from packaging companies, this ecosystem's main actors are governmental, standardizing organizations, food and beverage industry brand owners (Ruippo et al., 2023), and Finland's National Technology Research Center.

Many organizations, including charities, governments, and industry, try to predict and prepare for the future. Analyzing the activities of ecosystems worldwide can provide valuable insights into emerging trends and collaboration opportunities. Tracking market trends in sustainable technologies and investment patterns in relevant companies can indicate the level of commercial interest and potential for future growth in specific technologies.

23.4 INTERNATIONAL AGREEMENTS, POLICIES, AND FRAMEWORKS

International agreements and national policy frameworks related to sustainability are sources of knowledge that can help to understand the regulatory environment and potential support for specific technologies. The United Nations has established several intergovernmental agreements and policy frameworks related to sustainability.

The 2030 Agenda for Sustainable Development, adopted in September 2015, is a plan of action for people, the planet, and wealth that aims to support universal peace and uproot poverty. It comprises 17 Sustainable Development Goals (SDGs) and 169 targets, which are characterized by seamless integration and indivisibility. It is a universal sustainable development roadmap that applies to developed and developing countries.

The Addis Ababa Action Agenda is a global framework for fostering sustainable growth. It was adopted at the Third International Conference on Financing for Development in July 2015 and recommended by the General Assembly in its resolution 69/313 of 27 July 2015. The Addis Agenda aims to support the application of the 2030 Agenda by providing a comprehensive agenda for financing sustainable undertakings. It aligns all financing flows, including domestic public resources, private finance, and international cooperation (United Nations, 2015).

The Paris Agreement is an international treaty implemented by 196 countries at the 21st Conference of the Parties (COP21) of the United Nations Framework Convention on Climate Change (UNFCCC) in December 2015. The agreement aims to constrain the rise in global temperatures to well below 2°C above pre-industrial levels and to restrict the increase to 1.5°C above pre-industrial benchmarks. Its objective is also to strengthen countries' ability to deal with the consequences of climate change and support them in their efforts (United Nations, 2016).

The Convention on Biological Diversity (CBD) is an international treaty that aims to maintain biological diversity, promote sustainable use of its elements, and ensure the fair and reasonable sharing of benefits from using genetic resources (Xu & Wang, 2023).

23.5 SUSTAINABLE TECHNOLOGICAL MEGATRENDS

Megatrends are the global shifts that shape our world and society. Technological innovations can potentially revolutionize the global economy, influencing the main sectors such as energy, transportation, manufacturing, and healthcare.

There are different approaches and strategies to address global megatrends in sustainability. There is no single way to achieve sustainable goals. Sustainability in technology can be achieved by combining three general strategies depending on local, regional, national, or global conditions:

- Substitution: Sustainable technology facilitates a transition from nonbiodegradable to biodegradable materials and replaces nonrenewable resources with renewables.
- Prevention: Sustainable technology avoids deterioration, pollution, and other negative environmental influences of its use and production.
- Efficiency: Sustainable technology is efficient in terms of consuming energy and resources. The trade-off between short-term cost-effectiveness and long-term sustainable goals should be balanced.

All the three mentioned strategies have their implications in the environmental, social, and economic dimensions. Table 23.1 summarizes various strategies to achieve technological sustainability.

This matrix provides a structured visualization of how each strategy aligns with ecological, economic, and social considerations. Technological innovation can be called sustainable if it can be described within the matrix framework, specifying the relevant strategy and its three sustainable aspects.

TABLE 23.1
Matrix of Sustainable Strategies in Ecological, Economic, and Social Dimensions

		Ecological Aspect	Economic Aspect	Social Aspect
Substitution Strategies	Materials shift	Biodegradable materials usage	Sustainable sourcing of raw materials	Ethical labor practices and fair wages
	Resource replacement	Renewable energy sources	Circular economy practices	Inclusive technology access
Prevention Strategies	Environmental impact mitigation	Pollution control measures	Cost-effective waste management	Community engagement and education
	Life cycle assessment (LCA)	Sustainable design and manufacturing	Resource efficiency and optimization	Stakeholder awareness and involvement
Efficiency strategies	Energy efficiency	Renewable energy	Cost-effective energy use	Job creation and skills development
	Resource optimization	Waste reduction and recycling	Lean manufacturing principles	Socially responsible supply chain

23.5.1 Clean Energy Production

According to the US Energy Information Administration's prediction, the global energy demand will increase by 50% in 2050 (EIA 2023). Exploiting fossil fuels causes high greenhouse gas emissions, contributing to air pollution and global warming (Sazal, 2020). According to Statista (2023), the fossil fuel share in global electricity production reached 61.27% in 2022, and it is decreasing slowly but consistently. Therefore, the immediate development of green energy solutions is a fundamental megatrend to face global energy demands and sustainable development.

A current report by the International Energy Agency (IEA, 2023) paints a promising future for renewable energy, projecting that its global production in 2024 will match that of fossil fuels. The prime movers of this green revolution are wind and solar power. These sustainable energy leaders are supposed to double their global capacity over the next five years, exceeding gas and coal. However, the path to clean energy has faced severe financial barriers in recent years. Material costs have been a significant challenge in deploying new renewable projects in recent years. The cost of fossil fuels and electricity has increased, making transport more expensive. This negatively influences new construction projects in remote areas. For instance, growing prices of PV-grade polysilicon, steel, and copper increased the production cost of solar PV modules, wind turbines, and biofuels in the global market. However, in spite of this difficult situation, the price of renewables still stays competitive in the market. This is because the cost of fossil fuel production has also increased due to the Russian invasion of Ukraine (Milewska & Milewski, 2023) and governments seeking to secure energy for winter (UK Parliament, 2023).

The clean energy trend is also going to contribute to the social sustainability aspect. According to predictions published by IRENA and the International Labor Organization, 38 million people will get new jobs within the renewable energy sector over the next decade and 43 million by 2050 (IRENA & ILO, 2021).

23.5.2 Sustainable Digital Transformation

Digital innovations are expected to reshape most businesses not only as a tool to lower costs and improve economic effectiveness but also to assist sustainable change. Many organizations are investing in AI and automation to support their sustainability goals. The latest report by Capgemini (Shishodia, 2023) has shown that businesses are deploying AI and automation in tandem, not only to obtain precise insights into their operations but also to measure the cross-functional impact on the environment.

Nowadays, blockchain technology, the Internet of Things (IoT), and artificial intelligence (AI) are gaining much interest as remarkable innovations that will disrupt various industries on many levels. These are general-purpose technologies; however, in this chapter, we will explore their impact on sustainability. These promising technologies can help mitigate climate change and help control and reduce CO_2 emissions.

23.5.2.1 Internet of Things (IoT)

In maintaining a sustainable environment, it is very important to control its parameters in order to manage them. The Internet of Things is a modern technology that is already revolutionizing many industries. IoT is a very broad concept. It involves the idea of connecting devices equipped with various types of sensors to a global network. These devices can communicate with each other via IoT platforms. An IoT platform is the IT infrastructure that enables analyzing data from different devices for the purposes of giving recommendations, detecting patterns, or instantly discovering potential risks.

Applications of IoT enable achieving sustainable goals and may be treated as strategic resources. According to the World Bank (2023), today, more than half of the world's inhabitants live in urban agglomerations. This trend will continue, and the cities' population is expected to double by 2050. Transport plays a crucial role in city development by providing inhabitants with access to markets, employment, education, recreation, and healthcare. Internet of Things systems currently play a very important role in pursuing transport sustainability.

Smart IoT-based traffic management systems enable a control center to monitor traffic throughout a city. Data about conditions such as traffic jams or congestion can be analyzed using IoT devices and enable dynamic traffic adjustments. With traffic light optimization and entry alarms, the algorithms find optimal routes, aiming at reducing travel time and fuel consumption by freight vehicles, which are among the most substantial energy users and carbon dioxide emitters. The IoT technology also facilitates communication between intelligent vehicles at intersections or other problematic road touchpoints. Intelligent traffic management systems help provide priority access to emergency vehicles such as ambulances, police vehicles, or fire trucks.

Another sustainable benefit of IoT is the reduction of waste. Many communities worldwide have already implemented intelligent IoT-based systems to better manage waste and help the environment. Modern garbage cans have sensors to detect fill levels. The real-time data is sent to waste management companies, which use it to optimize truck routes in areas where urgent priority is needed.

IoT can also contribute to waste management across the food supply chain. It enables real-time tracking of the complete journey of goods, from when they are first produced or acquired to when they are delivered to the final customer. Food temperature in the so-called cold chain monitoring systems allows suppliers, manufacturers, and shipping companies to monitor the temperature of their food storage facilities accurately. Besides waste reduction, using IoT systems during food transport, storage, and serving contributes to socially sustainable goals by minimizing the health risks to end-consumers.

IoT technology can also improve work safety, which is a crucial concern of every industrial company. The networked sensors installed around a factory or warehouse and embedded in workers' clothes provide real-time information about employees' activity and environmental conditions such as air composition, temperature, pressure, light, or noise. Data analysis allows for the quick detection of any hazardous circumstances or behavior.

IoT devices can facilitate greenhouse gas emissions calculation, tracking, and reporting. Sensors are installed in the components of the manufacturing lines, vehicles, and other facilities to measure the running time of equipment. Based on this data, energy usage can be calculated and reported in real time.

These few examples illustrate the use of modern IoT technology for sustainable purposes in an ecological, social, and economic context. However, connected sensor networks would be of little use without the analytical capabilities of large data sets supported by artificial intelligence algorithms.

23.5.2.2 Artificial Intelligence

Artificial intelligence has recently stepped into a new level of maturity, becoming a general-purpose solution able to positively impact the productivity of many technologies and systems. AI also provides countless opportunities and future applications for sustainable development.

Artificial intelligence aims to simulate human intelligence quickly and accurately. AI's predictive capabilities enable environmental monitoring systems to identify patterns and trends that may be invisible to the human eye. This ultimately helps to predict and prevent pollution, prepare for natural disasters, and support wildlife conservation efforts.

A vital component of modern artificial intelligence (AI) systems is the ability to discover patterns in large amounts of data. The following examples are intended to illustrate the usefulness of AI from an ecological perspective.

AI systems are being used to examine data from IoT sensors in the soil to help farmers make more informed decisions about irrigation and fertilization. Farmers may reduce waste and save money by predicting soil moisture levels, optimizing irrigation, and detecting nutrient deficiencies.

Another exciting technology coupled with AI is image-processing drones. Combining real-time machine learning technology with the mobility and aerial imagery capabilities of autonomous drones, farmers perform their everyday tasks in agriculture. AI algorithms are also used for image processing in agriculture (Chen & Kuo, 2022). Drones can now process the images they capture and report back in real time. For example, in Brazil, AI image-processing drones are utilized for pest control and plant disease prevention. By analyzing drone-captured images, AI algorithms can identify early signs of pest invasions and plant diseases, allowing farmers to take quick, targeted actions to mitigate risks (Silva, 2022). This technology increases farms' productivity, reduces the need for toxic chemical pesticides, and saves costs associated with their purchase.

AI-powered drones also contribute to sustainable energy production by improving the efficiency of solar farms and wind farms. The efficiency and reliability of these modern energy farms are highly dependent on maintenance, which traditionally was slow, laborious, and expensive. Drones are used to capture visual data, which can be analyzed for hotspot detection and the flagging of anomalies.

AI can also contribute to achieving sustainable goals in other ways unrelated to environmental monitoring. Analyzing large data sets about customers' shopping preferences and online behavior can contribute to developing more sustainable production strategies. The problem that many companies face is the uncertainty about what products people will want to buy. This dilemma touches on the economic performance of a company and the environment as well. The production resources, such as water and energy, are wasted on the products that soon become waste, which is often hard to utilize. Therefore, there is a need to use AI in Life-Cycle Assessment (LCA) models to measure the environmental impacts of a product or service. As data collection is an essential and time-consuming part of the life cycle assessment, AI algorithms can help in all four LCA steps: the Goal and Scope phase (e.g., definition of system boundaries), the life cycle inventory analysis (e.g., data collection of input and output data), the life cycle impact assessment (conversion into environmental impacts), and the interpretation (evaluation of results) (Köck et al., 2023).

23.5.2.3 Blockchain

Blockchain is the next disruptive technological innovation that can significantly enhance sustainability efforts in various domains and industries.

A blockchain is a type of distributed database (or ledger) shared among a network's nodes. It stores data in blocks that are connected together using cryptography. The key features of blockchain are:

- Immutability: The data that is once recorded in a distributed database cannot be altered due to its unique design and the use of cryptographic principles.
- Decentralization: Because of blocks distribution, no single person has control over the entire blockchain. Instead, all users collectively retain control. This unique feature eliminates the requirement for central governance and decreases the risk of manipulation.
- Transparency: All network participants have access to the distributed ledger, which provides the ability to check the transactions and can increase trust among participants.
- Smart Contracts: These are self-executing computer programs in which the terms of the agreement are coded as if–then instructions. They can automate data processing, eliminating the need for intermediaries.

Blockchain is best known for its implementation in cryptocurrency systems (e.g., Bitcoin), but the potential of this technology goes far beyond that. Blockchains can be used in any setting where secure and verifiable records are essential. An example of blockchain areas of application for sustainable practices is ensuring supply chain transparency and traceability. This promotes fair supplier and customer relationships and trust between the supply chain stakeholders.

Implementing blockchain and Internet of Things devices, such as smart sensors and RFID tags, enables more effective goods tracking through the various stages of the supply chain. Blocks contain the product's status and condition records, e.g., location, temperature, and humidity, at any delivery stage. Since transactions are always up-to-date and marked with the timestamp, contractors can check the status of shipments in real time. Demand for blockchain applications for logistic services is also motivated by customers. Society becomes more aware of how important it is to know the origins of the materials and ingredients and whether they were produced or acquired ethically.

Another application of blockchain technology related to environmental protection is tracking and verifying carbon emissions and other environmental metrics, holding companies accountable for their sustainability claims. Blockchain provides a platform for hosting and executing smart contracts, which work autonomously without human intervention. These contracts can be used to enforce sustainability commitments.

Blockchain technology can also be used to support socially sustainable goals by improving governance systems, enabling e-voting, and reducing corruption by providing transparency.

However, it's important to note that while this new technology has the potential to contribute to sustainability, it also faces challenges. For instance, the energy consumption of blockchain, particularly in the case of cryptocurrencies like Bitcoin, can be quite high. Therefore, it's crucial to continue exploring ways to mitigate these challenges as we harness blockchain for sustainable development.

23.5.3 Circular Economy and Sustainable Design

The traditional linear economy relies on extracting resources, making products, using them, and then disposing of them as garbage. On the other hand, the circular economy vision is to create a closed-loop system where resources are continuously reused and cycled back into the economy. This approach seeks to minimize waste generation and environmental deterioration while promoting sustainable resource use.

Implementing the assumptions of the circular economy would not be possible without a fundamental shift in our economic mindset. A collaborative effort from governments, businesses, consumers, and individuals is crucial to transition from a linear, take–make–dispose model to a restorative one. This shift requires a holistic view of the connections between economic, social, and environmental systems.

Governments play a central role in promoting a circular economy by shaping the policies, regulations, and incentives for businesses and individuals. They can dictate extended producer responsibility (EPR) and invest in waste collection, recycling, and remanufacturing infrastructure. Another way is to offer tax breaks and subsidies for eco-friendly businesses. As research shows, many countries (Ivanova & Laptiev, 2019) use incentives to stimulate the efficient use of energy resources when collecting corporate income tax. Such incentives include, for example, tax reductions for energy-efficient commercial buildings, tax credits for qualified electric motor vehicles, and residential clean energy credits.

The role of governments and educational institutions is also to raise consumer awareness and highlight the economic and ecological benefits associated with everyday choices.

Individuals can adopt practices in their daily lives that reduce waste and promote sustainability, such as composting food scraps, repairing and reusing items, and choosing products that are manufactured from recycled materials or have minimal environmental impact.

A perfect example of such an approach is the French government's initiative to offer subsidies for anyone who repairs shoes or clothing. This idea encourages consumers to visit cobblers and tailors instead of throwing away old shoes and clothes.

Consumers can make informed choices about their purchases, prioritizing durability, repairability, and recyclability. They can also support circular businesses and advocate for policies that promote a circular economy.

On the other hand, companies can help the circular economy by redesigning their products, extending their lifespans, and enabling repair and maintenance. They can also collaborate within sustainable innovation ecosystems to share knowledge, develop new technologies, and scale up circular solutions. Sustainable design is a vital concept of the circular economy. It encompasses the principles of environmental stewardship, resource efficiency, and human well-being, aiming to create products, services, and systems that minimize their negative environmental impacts.

Sustainable design principles are applied throughout the product lifecycle, from conception to disposal. The first step in this cycle is material selection aimed at choosing materials with low environmental impact. Examples of such materials are recycled polyester (rPET) produced from recycled plastic bottles and bio-based materials, such as Mycofoam, used instead of polystyrene that can be easily recycled or reused.

Further considerations of the product lifecycle development encompass designing products in a way that makes them easy to take apart and reassemble, facilitating repair, refurbishment, and recycling. Sustainable design practices, such as modularity and easy disassembly, also facilitate the recovery of valuable materials from discarded products, reducing the volume of waste sent to landfills and incinerators.

Sustainable design practices should aim to significantly extend the lifetime of products, reducing the need for new production, which usually involves resource extraction and manufacturing emissions.

A separate challenge is designing products that minimize the consumption of resources such as electricity, fuel, or water during the product's exploitation. An example of an energy-saving product is LED lamps that replaced the previously used energy-saving fluorescent lamps. The problem with the latter was difficult and expensive disposal. Moving further towards energy-saving and fully biodegradable light sources, organic light-emitting devices (OLEDs) have recently become the leading technology (Pode, 2020).

Sustainable design is also a growing trend in architecture, further underlining smart city trends that aim to create buildings and urban facilities that are environmentally friendly, energy-efficient, and socially sustainable. As the construction industry is one of the major contributors to global energy consumption, sustainable architectural design incorporates features such as green roofs, natural ventilation, and rainwater harvesting, which can enhance the environmental resilience of buildings and their surroundings.

23.6 CONCLUSIONS AND ETHICAL CONSIDERATIONS

Technology is now an inevitable part of social and economic development. Exploring technological trends, their interconnections, and mutual impacts is a crucial step toward creating a more harmonious and sustainable future. The creation and application of green technologies often raise ethical concerns. In recent years, we have faced challenges related to data privacy, algorithmic bias, potential misuse, social inequality, and exclusion. Anticipating these issues early on allows us to incorporate ethical frameworks and safeguards into developing and deploying these technologies.

An excellent example of a sustainable technology that raises social controversy is the rapid development of Smart Grids (SGs) technologies to equip electricity networks with ICT solutions. Smart energy systems can detect changes in the network such as power losses, frauds, or backouts. SGs react and respond to detected events. They can also contribute to the efficiency and reliability of energy providers and facilitate the integration of renewable energy infrastructures. SGs enable electricity suppliers to obtain near real-time reports on the energy consumption patterns of consumers. This raises concerns about privacy due to the possibility of inferring sensitive consumer data. Therefore, social aspects in SGs are principal in guaranteeing their widespread and successful deployment.

Another controversial and very up-to-date aspect of sustainable innovations and policies is the ban on the use of vehicles with combustion engines. Rising air pollution is becoming a problem, prompting some countries to plan to ban further production of petrol cars.

Many countries, including the UK and France, are going to ban the sale of new petrol or diesel vehicles by 2040. However, without an available and affordable alternative, the ban on combustion engine vehicles may be associated with extremely negative social and economic consequences. History shows that technology bans are effective when the required technology is readily available and affordable. For example, the 1987 Montreal Protocol, which aimed to eliminate ozone-depleting chlorofluorocarbons (CFCs), was effective because alternatives to CFCs were already being developed and implemented. Inappropriate dates for the prohibition of new combustion engine cars could lead to public backlash against the clean energy transition and even against the European Union.

The switch to using only electric vehicles is also not without controversy. The problem is the production of electric vehicle batteries, which rely on rare minerals. In particular, using lithium, cobalt, nickel, and other metals in electric vehicle battery packs has raised discussions about human rights violations and worker protection laws in countries where these materials are mined. This leads to concerns about environmental and labor exploitation in mining operations. As electric vehicles increase in market share, they will be more responsible for the impact of battery production. It is worth noting that these effects are not solely due to electric vehicles. They are also driven by global demand for cell phones, laptops, and other portable electronic devices, which are an integral part of social activity.

These few examples illustrate that balancing the ethical implications of sustainable technology requires careful consideration. The most important goal is for sustainable innovations to be created in a sustainable way. Anticipating future trends, we can focus on developing innovations that accelerate economic growth, minimize their ecological footprint, and maximize their positive impact on society.

REFERENCES

Acar, C., & Dincer, I. (2018). Hydrogen production. *Comprehensive Energy Systems*, 3(1), 1–40. doi:10.1016/j.egyr.2023.11.060

Anand, A., & Bhambri, P. (2018). Orientation, scale and location invariant character recognition system using neural networks. *International Journal of Theoretical & Applied Sciences*, 10(1), 106–109.

Bhambri, P., Sinha, V. K., & Jaiswal, M. (2019). Change in iris dimensions as a potential human consciousness level indicator. *International Journal of Innovative Technology and Exploring Engineering*, 8(9S), 517–525.

Chen, S. F., & Kuo, Y. F. (2022). Artificial intelligence for image processing in agriculture. In S. Ma, T. Lin, E. Mao, Z. Song, & K. C. Ting (Eds.), *Sensing, Data Managing, and Control Technologies for Agricultural Systems. Agriculture Automation and Control.* Springer, Cham. https://doi.org/10.1007/978-3-031-03834-1_7

Costa, J., & Matias, J. C. O. (2020). Open innovation 4.0 as an enhancer of sustainable innovation ecosystems. *Sustainability*, 12, 8112. https://doi.org/10.3390/su12198112

EIA. (2023). *Today in Energy.* U.S. Energy Information Administration. https://www.eia.gov/todayinenergy/detail.php?id=49876#

EISMEA. (2023). European innovation ecosystems enabling innovation ecosystem actors to work together across Europe. https://eismea.ec.europa.eu/programmes/european-innovation-ecosystems_en

EPO & IRENA. (2022). Patent insight report. Innovation trends in electrolysers for hydrogen production, EPO, Vienna. https://www.irena.org/publications/2022/May/Innovation-Trends-in-Electrolysers-for-Hydrogen-Production

Granstrand, O., & Holgersson, M. (2020). Innovation ecosystems: A conceptual review and a new definition. *Technovation*, 90, 102098.

IEA. (2023). Renewable energy market update - June 2023.https://www.iea.org/news/renewable-power-on-course-to-shatter-more-records-as-countries-around-the-world-speed-up-deployment

IRENA & ILO. (2021). *Renewable Energy and Jobs – Annual Review 2021.* International Renewable Energy Agency, International Labour Organization.

Ivanova, O., & Laptiev, V. (2019). Tax incentives for innovation in the energy sector. *Acta Innovations*, 32, 20–28.http://dx.doi.org/https%3A//doi.org/10.32933/ActaInnovations.32.3

Köck, B., Friedl, A., Serna Loaiza, S., Wukovits, W., & Mihalyi-Schneider, B. (2023). Automation of life cycle assessment—a critical review of developments in the field of life cycle inventory analysis. *Sustainability*, 15, 5531. https://doi.org/10.3390/su15065531

Kumar, S. S., & Lim, H. (2022). An overview of water electrolysis technologies for green hydrogen production. Energy Reports, 8, 13793–13813.

Mikova, N., & Sokolova, A. (2019). Comparing data sources for identifying technology trends. *Technology Analysis & Strategic Management*, 31(11), 1353–1367. https://doi.org/10.1080/09537325.2019.1614157

Milewska, B., & Milewski, D. (2023). The impact of energy consumption costs on the profitability of production companies in poland in the context of the energy crisis. *Energies*, 16, 6519.https://doi.org/10.3390/en16186519

Moore, J. (1993). Predators and prey: A new ecology of competition. *Harvard Business Review*, 71, 75–86.

Pode, R. (2020). Organic light-emitting diode devices: An energy-efficient solid-state lighting for applications. *Renewable and Sustainable Energy Reviews*, 133, 110043. https://doi.org/10.1016/j.rser.2020.110043

Ruippo, L., Koivula, H., Korhonen, J., et al. (2023). Innovating for sustainability: Attributes, motivations, and responsibilities in the finnish food packaging ecosystem. *Circular Economy and Sustainability*, 3, 919–937. https://doi.org/10.1007/s43615-022-00217-2

Satiapal, S. (2017). *Hydrogen: A Clean, Flexible Energy Carrier 2017.* Office of Energy Efficiency & Renewable Energy.https://www.energy.gov/eere/articles/hydrogen-clean-flexible-energy-carrier

Sazal, N. (2020). Emerging technologies by hydrogen: A review. *International Journal of Hydrogen Energy*, 45, 18753–18771.https://doi.org/10.1016/j.ijhydene.2020.05.021

Segev, A., Jung, S., & Choi, S. (2015). Analysis of technology trends based on diverse data sources. *IEEE Transactions on Service Computing*, 8(6), 903–915.

Shishodia, P. (2023). Creating a circular economy through AI. *Capgemini*.https://www.capgemini.com/insights/expert-perspectives/creating-a-circular-economy-through-ai/

Silva, F. (2022). Brazilian farmers use AI image processing drones for pest control and disease prevention. *Brazil AgriNews*, August 15, 2022.

Statista. (2023). Fossil fuel share in electricity production worldwide from 2000 to 2022.https://www.statista.com/statistics/1303803/global-fossil-fuel-share-in-power-generation/

Tansley, A. G. (1935). The use and abuse of vegetational concepts and terms. *Ecology*, 16, 284–307.

Tao, F., Zuo, Y., Xu, L. D., Lv, L., & Zhang, L. (2014). Internet of Things and BOM-Based life cycle assessment of energy-saving and emission-reduction of products. *IEEE Transactions on Industrial Informatics*, 10, 1252–1261.

UK Parliament. (2023). Preparing for the winter: Government response to the committee's first report of session 2022–23. https://publications.parliament.uk/pa/cm5804/cmselect/cmesnz/401/report.html

United Nations. (2015). Addis ababa action agenda of the third international conference on financing for development.https://sustainabledevelopment.un.org/content/documents/2051AAAA_Outcome.pdf

United Nations. (2016). The Paris agreement.https://unfccc.int/process-and-meetings/the-paris-agreement

World Bank. (2023). Urban development.https://www.worldbank.org/en/topic/urbandevelopment/overview

Xu, J., & Wang, J. (2023). Analysis of the main elements and implications of the Kunming-Montreal Global Biodiversity Framework. *Biodiversity Science*, 31(4), 23020.https://doi.org/10.17520/biods.2023020

Zapata-Sierra, A., Cascajares, M., Alcayde, A., & Manzano-Agugliaro, F. (2022). Worldwide research trends on desalination. *Desalination*, 519, 115305. https://doi.org/10.1016/j.desal.2021.115305

24 Unveiling South Africa's Path to a Sustainable Future

A Closer Look at Its Evolving Renewable Energy Sector

Privilege Cheteni and Ikechukwu Umejesi

24.1 INTRODUCTION

The global pursuit of mitigating climate change and achieving sustainability milestones has seen a slow but persistent march. Climate change is still a major concern, even after landmark international accords such as the 1997 Kyoto Protocol and the 2016 Paris Agreement. Within this framework, international organizations and governments work to keep global warming well below 1.5°C above pre-industrial levels; however, there is significant regional and national variation in the rate of progress towards this goal. Swapping out traditional fossil fuels for renewable energy is a major front in this fight (Ram et al., 2022). Various factors, including a lack of technical capability and reluctance to invest in new energy sources, cause the rate of this transition to vary globally. When it comes to slow adoption of sustainable technologies, nowhere is this more apparent than in Africa. The continent has suffered from a combination of a lack of investment in renewable energy and unclear policy goals. This chapter delves into the unique case of South Africa, a nation that has emerged as a pioneer in the African continent by actively developing policies on renewable energy.

24.1.1 Importance of Case Studies

Understanding the intricacies and complexities involved in the implementation of renewable energy initiatives demands a nuanced approach. Case studies play a pivotal role in unraveling the multifaceted journey toward sustainable energy adoption. The South African experience, serving as a rich tapestry of challenges and triumphs, offers a lens through which we can discern valuable insights applicable to the global pursuit of a greener future.

DOI: 10.1201/9781003475989-28

Examining South Africa's renewable energy landscape allows us to uncover the challenges faced by a nation striving for a sustainable future. From initial hurdles in policy formulation to the practical difficulties of integrating renewable sources into an existing energy infrastructure, the case study illuminates the roadblocks encountered along the way (Hanto et al., 2022). These challenges serve as cautionary tales for other nations, offering a realistic portrayal of the obstacles to be overcome in the pursuit of a cleaner and more sustainable energy sector.

South Africa's journey to cleaner sources of energy has been underlined by a narrative of "successes" and "achievements". The dedication of South Africa to renewable energy has led to the effective implementation of various technologies, which have had a substantial impact on its energy mix (Hanto et al., 2022). The successes documented in the case study become beacons of inspiration for other countries in Africa and beyond seeking viable models for transitioning away from fossil fuels. Therefore, by dissecting these successes, one can gain insights into effective strategies and best practices that can be replicated on a global scale.

An essential part of the case study is South Africa's intentional policymaking to encourage and direct the shift to renewable energy sources. The examination of these policy frameworks reveals the importance of proactive government intervention in fostering a conducive environment for sustainable energy development (Ibrahim et al., 2021). Policymakers globally can draw lessons from South Africa's experience to tailor effective strategies that align with their unique socio-economic contexts, thus accelerating the adoption of renewable energy initiatives.

Technological innovation is at the heart of any successful renewable energy transition (Akinbami, Oke, & Bodunrin, 2021). South Africa's case provides a comprehensive exploration of the technological advancements driving its sustainable energy sector. From large-scale solar farms to innovative grid management solutions, the technological trajectory becomes a roadmap for other nations seeking to harness the full potential of renewable resources. Lessons learned from South Africa's technological journey can inform strategies for overcoming technological barriers and maximizing the efficiency of sustainable energy solutions.

Beyond its national borders, South Africa's case study underscores the importance of global collaboration in the pursuit of a greener future. Insights gained from this case study can inspire collaborative efforts between nations, encouraging the exchange of knowledge, expertise, and resources (Uhunamure & Shale, 2021). At a time when the entire world is trying to figure out how to deal with climate change, South Africa's story shows how international cooperation can speed up the switch to renewable energy.

24.1.2 Objectives of the Study

This chapter endeavors to achieve the following objectives:

- Uncover the intricacies of South Africa's sustainable energy sector, including its policy framework and technological advancements.

- Explain how case studies help us understand the dynamics of renewable energy implementation.
- Evaluate the contribution of South Africa's renewable energy sector to the nation's broader goal of attaining a well-balanced and sustainable energy mix.

24.2 SOUTH AFRICA'S RENEWABLE ENERGY LANDSCAPE: A COMPREHENSIVE OVERVIEW

24.2.1 Policy Framework and Legislative Initiatives

South Africa's unwavering commitment to sustainable energy is underpinned by a robust policy framework that serves as the backbone of its renewable energy initiatives. The government is taking a strategic approach by carefully crafting and enforcing policies that encourage and support the broad use of renewable energy sources. The evolution of these policies reflects a dynamic response to the changing landscape of energy needs and environmental concerns.

In terms of renewable energy policy, the Renewable Energy Independent Power Producer Procurement Program (REIPPPP) is spearheading South Africa's efforts. Launched in 2011, this pioneering initiative represents a paradigm shift by introducing private sector participation in the country's energy sector. By allowing independent power producers to generate renewable energy and sell it to the national grid, REIPPPP has been a game-changer, fostering competition, driving down costs, and accelerating the deployment of renewable technologies (Uhunamure & Shale, 2021).

One of the key strengths of South Africa's sustainable energy policy lies in its ability to attract private investment (Akinbami, Oke, & Bodunrin, 2021). There has been an influx of capital into renewable energy projects thanks to the government's efforts to foster an atmosphere that encourages private sector involvement (Kuzhaloli et al., 2020). The competitive bidding process introduced by REIPPPP ensures that projects are not only economically viable but also financially attractive to private investors, creating a win–win scenario that propels the nation toward a sustainable energy future.

Figure 24.1 displays the 2020 renewable energy status in South Africa. Coal + others represent a larger proportion of 75%, followed by oil with a 14% contribution to the national grid. Renewables stand at 6%, with gas contributing 3% and nuclear contributing 2%.

A distinctive feature of South Africa's policy framework is its emphasis on fostering innovation within the renewable energy sector. Encouraging a diverse range of technologies, from solar and wind to biomass and hydroelectric, the policies create a fertile ground for experimentation and advancement (Usman, Akadiri, & Adeshola, 2020). The incentivized competition among independent power producers has driven innovation, pushing the boundaries of what is technologically achievable in the realm of renewable energy.

Increasing the variety of energy sources used by the country is a primary objective of South Africa's green energy policies. After years of relying on coal, South

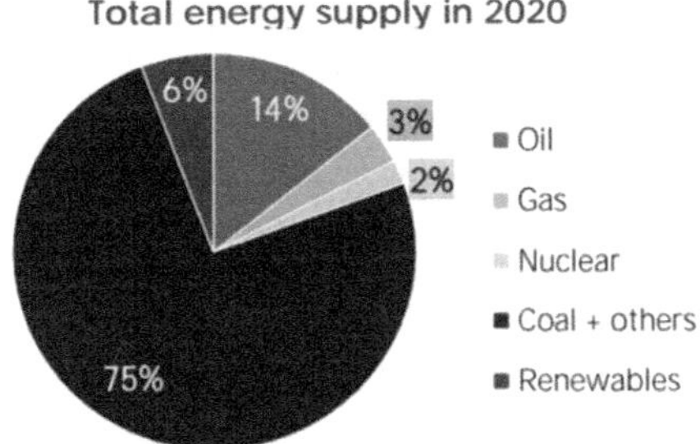

FIGURE 24.1 South Africa's renewable energy landscape.

Africa has been able to diversify its energy sources and become less reliant on fossil fuels thanks to programs like REIPPPP. Iorember et al. (2021) highlighted that this diversification not only mitigates environmental impact but also enhances energy security, making the nation more resilient to global energy market fluctuations.

A notable strength of South Africa's sustainable energy policies lies in their long-term vision and adaptability. The government has demonstrated a commitment to regularly reviewing and updating policies to align with evolving technological trends, global best practices, and emerging environmental challenges. This adaptability ensures that the nation remains at the forefront of sustainable energy innovation, positioning itself as a dynamic player in the global transition toward a greener future (Hanto et al., 2022).

24.2.2 Technological Advancements Driving the Transition

South Africa's strides in renewable energy are intricately woven with technological advancements that form the bedrock of its transition (See Figure 24.2). The nation's commitment to diversifying its energy mix has led to the embrace of a diverse array of renewable technologies, positioning South Africa as a beacon of innovation in the global quest for sustainable energy solutions (OIkuski, 2023).

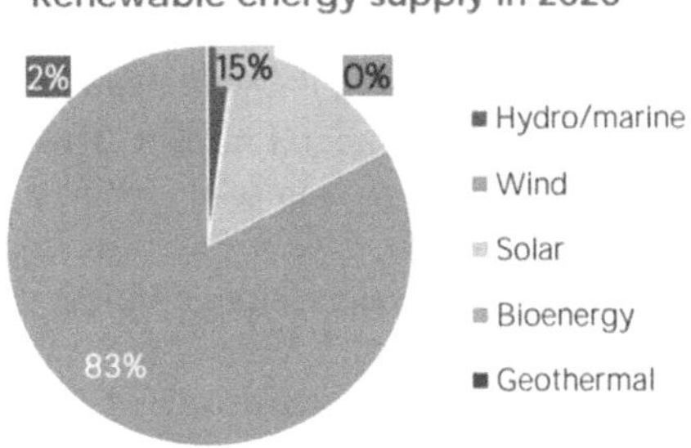

FIGURE 24.2 Renewable energy supply in 2020. Source: Akinbami, Oke & Bodunrin, 2021.

Beyond solar, wind, and hydroelectric power, South Africa has also ventured into biomass and bioenergy initiatives (Bakshi et al., 2021). These technologies utilize organic materials to generate energy, offering a sustainable solution for both power generation and waste management, which contributes 83% to the national grid (Ibrahim et al., 2021). The integration of biomass technologies demonstrates South Africa's holistic approach to renewable energy, addressing multiple environmental challenges while contributing to a more circular and sustainable economy.

Among the standout contributors to South Africa's renewable energy success is the significant growth in solar power. The nation's abundant sunlight resources have been harnessed through the development of large-scale solar farms, contributing 15% of renewable energy to the national grid (Akinbami, Oke, & Bodunrin, 2021). These solar farms, equipped with state-of-the-art photovoltaic technology, not only harness energy efficiently but also underscore South Africa's commitment to tapping into its natural resources for clean energy production.

South Africa's pursuit of a balanced energy mix extends to harnessing hydroelectric power. Strategically utilizing the nation's water resources has led to the implementation of hydroelectric projects, which contribute 2% to the national grid, although they are not as prominent as solar power (Iorember et al., 2021). South Africa's renewable energy strategy is flexible enough to incorporate advanced hydroelectric technologies, drawing on a variety of renewable resources to satisfy the country's increasing energy needs.

New developments in energy storage and control of the grid are essential to the sustainability of renewable power. South Africa has invested in cutting-edge energy storage solutions, such as advanced battery technologies, to address the intermittent nature of solar and wind power. Improved system stability and efficiency are additional benefits of renewable energy sources that can be easily integrated into the current power grid thanks to developments in grid management technologies (Akinbami, Oke, & Bodunrin, 2021).

24.3 BACKGROUND ON RENEWABLE ENERGY

24.3.1 Overview of Renewable Energy and Its Principles

A new age of environment-friendly power generation is about to begin, and renewable energy is at the vanguard of this revolution. At its essence, the principles of renewable energy revolve around harnessing power from sources that are inherently replenishing and have minimal environmental impact. This departure from conventional fossil fuels is rooted in the recognition of their finite nature and the imperative to address climate change (Oikuski, 2023). The principles governing renewable energy underscore a commitment to conservation, efficiency, and a reduced ecological footprint.

Renewable energy encompasses a diverse array of sources, each with unique characteristics and applications. Photovoltaic cells and solar thermal systems convert sunlight into electricity and heat, making solar energy one of the most pervasive sources of renewable energy (Iorember et al., 2021). Wind energy harnesses the kinetic energy of the wind, turning it into electrical power through wind turbines. Both geothermal energy and hydropower harness the mechanical energy of water in motion to create electrical currents, but geothermal energy also draws on the Earth's natural heat. Ocean energy, which includes tidal and wave power, and biomass, which is made of organic materials, are two more components of the diverse portfolio of renewable resources.

Renewable energy is gaining popularity due to its many advantages. In addition to cutting down on emissions of greenhouse gases, renewable energy helps clean up the air and water that conventional power plants pollute. The diversity of renewable sources ensures a resilient and adaptable energy infrastructure, less susceptible to geopolitical uncertainties (Hanto et al., 2022). Job creation is a notable byproduct, spanning manufacturing, installation, and maintenance of renewable energy systems, fostering economic growth and social development.

Nevertheless, there will be obstacles along the way to a renewable future. Innovations in energy storage are required because renewable power sources like solar and wind are intermittent. Integrating renewable energy into existing grids requires substantial investments in infrastructure, highlighting the importance of grid modernization. Large-scale projects may impact ecosystems and necessitate careful consideration of land use. Overcoming technological limitations and addressing initial costs are ongoing challenges that require sustained innovation and supportive policies.

In navigating this transition, public perception and policy support play pivotal roles. Widespread awareness and understanding of the benefits of renewable energy are essential for garnering public support (Ibrahim et al., 2021). Governments and policymakers must enact consistent and supportive policies that incentivize the adoption of renewable technologies. This intersection of public awareness and policy frameworks forms the bedrock for a successful and sustainable transition to renewable energy, aligning with the principles of environmental stewardship and long-term resource conservation.

24.3.2 Benefits and Challenges of Adopting Renewable Energy

24.3.2.1 Benefits

One way renewable energy helps fight climate change is by drastically cutting emissions of greenhouse gases (Pillot et al., 2019). It also reduces pollution of both air and water, which is another benefit mentioned by Panos et al. (2023) in relation to traditional energy sources. Reduced vulnerability to shortages caused by over-reliance on fossil fuels is another benefit of a diverse portfolio of renewable energy sources. A more robust and flexible energy infrastructure is guaranteed by this variety.

Many people find work in the renewable energy industry, which includes production, installation, maintenance, and R&D (Ram et al., 2022). Employment opportunities like these boost the economy and improve people's quality of life. Relying on renewable, indigenous resources helps countries become more energy independent and less susceptible to the geopolitical conflicts that surround fossil fuels (Ibrahim et al., 2021). Renewable energy has become much more affordable due to technological advancements and economies of scale, according to Panos et al. (2023). In the long run, renewable energy typically has cheaper operational costs than traditional energy.

24.3.2.2 Challenges of Renewable Energy

A reliable power supply is made more difficult by the fact that renewable energy sources like solar and wind are intermittent. Hybrid systems and energy storage solutions are crucial for dealing with this intermittent power (Iorember et al., 2021). It will take a lot of investment in infrastructure to connect renewable energy to the grid. To ensure smooth integration, grid modernization and smart grid technology development are of the utmost importance.

Large-scale renewable energy projects, particularly in solar and wind, may require significant land use. Balancing the need for renewable energy with minimizing environmental impact is a complex challenge. Current technologies have limitations, including the efficiency of energy conversion and energy storage capacity. Ongoing research and development are necessary to overcome these limitations (Ram et al., 2022).

Although renewable energy has lower operating costs over time, it may require a hefty investment to set up the necessary infrastructure. To overcome this obstacle, we need policies that support the cause and financial incentives. Furthermore, renewable energy adoption is highly dependent on public perception and policy support. Overcoming resistance and garnering widespread support often requires effective communication and education initiatives. Additionally, consistent and supportive government policies are essential for creating an environment conducive to renewable energy adoption.

24.4 METHODOLOGY

24.4.1 Research Approach

To comprehensively explore South Africa's evolving sustainable energy sector, a multifaceted research approach was employed. The methodology integrates both a thorough literature review and a meticulous analysis of case studies. The literature review serves as the foundation, providing a comprehensive understanding of the theoretical frameworks, policy landscapes, and technological advancements in the field of sustainable energy. This foundational knowledge is then contextualized and enriched through the analysis of pertinent case studies, allowing for a nuanced exploration of real-world applications and experiences.

24.4.2 Selection Criteria for Case Studies

The selection of case studies was purposively guided by a set of predefined criteria aimed at ensuring a representative and insightful exploration of South Africa's sustainable energy landscape. Criteria included geographical diversity to capture regional nuances, a range of technological applications, and a mix of urban and rural contexts. Emphasis was placed on including projects that have demonstrated significant impact, innovation, and resilience in the face of challenges. The selected case studies thus form a mosaic that reflects the complexity and diversity of South Africa's sustainable energy transition.

24.4.3 Data Collection Methods

To gather a holistic understanding of South Africa's sustainable energy journey, a desktop approach was employed. Articles, bulletins, reports, strategic plans, and other documents were utilized to gather information on renewables in South Africa. These documents provided valuable insights into the policymaking process, challenges faced, and the strategic vision driving sustainable energy initiatives.

24.4.4 Integration of Stakeholder Perspectives

Recognizing the importance of incorporating diverse perspectives, the research methodology prioritized the integration of stakeholder input. Stakeholder engagement workshops were organized to facilitate discussions, share knowledge, and validate findings. This participatory approach ensures that the research not only captures the macro-level policy and technological aspects but also resonates with the experiences and aspirations of the communities directly impacted by sustainable energy projects (Akinbami, Oke, & Bodunrin, 2021).

24.4.5 Comparative Analysis and Synthesis

This research methodology includes a rigorous comparative analysis of the case studies, synthesizing findings to draw overarching insights. By comparing and contrasting the various projects, technologies, and policy frameworks, the research aims to distill key success factors, challenges, and transferable lessons. This synthesis is crucial for extrapolating broader implications applicable not only to South Africa but also to other nations navigating their unique paths toward sustainable energy adoption.

24.5 CASE STUDY 1: KWAZULU-NATAL PROVINCE

With the exception of ocean power and hydropower, the province of KwaZulu-Natal (KZN) has a 45 GW renewable resource potential that could be used to generate electricity. This amount is distributed as follows: 53.63% of global normal irradiance (GHI), 23.28% of direct normal irradiance (DNI), 13.52% of wind energy, 9.51% of geothermal, and 0.06% of biomass energy. However, with the results exposed in this case study, further investigations can be done to determine the net electrical energy that can be produced from those sources for the design of renewable energy systems and energy harvesting as well as the appropriate areas to install renewable energy systems.

24.5.1 Description of the Selected Case Study

Perched on the sunny eastern coast of South Africa, KwaZulu-Natal ranks second in both population and economy among the provinces. Featuring an impressive 45 GW of potentially exploitable renewable resources, the South African province of KwaZulu-Natal (KZN) stands out as an interesting case study in the renewable energy industry (Panos et al., 2023). Energy from wind, geothermal, and biomass accounts for a small fraction of this potential, while global normal irradiance (GHI) accounts for 53.63%, direct normal irradiance (DNI) for 23.28%, and the remaining 1.352% comes from other renewable sources (TotalEnergies, 2023). Noteworthy to mention are the diverse and untapped renewable energy possibilities in the region, as this potential does not include ocean energy and hydropower.

24.5.2 Overview of Renewable Energy Used and Its Sustainability Features

KZN's renewable energy landscape encompasses a spectrum of sources, each contributing to the sustainability features of the region. Solar energy, represented by both global normal irradiance and direct normal irradiance, holds significant promise. The abundance of sunlight, especially in a province like KZN, positions solar power as a reliable and sustainable source. Wind energy, constituting 13.52% of the renewable potential, further diversifies the energy mix. The inherent sustainability of wind power lies in its low environmental impact and minimal greenhouse gas emissions during operation (Panos et al., 2023).

Geothermal energy, though a smaller fraction of the potential, adds another layer of sustainability (Ibrahim et al., 2021). Harnessing the Earth's internal heat can provide a consistent and baseload source of energy, contributing to the reliability of the overall energy system. Additionally, the presence of biomass energy resources, though a smaller percentage, offers opportunities for decentralized and community-based energy solutions.

24.5.3 Evaluation of the Impact of Renewable Energy on the Environment and Society

The potential of renewable energy in KZN has a multifaceted impact on both the environment and society. Environmentally, the reliance on abundant solar and wind resources minimizes the carbon footprint associated with traditional energy sources, reducing greenhouse gas emissions and mitigating climate change (Ram et al., 2022). The emphasis on geothermal and biomass resources adds to the sustainability quotient by tapping into sources that have lower environmental impacts compared to conventional fossil fuels.

Societally, the integration of renewable energy systems presents economic opportunities and enhances energy security. The development of renewable energy projects in KZN not only fosters job creation but also promotes local economic development. Furthermore, by decentralizing energy production through renewable sources, the region becomes more resilient to external energy supply shocks, ensuring a consistent and reliable energy supply for communities (TotalEnergies, 2023).

24.5.4 Future Implications and Further Investigations

The preceding discussion on the renewable energy potentialities of South Africa lays the groundwork for future investigations and actions. The identification of renewable resource potentials provides a foundation for designing renewable energy systems and energy harvesting strategies. Discovering the potential net electrical energy production from these sources, as well as improving the planning and execution of renewable energy projects in KZN, can be the subject of future research. The study also emphasizes the need for further exploration to identify appropriate areas for the installation of renewable energy systems. Considering the geographical diversity and the specific characteristics of each energy source, strategic placement becomes crucial for maximizing efficiency and harnessing the full potential of KZN's renewable resources.

In essence, the case study of renewable energy potential in KwaZulu-Natal serves as a microcosm of South Africa's broader journey towards a sustainable energy future. It highlights the untapped potential within the region and underscores the importance of strategic planning, technological innovation, and community engagement in unlocking the full spectrum of benefits offered by renewable energy sources.

24.6 CASE STUDY 2: GAUTENG PROVINCE

Gauteng is the most populous province and the economic nerve center of South Africa. In the case of Gauteng Province, one major challenge being faced is the limited mandate that provinces have around energy generation and the fact that they are reliant on national developments in terms of formulating and implementing strategies and frameworks (Hanto et al., 2022). The Gauteng Integrated Energy Strategy of 1998 states that Gauteng Province is heavily reliant on non-renewable resources for energy. This presents a challenge in terms of promotion and investment in renewable energy when there is an abundance of, for example, cheap coal. Unless this challenge is properly addressed in the strategy, all fuel-switching attempts from coal to cleaner forms of energy will be compromised.

24.6.1 Description of the Selected Case Study

Gauteng Province emerges as a distinctive case study within the South African context, marked by its complex energy landscape and the challenges inherent in transitioning to renewable energy sources. The province faces a major challenge due to the limited mandate that provinces hold concerning energy matters. Reliant on national developments and frameworks, Gauteng grapples with the dominance of non-renewable resources in its energy portfolio. The Gauteng Strategy for Sustainable Development State of Play Report underlines the province's heavy dependence on non-renewable resources, particularly cheap coal, posing a significant hurdle to the promotion and investment in renewable energy.

24.6.2 Explanation of Renewable Energy Practices Employed

Despite the challenges, Gauteng Province is exploring avenues for incorporating renewable energy practices. Energy sources that are less harmful to the environment, such as solar, wind, and biomass, are gradually replacing non-renewable ones. The plan is to wean ourselves off cheap coal as much as possible through fuel-switching initiatives. One prominent aspect of the renewable energy practices is solar energy, which is abundant in the area. Utilizing sustainable power sources in Gauteng, renewable energy is generated by installing solar panels and wind turbines.

24.6.3 Assessment of Economic and Environmental Benefits

There are ecological and financial benefits to switching to renewable energy sources in Gauteng. From an economic perspective, switching to renewable energy sources boosts local businesses and generates new job openings. Building and sustaining renewable energy infrastructure helps the local economy thrive by creating more opportunities for skilled workers (Ram et al., 2022). In addition, Gauteng hopes to improve energy security and lessen economic vulnerabilities caused by changes in the price of fossil fuels by decreasing reliance on non-renewable resources.

To lessen the negative effects of energy production on the environment and cut down on carbon emissions, the shift to renewable energy sources is crucial. It is Gauteng's hope that its use of renewable energy sources like solar and wind will bolster international and national initiatives to slow the rate of climate change. Improving air quality and addressing public health concerns related to respiratory diseases are both achieved through the reduction of air pollutants associated with burning coal.

24.6.4 Overcoming Challenges

The identified challenge of limited provincial mandate and dependence on national developments poses a critical obstacle. To address this, the strategy for sustainable development in Gauteng must advocate for a more decentralized decision-making process, empowering provinces to tailor energy solutions to their specific needs. Collaboration between provincial and national entities is crucial to synchronize efforts and align strategies for a cohesive and effective renewable energy transition.

24.7 CASE STUDY 3: EASTERN CAPE PROVINCE

The Eastern Cape Province consumes approximately 9,538 (GWh) per annum, which is 4.4% of the national total. With the introduction of the capacity procured under the REIPPPP, approximately a third of the province's power needs will be produced locally from renewable energy sources. The province identified the opportunity to enable development for local electricity generation to exceed local demand within 15–20 years (Hanto et al.). Eastern Cape's vision is to create an enabling environment for sustainable energy investments and implementation in the country. In support of this vision, various South African government departments and international organizations have become involved in the development and implementation of innovative renewable energy programs and encourage the use of the energy created (Ram et al., 2022).

24.7.1 Description of the Selected Case Study

Sitting on the eastern seaboard of South Africa, the Eastern Cape Province is the fourth most populous province in South Africa and its sixth largest economy in terms of GDP. The Eastern Cape Province stands as a compelling case study in the South African context, with a distinctive focus on renewable energy. Consuming approximately 9,538 GWh annually, which accounts for 4.4% of the national total, the province is undergoing a significant transformation with the introduction of capacity procured under the Renewable Energy Independent Power Producer Procurement Program (REIPPPP). This initiative marks a substantial shift toward local renewable energy production, with the goal of exceeding local demand within 15–20 years (Uddin et al., 2023). The Eastern Cape has identified this opportunity to create a conducive environment for sustainable energy investment, aligning with the broader vision of fostering local electricity generation.

24.7.2 Discussion of Renewable Energy in the Province

The Eastern Cape's embrace of renewable energy, facilitated by the REIPPPP, signifies a paradigm shift in the province's energy landscape. With a third of its power needs set to be produced locally from renewable sources, the province positions itself at the forefront of the national drive toward a sustainable energy future. The case study underscores the commitment of the Eastern Cape to not only meet its energy demands but also to contribute substantially to the national goal of diversifying the energy mix and reducing dependence on non-renewable resources.

24.7.3 Analysis of Energy Efficiency and Environmental Impact

A domino effect of beneficial impacts on energy efficiency and the environment results from the Eastern Cape's adoption of renewable energy projects. Compared to centralized power generation, which results in transmission and distribution losses, locally generated renewable energy improves energy efficiency (Ram et al., 2022). Improving energy efficiency by means of distributed, environmentally friendly power sources is in line with the provincial government's plans to foster an atmosphere conducive to long-term energy investments.

In terms of environmental impact, switching to renewable energy sources greatly reduces the amount of carbon dioxide emissions caused by power generation (Panos et al., 2023). As part of its efforts to combat climate change, the Eastern Cape has pledged to reduce emissions of greenhouse gases by increasing its use of renewable energy sources to meet or exceed local demand. Traditional energy sources are known to cause air pollution and resource depletion; however, these problems can be mitigated through the use of renewable energy.

24.7.4 Government and International Partnership in Renewable Energy

The vision of creating an enabling environment for sustainable energy investment in the Eastern Cape has garnered support from various South African government departments and international organizations. This collaborative effort emphasizes the importance of cross-sectoral cooperation in advancing renewable energy initiatives (Uddin et al., 2023). The involvement of international organizations underscores the global significance of such endeavors and highlights the Eastern Cape's commitment to drawing on international expertise and resources for the successful implementation of innovative renewable energy programs.

24.7.5 Future Prospects and Community Development

Looking ahead, the Eastern Cape Province is poised for significant advancements in its renewable energy sector. The province's goal to exceed local demand within 15–20 years not only ensures energy security but also presents opportunities for community development. Local electricity generation initiatives create jobs, foster economic growth, and empower communities to actively participate in the renewable energy landscape (Uddin et al., 2023). Future prospects involve the continued collaboration between the government, international organizations, and local communities to ensure a sustainable, inclusive, and resilient energy future for the Eastern Cape.

24.8 COMPARATIVE ANALYSIS OF CASE STUDIES

24.8.1 Identification of Commonalities among the Case Studies

All three provinces, KwaZulu-Natal, Gauteng, and the Eastern Cape, share a common commitment to transitioning to renewable energy sources. South Africa is working toward a cleaner, more sustainable energy future, and renewable resources are an important part of that mix.

A noteworthy commonality across the case studies is the recognition of the vast renewable potential within each province. Whether it be the solar and wind resources in KwaZulu-Natal, the challenges and opportunities in Gauteng, or the transformative vision of local electricity generation in the Eastern Cape, all provinces understand the importance of harnessing their unique renewable resources. This recognition serves as a foundation for strategic planning and the development of policies to unlock the full potential of renewable energy (Uhunamure & Shale, 2021).

24.8.2 Identification of Differences among Case Studies

Gauteng Province is facing a number of challenges, including the the limited mandate to turn around the energy situation, moving towards renewable energy sources, and the fact that they are reliant on national developments in terms of formulating and implementing strategies.

KwaZulu-Natal's case study highlights the province's wealth of renewable potential, including solar, wind, and biomass resources. The emphasis on harnessing these resources positions KwaZulu-Natal as a potential leader in diversified renewable energy solutions. The challenge here lies in conducting further investigations to determine the net electrical energy that can be produced from these sources and strategically planning for their optimal utilization.

The Eastern Cape distinguishes itself with a transformative vision focused on local electricity generation. The province aims to exceed local demand within a specific timeframe, showcasing a proactive approach to renewable energy adoption. This ambitious vision sets the Eastern Cape apart, emphasizing a decentralized energy model that empowers local communities and contributes significantly to the province's energy needs.

24.8.3 Evaluation of Successes and Challenges

24.8.3.1 Successes

The success in KwaZulu-Natal goes beyond the identification of renewable resource potential; it represents a pioneering step in understanding the intricate dynamics of renewable energy systems. By quantifying the renewable potential in solar, wind, and biomass, KwaZulu-Natal lays the groundwork for informed decision-making. This success not only serves as a blueprint for the province's sustainable energy journey but also contributes valuable insights for other regions contemplating similar transitions.

Gauteng's successes are notable in the face of challenges related to provincial mandates and coal dependence. The province's adoption of solar, wind, and biomass use signals a shift toward a diversified energy portfolio. This success underscores the resilience and adaptability of Gauteng in navigating the complexities of transitioning from non-renewable to cleaner energy sources. Gauteng's initiatives provide a model for other regions striving for a more sustainable and balanced energy mix.

The Eastern Cape's achievements are monumental, with a substantial shift toward local renewable energy production. The province's proactive approach, facilitated by the REIPPPP, not only meets a third of its power needs sustainably but also exemplifies leadership in the national transition to sustainable energy. The Eastern Cape's successes highlight the transformative power of strategic planning and collaboration, setting a precedent for other provinces to follow in achieving ambitious renewable energy goals.

24.8.3.2 Challenges

The challenge in KwaZulu-Natal underscores the need for sustained efforts in further investigations to determine the net electrical energy that can be harnessed from identified renewable sources. Strategic planning for the optimal placement and integration of renewable energy systems remains crucial. KwaZulu-Natal's challenge serves as a reminder that the journey to sustainable energy requires continual refinement and adaptability to evolving technological and environmental factors.

Gauteng's key challenge of limited provincial mandates and the attractiveness of cheap coal demands a strategic, multifaceted approach. Advocating for decentralized decision-making and fostering collaboration between provincial and national entities are essential steps toward overcoming these challenges. Gauteng's experience serves as a lesson in the importance of navigating economic considerations while prioritizing the long-term benefits of renewable energy adoption.

While the Eastern Cape shows a strong commitment to renewable energy, challenges may arise in achieving the ambitious vision of exceeding local demand within a specific timeframe. Continuous collaboration and strategic planning are essential to address potential obstacles. The Eastern Cape's challenges underscore the importance of aligning ambitious goals with realistic timelines, emphasizing the need for flexibility in the pursuit of sustainable energy objectives.

24.8.4 Discussion of Lessons Learnt

The case studies collectively emphasize that strategic planning and collaboration are foundational to successful renewable energy transitions. Proactive strategic planning not only addresses current challenges but also anticipates future shifts in energy demand and technology. Collaboration ensures synchronized efforts and aligned strategies, facilitating a cohesive and effective renewable energy transition that considers both short-term milestones and long-term sustainability.

Community involvement and economic development emerge as integral components of successful renewable energy projects. Gauteng and the Eastern Cape's experiences highlight the need for inclusive approaches that consider the socio-economic dynamics of each region. Beyond environmental benefits, renewable energy projects can stimulate local economies, create employment opportunities, and empower communities, showcasing the broader societal impact of sustainable energy initiatives. Additionally, the ability to adapt to evolving technological trends and innovations is critical for sustainable energy transitions. The case studies underscore the importance of embracing diverse renewable technologies, such as solar, wind, biomass, and geothermal. This adaptability ensures a resilient and versatile energy infrastructure capable of meeting evolving energy needs while staying at the forefront of technological advancements in the renewable energy sector.

Advocacy for supportive policies and the empowerment of provincial mandates are vital lessons drawn from the challenges faced, particularly in Gauteng. The need for provinces to have more authority in formulating and implementing strategies and frameworks becomes evident. Effective policy advocacy ensures that provinces can tailor solutions to their specific energy landscapes, fostering a conducive environment for sustainable energy investments and implementation.

24.9 CONCLUSION

24.9.1 Summary of Key Findings

The exploration of South Africa's evolving sustainable energy sector, with a focus on case studies in KwaZulu-Natal, Gauteng, and the Eastern Cape, has yielded valuable insights into the complex dynamics of renewable energy adoption. Key findings include the identification of significant renewable resource potential in KwaZulu-Natal, Gauteng's commendable initiatives to shift towards cleaner energy sources, and the Eastern Cape's proactive approach resulting in a substantial shift toward local renewable energy production.

The successes and challenges encountered in these case studies underscore the importance of tailored strategies, strategic planning, and collaboration between provincial and national entities. KwaZulu-Natal's commitment to further investigations, Gauteng's proactive shift despite challenges, and the Eastern Cape's ambitious vision collectively contribute to the broader narrative of South Africa's transition to a sustainable energy future.

24.9.2 Implications for Future Research and Practice in Renewable Energy

The findings from these case studies have significant implications for future research and practice in the field of renewable energy. Firstly, the need for continued research in quantifying and harnessing renewable resource potential, as demonstrated in KwaZulu-Natal, is crucial for effective strategic planning and resource utilization. Understanding the economic and policy landscape, as exemplified by Gauteng, emphasizes the importance of addressing challenges related to provincial mandates and economic factors for successful renewable energy adoption.

The Eastern Cape's success in local renewable energy production highlights the potential benefits of decentralized models and serves as a case for scaling such initiatives nationally. Future research should delve into refining strategies for achieving ambitious renewable energy goals while ensuring alignment with realistic timelines and adaptable approaches.

24.9.3 Final Thoughts on the Importance of Case Studies

Case studies play a pivotal role in advancing renewable energy research and practice. They provide real-world insights into the complexities, successes, and challenges of sustainable energy transitions. The nuances revealed in KwaZulu-Natal, Gauteng, and the Eastern Cape showcase the importance of case studies in capturing the contextual intricacies that can inform more effective renewable energy policies and practices. Additionally, case studies serve as living examples of the intersection between theory and practical implementation, offering lessons that extend beyond borders. They enable policymakers, researchers, and practitioners to draw from tangible experiences, fostering a more nuanced understanding of what works and what challenges may be encountered in the journey toward a sustainable energy future.

REFERENCES

Akinbami, O. M., Oke, S. R., & Bodunrin, M. O. (2021). The state of renewable energy development in South Africa: An overview. *Alexandria Engineering Journal*, 60(6), 5077–5093.

Bakshi, P., Bhambri, P., & Thapar, V. (2021). A review paper on wireless sensor network techniques in Internet of Things (IoT). In International Conference on Contemporary Issues in Engineering & Technology.

Hanto, J., Schroth, A., Krawielicki, L., Oei, P. Y., & Burton, J. (2022). South Africa's energy transition–unraveling its political economy. *Energy for Sustainable Development*, 69, 164–178.

Ibrahim, I. D., Hamam, Y., Alayli, Y., Jamiru, T., Sadiku, E. R., Kupolati, W. K., & Eze, A. A. (2021). A review on Africa energy supply through renewable energy production: Nigeria, Cameroon, Ghana and South Africa as a case study. *Energy Strategy Reviews*, 38, 100740.

Iorember, P. T., Jelilov, G., Usman, O., Işık, A., & Celik, B. (2021). The influence of renewable energy use, human capital, and trade on environmental quality in South Africa: multiple structural breaks cointegration approach. *Environmental Science and Pollution Research*, 28, 13162–13174.

Kuzhaloli, S., Devaneyan, P., Sitaraman, N., Periyathanbi, P., Gurusamy, M., & Bhambri, P. (2020). *IoT Based Smart Kitchen Application for Gas Leakage Monitoring*. IN Patent App. 202,041,049,866 A.

OIkuski, T. (2023). The modern status and prospects for further development in the Australian energy sector: Transformation, external economic relations, investment climate. *Polityka Energetyczna*, 26(3), 65–80.

Panos, E., Glynn, J., Kypreos, S., Lehtilä, A., Yue, X., Gallachóir, B. Ó., & Dai, H. (2023). Deep decarbonisation pathways of the energy system in times of unprecedented uncertainty in the energy sector. *Energy Policy*, 180, 113642.

Pillot, B., Muselli, M., Poggi, P., & Dias, J. B. (2019). Historical trends in global energy policy and renewable power system issues in Sub-Saharan Africa: The case of solar PV. *Energy Policy*, 127, 113–124.

Ram, M., Bogdanov, D., Aghahosseini, A., Gulagi, A., Oyewo, A. S., Mensah, T. N. O., & Breyer, C. (2022). Global energy transition to 100% renewables by 2050: Not fiction, but much needed impetus for developing economies to leapfrog into a sustainable future. *Energy*, 246, 123419.

TotalEnergies, S. E. TotalEnergies Energy Outlook. (2023). The energy transition: challenges and opportunities-13 november 2023. Analysis by sector and by scenario-13 November 2023. 2023 Strategy and outlook: More energy, less emissions, growing cash flow-September 27, 2023. Strategy, Sustainability and Climate: global strengths, global results, more energy, less emissions-March 21, 2023. 2022 Results and 2023 Objectives: Global strengths, global results-February 8, 2023.

Uddin, G. S., Sahamkhadam, M., Yahya, M., & Tang, O. (2023). Investment opportunities in the energy market: What can be learnt from different energy sectors. *International Journal of Finance & Economics*, 28(4), 3611–3636.

Uhunamure, S. E., & Shale, K. (2021). A SWOT Analysis approach for a sustainable transition to renewable energy in South Africa. Sustainability, 13(7), 3933.

Usman, O., Akadiri, S. S., & Adeshola, I. (2020). Role of renewable energy and globalization on ecological footprint in the USA: Implications for environmental sustainability. *Environmental Science and Pollution Research*, 27(24), 30681–30693.

25 Lean Management, Kaizen, and TIM WOODS Reduction in Logistics

Dorota Klimecka-Tatar

25.1 INTRODUCTION

The need for improving the quality of services arises from the ever-increasing requirements of subcontractors, customers, and clients, as well as the constantly growing competitiveness of the contemporary service sector. The improvement of quality is also associated with implementing changes aimed at increasing sustainability at every stage of logistics, business, or production processes. In the current favorable economic and social conditions, service providers compete using various methods to acquire new customers, directly leading to excellent results and profits (Dwyer and Wright, 2019). There are numerous methods, techniques, and concepts for process improvement that do not require significant financial outlays. Improvement tools that enhance existing processes without necessitating a complete reorganization are the most desirable ones (Chioatto and Sospiro, 2023; Chioatto et al., 2023).

25.1.1 LIMITATIONS AND CONSTRAINTS IN PROCESS AND TECHNOLOGICAL SUSTAINABILITY IMPROVEMENT

The theory of constraints (TOC) is the most popular theory indicating the limitations that, in the enterprise, are the sources of failures (delays, errors, waste, etc.) (Dabholkar, 2015). In the course of the process, such limitations are often called bottlenecks. The bottleneck limits development because it inhibits actions that must be taken at later stages of the process flow. In extreme cases, the bottleneck can even lead to process suspension (Bisogno et al., 2017).

On this basis, the main assumption of TOC is that the value and effectiveness of any enterprise should be assessed based on its weakest link, not the strongest one (Singh et al., 2004). Thus, the company's efficiency can be underestimated due to the appearance of limitations. In this case, limitations are all the elements that are part of the process and reduce the system's ability to achieve a goal.

 DOI: 10.1201/9781003475989-29

The main principles of the theory of constraints include (Tao et al., 2017; Ben-Ammar et al., 2020; Orue et al., 2023).

- Each company has only a few constraints, the removal of which will result in a radical change in results.
- Continuous improvement of systems is manifested through a process consisting of actions.
- Technological sustainability and process improvement based on questions: What to change? What to change for? How to make changes?
- Service quality development involves recognition, location, and identification of the constraint, determination of resources and activities, reconstruction, and subsequent loops of the continuous improvement process.

25.1.2 Constraints Identification in Process Improvement

The identification of constraints (constraints in ensuring process sustainability) is the starting point because it provides information about the scope of the service, the ability of the service, and errors made during its implementation (Orue et al., 2023) At this stage, a bottleneck analysis is usually performed, and ways to maximize bottleneck exploitation are sought. Typically, each restriction is assigned operations and activities resulting from the course of the process (Lizarralde-Aiastui et al., 2020). It may be associated with the purchase of materials, information flow, shipment of finished products, etc. Correctly identifying the constraint enables the introduction of changes that will either eliminate the constraint completely or reduce its impact on the further course of the process. The cyclical nature of constraint identification is part of the continuous improvement process (repetition of the procedure for the next, new constraint)

Within the realm of services, various constraints may appear at every stage of the system, both at the input, output, and in the transformation area. These constraints can be associated with difficulties caused by the human factor, as well as by machines. Concerning the human factor, it should be noted that employees should have specific activities and tasks to perform, while the task of superiors is to ensure control and coordination of these activities. Failure to maintain an appropriate hierarchy may result in inefficiency at a given workplace and failure to achieve consistency due to mistakes made.

On the other hand, constraints in the operation of machinery and equipment may be caused by the improper organization of technological operations, downtime, excessive equipment overload, improper capacity, damage, failures, etc. (Sarkar et al., 2015).

25.1.3 Areas of Process Improvement

The most crucial areas for improvement are human resource management and machine maintenance. However, it is essential not to overlook other connection networks created in each process, including information and material connection networks or management systems (Trojanowska et al., 2018; Trojanowska & Dostatni, 2017; Wolniak et al., 2018).

A significant area for improvement during the process is the information stream. Organizational competitiveness is often linked to well-managed information and key knowledge. The bottleneck's essence lies in the flow of substantial information through one node, leading to its excessive load, making the information illegible or transformed. Communication congestions can arise when information is not effectively transferred between team members or subgroups. There are also situations in which intentional bottlenecks block certain information, limiting or completely depriving employees of access to it – which makes the process ineffective.

When analyzing the company as an entire system, attention should be paid to inventory management – an area that can be improved to reduce unnecessary costs. In the realm of resource, material, and inventory management, the introduction of practices like Just in Time (JIT) and Kanban is often necessary. Extended storage raises questions such as: Are stocks required? What volume/level should they be? How often should deliveries be made? What storage space is required? (Pacana & Czerwińska, 2019). Inventories may serve as a production buffer to ensure process continuity in case of uncertainty about actual demand or to provide security in case of delays and compliance issues with deliveries. Problems can arise from excessive stocks (a greater volume of non-rotating stocks than buffer stocks) and unnecessary stocks (stocks of non-rotating goods).

Another area requiring improvement in the process is environmental management. Issues may include problems with waste precipitation, their scale, lack of employee training, inability to segregate waste, or poorly organized waste storage spaces (no waste containers around the workplace). A poorly organized waste management system may result in significant direct and indirect costs. Direct costs occur when enterprises lack supervision over waste and pay high fees for disposal. Indirect costs arise when the company, due to poor organization of the environmental management system, cannot reduce operating time, which makes the process unsustainable (Puche et al., 2016). Failure to understand the relationship between sustainable management, customer requirements, and the company's needs can occur due to a lack of communication between departments with different competencies, such as the financial department in charge of the situation, but its role at the quota stage may not be clearly defined. Therefore, regardless of the type of process being carried out, identifying constraints in the course of the process is as important as analyzing the problems that have already arisen (Ingaldi & Ulewicz, 2019).

25.2 THE MOST IMPORTANT CONCEPTS OF PROCESS IMPROVEMENT

25.2.1 Kaizen Philosophy

The necessity to enhance the sustainability and quality of processes by eliminating bottlenecks and improving business processes is closely linked to the constantly growing economy and the trends in the sustainable global economy. Numerous methods, techniques, and concepts are available to improve processes while addressing errors and problems within them. One notable solution for enterprise improvement is the Kaizen philosophy, originating from Japanese enterprises. This philosophy revolves around effective management, integrating both philosophy and tools to solve problems with a focus on the continuous improvement of customer-/environment-oriented activities (Singh et al., 2005). Kaizen aims to identify problems and provides tools for their recognition, addressing improvements in the human factor, operating system, and general organization. Therefore, it emphasizes the importance of integrating activities throughout the entire organization.

By implementing small but continuous changes in the improvement process, uncertainty can be limited without the need for radical changes, ensuring success in a gradual yet effective manner. These changes apply to the entire organization, and the Kaizen method does not necessitate substantial investments.

25.2.2 Lean Management

The most popular technique in recent years – lean management (LM) – derives from the Kaizen approach and continuous improvement in the PDCA cycle. Lean management emphasizes the elimination of waste (*muda*) (Larina, et al., 2021; Mazur & Momeni, 2018; Šukalová & Ceniga, 2015).

LM refers absolutely to the entire organization and employees at all levels in the structure of the organization. LM is focused on seeking consensus among three variables: time, costs, and quality, which have a strong impact on the company's sustainability. Thus, LM aims to provide high (required) quality at the lowest (optimal) costs while maintaining delivery times (reduced cycle time). In LM, the needs of customers are carefully considered and analyzed, and the value of products or services is clearly defined from the customer's point of view (Suchacka et al., 2023). Organization in accordance with the Lean Management technique can guarantee a transparent stream of service/product value building through:

- recognition of value-adding (VA) and non-value-adding (NVA) activities;
- use of small batches of products/services;
- organization of the service process in order of operations;
- reduction of inter-operational stocks.

The LM concept emphasizes operational management, recognizing problems throughout the entire service and production process, and focusing attention on the maximum flow of streams. LM assumes that the sustainable functioning and effectiveness of the entire system is determined by waste – *muda*. The bottleneck downtime, excess inventory, misuse of people skills, equipment, and waste of time create unbelievable costs and have consequences throughout the system. The essence of this method is planning backwards for a given period, identifying bottlenecks, modifying procedures, scheduling for optimal use of resources, and developing a plan for non-critical resources (Conger, 2010; Klimecka-Tatar, 2021).

Process improvement can make it more sustainable. The adjustments and optimizations introduced in the process can contribute to waste reduction, more efficient resource utilization, and a minimized negative impact on the environment. Striving for a sustainable process often involves increasing operational efficiency, which can help achieve a balance between economic, social, and environmental aspects.

25.3 EXPERIMENTAL

One of the most commonly used methods for identifying stream flow (process flow) is visual management using typical imaging tools and process map creation, i.e., value stream mapping (VSM), which enables processing of Big Picture analysis (Klimecka-Tatar, 2018; Zwolińska, 2016). Process improvement is based on a three-level analysis:

Stage 1: creation of a current state map (CSM) of the process.
Stage 2: identification of areas for improvement (implementation of the plan according to the Kaizen philosophy with planned changes in areas potentially adversely affecting the flow of the process).
Stage-3: creation of a future state map (FSM) and analysis of the newly designed stream flow.

Value stream mapping is derived from the Lean philosophy, i.e., a lean approach, allows for improved processes. This method identifies sources of waste and areas for improvement. On this basis, it is possible to solve problems and improve the indicated activities. Value stream mapping (VSM) is performed using three steps:

- VSA – Value Stream Analysis – identification of the current state with the current flow of value streams.
- VSD – Value Stream Designing – creation of future state map concept.
- VSWP – Value Stream Work Plan – creation of an improvement plan.

The basis for analyzing necessary changes is the current state map (CSM), which should comprehensively present all material and information dependencies relevant to the course of the process. Creating a process map refers to a series of information that forms the basis for process visualization. As known, a typical Big Picture created during value stream mapping (VSM) analysis consists of five interdependent areas. Each of these areas is equally important for creating a map, and a thorough and precise understanding of these areas is essential for effectively improving the process. In line with the methodology presented in the work (Klimecka-Tatar, 2021) during the visualization of service processes, several questions should be posed. When mapping service processes, a logical map of information flow also becomes a valuable tool. The process of mapping information flow helps identify the departments responsible for various decisions made.

In this chapter, the procedural scheme during the implementation of the lean concept in the field of process improvement is presented. An analysis of the process in a large company specializing in the distribution of goods is also provided. The distribution company operates as a global retail network that collects, stores, and subsequently sells home furnishings. The recipients are stores within the distribution company group and, subsequently, retail customers. The service process analysis includes creating a logical map of the decision process, value stream mapping, identifying restrictions (bottlenecks), and analyzing operations adding value (AV) and not adding value (NAV) to the logistic service.

25.4 RESULTS AND DISCUSSION

25.4.1 Logic Map in Distribution Service Process

The logical map of the process visualizes the current state of the distribution process within the enterprise. Figure 25.1 presents the logic map for the process, allowing for the organization of the sequence of information flow between company departments. The diagram has been created based on the successive steps, determining:

- Mapping range (considering the beginning and end of the process).
- Activities included in the process.
- Hierarchy of operations performed.
- Connections between subsequent activities.
- Implementers of activities.

Management, operational, and support processes are distinguished on the logical map. The group of management processes includes planning, order acceptance, as well as the transfer of responsibility to operators. In the case of the discussed service process, the main operational processes encompass all activities contributing to the final goal of the service: unloading, acceptance of delivery, storage, packaging, and loading. The group of support processes includes inter-operational transport, control activities, protection of goods against dispatch, and proper maintenance.

The service process logical map serves as an extremely important tool when searching for waste areas, as it presents the current state of the organization along with the flow of value streams, primarily information streams. It also characterizes key system operators (departments). In this case, after accepting the order, the information passes through the departments of planning, acceptance, transport, completion, foiling, and loading.

Having a complete picture of the flow of the information stream through all departments, it is possible to create an improvement plan, as well as define customer service standards, plan the ordering of goods, coordinate inventory, and change/improve work organization in the enterprise.

25.4.2 Value Stream Mapping and Big Picture

In contrast to the logical map, the value stream mapping (VSM) map provides an opportunity to visualize the flow of both the information and material streams. The

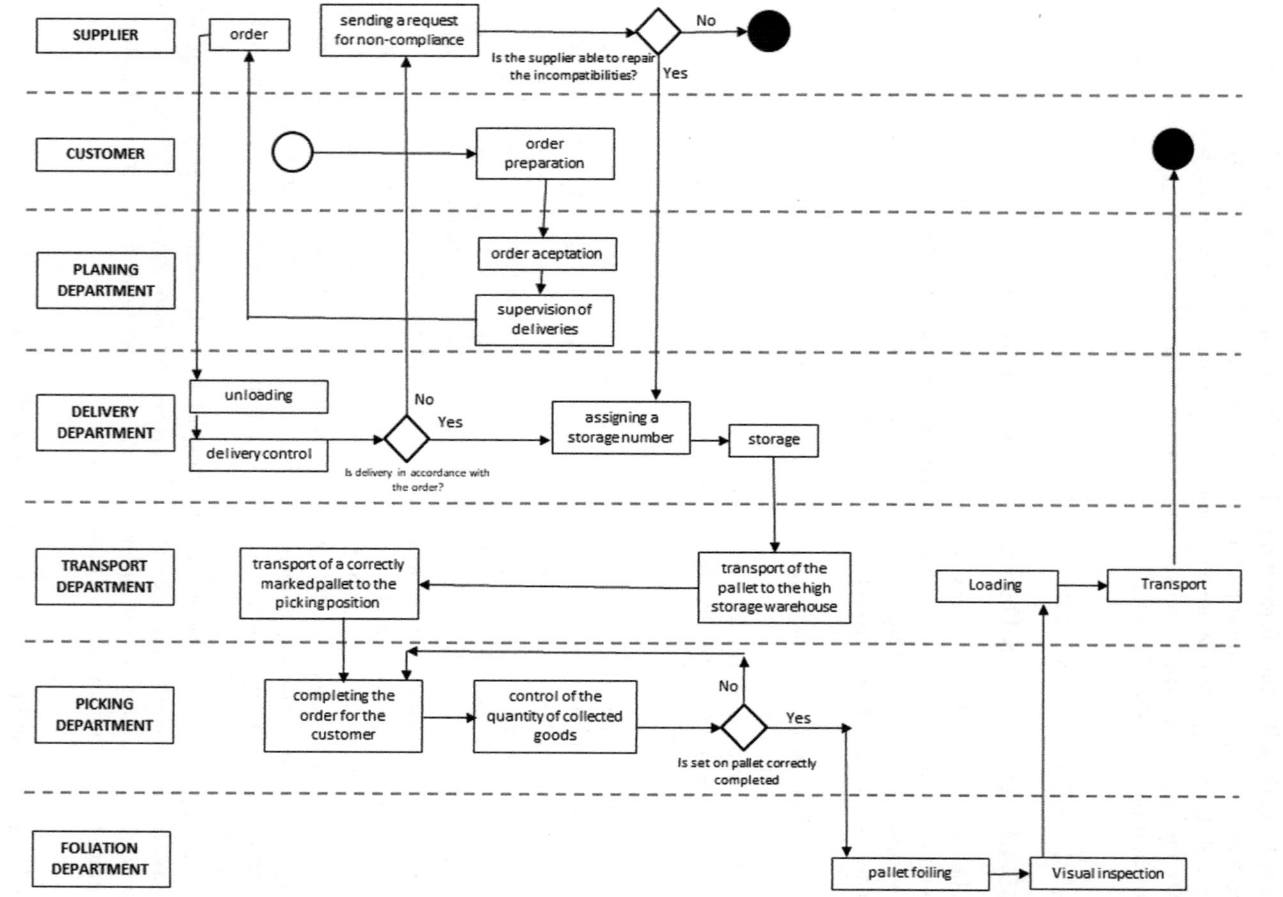

FIGURE 25.1 Logic map of the distribution process – flow of information streams

VSM map also enables the visualization of the coordination of the streams and the creation of a correlation between them over time. The Current State Map (CSM) in the distribution company is presented in Figure 25.2. The map illustrates numerous processes necessary to define the service as complete, ready, and full-valued in the eyes of the customer. It visualizes both the flow of goods and information, depicting the situation of the enterprise at the time of the research. Information about customers, suppliers, and warehouse stocks has also been included. Subsequently, the course of the distribution process with inter-station transport has been presented. The deployment of all operations over time is an important complement to the current state map.

The upper part of the map outlines the information flow starting from the customer through the enterprise planning department, creating order scheduling, and the picking department up to suppliers. The bottom part of the map presents the individual operations needed to perform the service, including goods receipt operations (unload, system receipt of delivery, definition of storage location), followed by storage, management of high storage warehouse (reservations of products needed for shipping, posting transport for picking up on warehouse elevators), packing/completing goods for a given store order, then foiling the goods, and the final stage – loading.

Operations marked as numbers 1, 4, and 5 are supplemented by qualitative and quantitative control. It is easy to observe that the longest operation is a storage operation, which can take up to 2 months. This operation is classified as a group of non-value-adding operations (NVA) – from the customer's perspective, it does not build the value of the service and should be modified. Each of the operations is performed by forklift operators with terminals controlled by the ERP system. The necessary information is displayed, directing the employee to the appropriate location of the goods. The lead time (L/T), i.e., the total process time of one distribution batch, is about 11 hour 10 min (excluding storage time of +2 months), while the value-adding operations are only 9 hour 40 min.

Based on the collected data and analysis of the value streams flow (materials and information flow) in a distribution company, it is possible to identify process bottlenecks. For this purpose, the types of waste (*muda*) have been indicated. In the Lean Six Sigma approach, production loss and waste are often called by the acronym TIM WOODS, which should be explained as (Conger, 2010):

T – Transport – excessive, unnecessary transport of materials, people and information.
I – Inventory – excessive deliveries, unnecessary stocks, collection of material, documents, and waste.
M – Motion – unnecessary movement.
W – Waiting – stopping the process, waiting for materials, information, documents, decisions.
O – Overproduction – making more than required.
O – Over-processing – excessive processing.
D – Defects – quality errors, incompatibilities, and deficiencies in production.
S – Skills – unused or misused employees' potential.

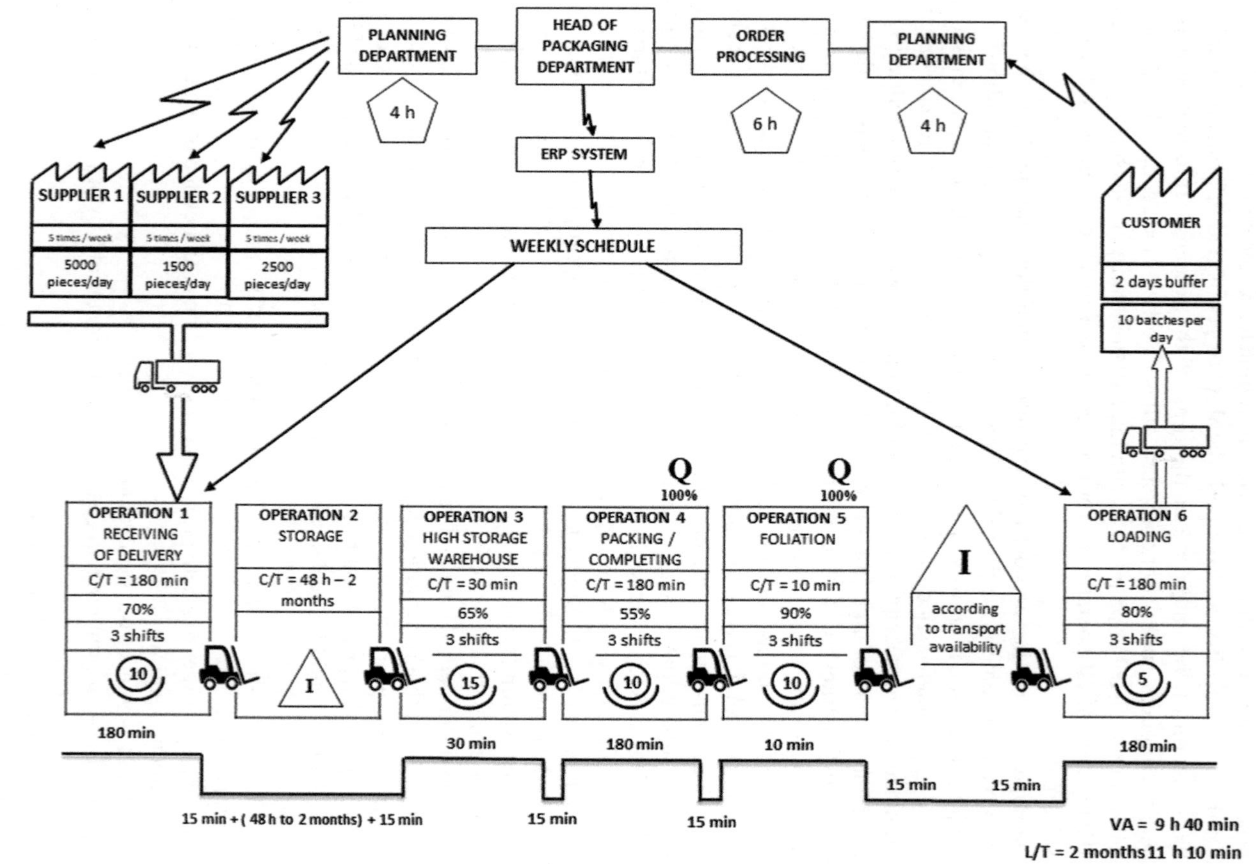

FIGURE 25.2 Current state map (CSM) for the distribution process in the research and observation period.

The pointed-out bottlenecks are presented in Table 25.1. The term *muda,* or waste, derived from the Toyota Production System (TPS), refers to all activities that do not add value, i.e., those for which the customer does not want to pay. They cause a waste of time and money for the organization and make the company less sustainable.

As evident from the data presented in Table 25.1, the primary types of waste in a distribution company include waiting, excessive processing, unnecessary inventory, unnecessary movements, and deficiencies/errors. The most significant losses and waste are observed in Operation 1 (receiving of delivery). Waiting occurs in a high storage warehouse when the goods carrier is reserved for shipment. The goods carriers change location from the high storage warehouse to the elevator. Due to elevator overload and unavailability, the carrier enters a waiting state, significantly extending the flow of the entire process. Such waiting causes delays in further stages of the process. Excessive processing is evident during the acceptance of deliveries. Often, due to haste, deliveries are unloaded without attention to their quality, necessitating improvement before insertion into the warehouse automation module. Excessive processing and unnecessary movements also occur in loading operations where incorrectly estimated loads result in excessive goods transport.

Storage times range from 48 hours to even 2 months, incurring unnecessary storage costs and extending the cycle time of the entire process. Errors, deficiencies, defects are an inevitable aspect of handling a large volume of goods, visible from the process's outset where pallets are incorrectly received and marked. The occurrence of such errors is a result of excessive haste at the beginning of the process. Unfortunately, these errors have a drastic impact on the process's correctness at later stages. Labeling errors contribute to delays necessitated to verify errors and resolve mislabeling (loss of goods, incorrect location of goods, incorrect storage levels). Visualization of the process's current state has clearly indicated areas for improvement – sources of waste (*muda*). The mapping process has defined the distribution process, its elements, and places where the stream flow was interrupted. The current state map outlines in detail the information and material flows.

25.4.3 Service Process Improvement Plans

Areas in need of improvement were identified at the outset of the process (Operation 1 – receiving of delivery). The problem arose from the inability to locate pallets, which, after being registered in the system, were transported to the wrong locations. Frequent mistakes, such as improper transport of pallets, were attributed to the presence of old system markings; obsolete barcodes were not removed after pallet recycling. Additionally, it was observed that many damaged pallets were transferred further in the workflow, leading to process suspension during loading. The replacement of damaged pallets with undamaged ones resulted in time loss, additional movements, operations, and unnecessary transport.

As a solution, the implementation of 5S practices has been proposed. In this case, 5S practices should focus on appropriately marking the storage positions of pallets, indicating their purpose, and prioritizing them. It has also been suggested to designate a specific area for sorting and selecting pallets. An employee should be

TABLE 25.1
Identification of Losses, Waste, and Bottlenecks in the process

	Sequence of operations in the process					
	Operation 1	Operation 2	Operation 3	Operation 4	Operation 5	Operation 6
Types of waste TIM WOODS	Receiving of delivery	Storage	Management in a high storage warehouse	Packing / completing	Foliation	Loading
T – Transprt	x					
I – Inventory		x	x			
M – Motion	x			x		
W – Waiting	x	x	x	x		x
O – Over-production	x					
O – Over processing						x
D – Defects	x			x		x
S – Skills	x					

appointed to conduct initial quality control of the pallets and ensure proper preparation, including the removal of old markings (obsolete barcodes). These changes aim to eliminate pallets that do not meet warehouse standards, and clear markings should prevent errors in warehouse transport. According to the 5S method, external and internal locations should be marked throughout the entire warehouse area, including storage areas. The storage area should be marked with colors, arrows, and location numbers, along with the assigned assortment type. The installation of information boards could be an additional advantage. Such solutions provide the warehouse operator with clear information on where to place a given assortment, facilitating the process of collecting goods for shipment.

Another problem identified for the enterprise is insufficient stock rotation – goods stored for too long. Excess inventory in a distribution company is wasteful due to incurred costs and the need for warehouse space. At the beginning of the improvement process, it was observed that storage times ranged from 48 hours to 2 months. The proposed solution involves implementing the first in, first out (FIFO) inventory organization technique and enhancing planning through the implementation of "go and check". These changes, besides reducing storage time, will support a greater assortment rotation. However, it should be noted that delivery planning should also consider seasonality, demand, and the inventory rotation ratio.

The next area identified for improvement was Operation 4 (packing/completing). Downtime at this stage resulted from the inability to locate the selected group of products or the physical absence of products. Although a store can check the availability of a given assortment in the warehouse when placing an order, what is available in the system may not always be physically available. This issue was mainly attributed to missing pallets during the registration stage (Operation 1), leading to subsequent deficiencies during packaging. Once again, the 5S method is proposed as the solution, aiming to eliminate problems in the further course of the process.

There was also waste identified at the end of the process in Operation 6 (loading). It was observed that after loading, parts of orders that did not fit in the trucks remained in the warehouse. The solutions to this problem are as follows:

- Planning should be based on the number of pallets, not load meters.
- Leave a load safety buffer in the car.
- Load the car with one order first (in the case of free space, top it up with other smaller orders).
- Take into account not only the number of pallets loaded but also the quality and utilization of loading space.
- Provide employee training in loading goods.
- Implement 100% control – verify all loaded pallets. The employee loading the delivery is responsible for it, and the manager is informed in case of irregularities.

Based on the observations and analysis of the value stream flow, process time axes have been constructed. The axes specify value-adding and non-value-adding activities. Figure 25.3 presents the process in relation to the current state map and the process in relation to the proposed changes.

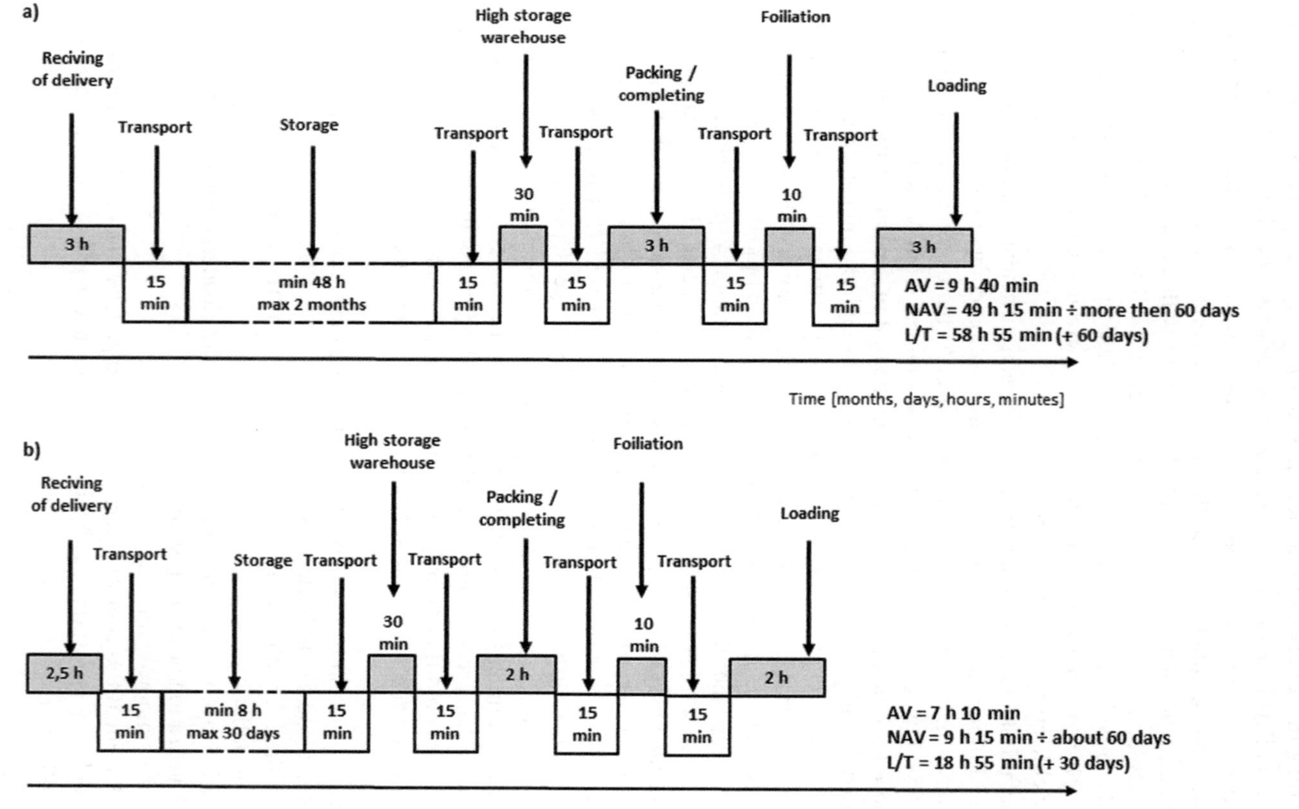

FIGURE 25.3 Structure of value-adding and non-value-adding activities: (a) according to the current state of the revised process; (b) after the implementation of the proposed changes.

Figure 25.3 illustrates all the activities required to perform the service, along with their durations. The process commences when the enterprise accepts the order and proceeds through the following activities: receiving the delivery, transporting to the warehouse, storage, transporting from the warehouse to separate automation lifts, movement in a high storage warehouse managed by the ERP system, picking/completing, packing goods, foiling, and loading. Some of these activities add value, meaning they are necessary from the customer's perspective to perform the service, while others, not deemed important or required by the recipient, do not add value to the finished product. Therefore, the presented improvement plan assumes that all changes will be aimed at eliminating bottlenecks in the flow and reducing non-value-adding activities.

The flow of value streams in the future state design should be enhanced by limiting the time of non-value-adding activities and eliminating bottlenecks in the flow. Such changes will make the service more effective and profitable while reducing the implementation process.

The implementation of Kaizen improvement in the distribution process has provided the opportunity to reduce the time of non-value-adding activities, mainly decreasing storage time and downtime. Inventory rotation in the warehouse has significantly increased (utilizing FIFO and "go and check"). Those responsible for purchasing supervision have determined the seasonal demand for goods, resulting in a significant freeing up of storage space.

In Operation 1 – acceptance of delivery, the time has been reduced by 30 min, benefiting Operation 6 – loading, where the control process was missing. However, the time for Operation 4 – packing/completing decreased by 1 hour due to the implementation of the 5S method at the beginning of the process.

25.5 CONCLUSIONS

The presented research results prove that the visualization of the value stream flow (creation of the current state map) makes it possible to easily identify the most important limitations of the problem-free process flow. The approach, in line with the Kaizen philosophy and the concept of Lean Management, also enabled the introduction of small changes, which consequently significantly improved the process. Eliminating problems related to various types of waste also contributes to making the process more sustainable. A sustainable service not only addresses environmental concerns but also enhances process efficiency. Through this, the entire process becomes more sustainable.

According to the presented analysis of the current state map (CSM), the total lead time of the process (L/T) was about 58 hours 55 min (+ 60 d), but while the introduction of improvement elements allowed a reduction of only 18 hours 55 min (+ 30 d).

The biggest effect has been brought by the introduction of pallet management principles in Operation 1 – receiving of delivery. Both the quality control of pallets and their storage in specially assigned places have significantly reduced downtime. The improvement of operations in the area of Operation 1 additionally positively affects operations in the further course of the process. 5S practices that have enforced segregation and supervision of supplies have proved to be a key element of improvement. In addition, certain assumptions of 5S mean that the availability of the range in the warehouse is more transparent, so proper inventory management is possible.

REFERENCES

Ben-Ammar, O., Bettayeb, B., & Dolgui, A. (2020). Integrated production planning and quality control for linear production systems under uncertainties of cycle time and finished product quality. *International Journal of Production Research*, 58(4), 1144–1160. doi: 10.1080/00207543.2019.1613580.

Bisogno, S., Calabrese, A., Ghiron, N. L., & Pacifici, A. (2017). Theory of constraints applied to scheduled and unscheduled patient flows: does it improve process performance? *International Journal of Services and Operations Management,* 26(3), 365. doi:10.1504/IJSOM.2017.081943.

Chioatto, E., Khan, M. A., & Sospiro, P. (2023). Sustainable solid waste management in the European Union: Four countries regional analysis. *Sustainable Chemistry and Pharmacy*, 33, art. no. 101037. doi:10.1016/j.scp.2023.101037.

Chioatto, E., & Sospiro, P. (2023) Transition from waste management to circular economy: The European Union roadmap. *Environment, Development and Sustainability,* 25(1), 249–276. doi:10.1007/s10668-021-02050–3.

Conger, S. (2010). Six Sigma and business process management. In J. vomBrocke & M. Rosemann (Eds.), *Handbook on Business Process Management 1. International Handbooks on Information Systems.* Springer, Berlin, Heidelberg, 127–146. doi:10.1007/978-3-642-45100-3_6.

Dabholkar, P. A. (2015). How to improve perceived service quality by increasing customer participation. In B. Dunlap (Eds.), *Proceedings of the 1990 Academy of Marketing Science (AMS) Annual Conference. Developments in Marketing Science: Proceedings of the Academy of Marketing Science.* Springer, Cham, 483–487. doi:10.1007/978-3-319-13254-9_97.

Dwyer, E., & Wright, O. (2019). Low-wage job growth, polarization, and the limits and opportunities of the service economy. *RSF: The Russell Sage Foundation Journal of the Social Sciences,* 5(4), 56. doi:10.7758/RSF.2019.5.4.02.

Ingaldi, M., & Ulewicz, R. (2019). How to make e-commerce more successful by use of kano's model to assess customer satisfaction in terms of sustainable development. *Sustainability,* 11(18), 4830. doi:10.3390/su11184830.

Klimecka-Tatar, D. (2018). Context of production engineering in management model of Value Stream Flow according to manufacturing industry. *Production Engineering Archives,* 21(21), 32–35. doi:10.30657/pea.2018.21.07.

Klimecka-Tatar, D. (2021). Analysis and improvement of business processes management – based on value stream mapping (VSM) in manufacturing companies. *Polish Journal of Management Studies*, 23(2), 213–23. doi: 10.17512/pjms.2021.23.2.13.

Kuhlang, P., Hempen, S., Sihn, W., & Deuse, J. (2013). Systematic improvement of value streams - fundamentals of value stream oriented process management. *International Journal of Productivity and Quality Management,* 12(1), 1. doi:10.1504/IJPQM.2013.054860.

Larina, I.V., Larin, A.N., Kiriliuk, O., & Ingaldi, M. (2021) Green logistics - Modern transportation process technology. *Production Engineering Archives*, 27(3), pp. 184–190. doi:10.30657/pea.2021.27.24.

Lizarralde-Aiastui, A., de Eulate, U. A.-P., & Mediavilla-Guisasola, M. (2020). A strategic approach for bottleneck identification in make-to-order environments: A drum-buffer-rope action research based case study. *Journal of Industrial Engineering and Management*, 13(1), 18–37. doi:10.3926/jiem.2868.

Mazur, M., & Momeni, H. (2018). LEAN production issues in the organization of the company - the first stage. *Production Engineering Archives,* 21(21), 36–39. doi:10.30657/pea.2018.21.08.

Naor, M., & Coman, A. (2017). Offshore responsiveness: theory of Constraints innovates customer services. *The Service Industries Journal,* 37(3–4), 155–166. doi:10.1080/02642069.2017.1303047.

Orue, A., Lizarralde, A., Apaolaza, U., & Amorrortu, I. (2023). Designing the process for implementing step three of the theory of constraints in a make-to-order environment: Integrating sales and operation planning. *Journal of Industrial Engineering and Management,* 16(2), 205–214. doi:10.3926/jiem.5127.

Pacana, A., & Czerwińska, K. (2019). Analysis of the causes of control panel inconsistencies in the gravitational casting process by means of quality management instruments. *Production Engineering Archives,* 25(25), 12–16. doi: 10.30657/pea.2019.25.03.

Puche, J., Ponte, B., Costas, J., Pino, R., & La Fuente, D. de (2016). Systemic approach to supply chain management through the viable system model and the theory of constraints. *Production Planning & Control,* 27(5), 421–430. doi:10.1080/09537287.2015.1132349.

Sarkar, B., Chaudhuri, K., & Moon, I. (2015). Manufacturing setup cost reduction and quality improvement for the distribution free continuous-review inventory model with a service level constraint. *Journal of Manufacturing Systems,* 34, 74–82. doi:10.1016/j.jmsy.2014.11.003.

Singh, P., Singh, M., & Bhambri, P. (2004, November). Interoperability: A problem of component reusability. In International Conference on Emerging Technologies in IT Industry (p. 60).

Singh, P., Singh, M., & Bhambri, P. (2005, January). Embedded systems. In Seminar on Embedded Systems (pp. 10–15).

Suchacka, M., Pabian, A. M., & Ulewicz, R. (2023). Industry 4.0 and socio-economic evolution. *Polish Journal of Management Studies,* 28(1), 311–329. doi:10.17512/pjms.2023.28.1.18.

Šukalová, V., & Ceniga, P. (2015). Application of the theory of constraints instrument in the enterprise distribution system. *Procedia Economics and Finance,* 23, 134–139. doi:10.1016/S2212-5671(15)00445–1.

Tao, X., Xia, T., & Xi, L. (2017). Dynamic opportunistic maintenance scheduling for series systems based on theory of constraints (TOC)–VLLTW methodology. *Journal of Manufacturing Science and Engineering,* 139(2), 257. doi:10.1115/1.4034474.

Trojanowska, J., & Dostatni, E. (2017). Application of the theory of constraints for project management. *Management and Production Engineering Review,* 8(3), 87–95. doi:10.1515/mper-2017–0031.

Trojanowska, J., Kolinski, A., Varela, M. L. R., & Machado, J. (2018). The use of theory of constraints to improve production efficiency – industrial practice and research results. *DEStech Transactions on Engineering and Technology Research* (icpr). doi:10.12783/dtetr/icpr2017/17667.

Wolniak, R., Skotnicka-Zasadzień, B., & Zasadzień, M. (2018). Application of the theory of constraints for continuous improvement of a production process-case study. *DEStech Transactions on Social Science, Education and Human Science (seme).* doi:10.12783/dtssehs/seme2017/18023.

Zwolińska, B. (2016). Use of the method VSM to the identify muda. *Research in Logistics and Production,* 6(6), 513–522. doi:10.21008/j.2083–4950.2016.6.6.3.

26 Catalysing Sustainable Progress

Empowering MSMEs through Tech Innovation for a Bright Future

S. Ruby, T. Biju, and Pankaj Bhambri

26.1 INTRODUCTION

Information technology (IT) is crucial in increasing the profitability of organisations. It fosters competition among small and medium-sized enterprises (SMEs), prompting them to prioritise research and development as well as access cutting-edge technology (Rana et al., 2024). IT is a system that allows small and medium-sized enterprises (SMEs) to effectively address customer complaints and needs by facilitating the transmission of information via a variety of means, including social networking sites, phone calls, and emails. Enterprises can utilise several IT service platforms for purposes such as education, promotion, sales, and trade, among numerous other applications of this technology. The diverse advantages of Information Technology for micro, small, and medium enterprises (MSMEs) are extensively recognised. It has the potential to improve the efficiency of MSMEs, decrease expenses, and broaden their market presence, both domestically and internationally. Given the significance of the MSME sector in the national economy, individual MSMEs collectively derive benefits that result in good outcomes such as job creation, revenue generation, and enhanced competitiveness of the country's enterprises. India is a rapidly developing country that is fiercely competing with global superpowers in all areas. India has experienced significant progress in various domains, including enterprises, defence, economics, sports, education, and other sectors, in recent years. SMEs and MSMEs are widely recognised as the primary drivers of economic growth on a global scale. Collectively, they constitute 90% of all businesses globally, or 70% of total employment. When discussing India, SMEs are expected to have a crucial impact on realising the goal of a $5 trillion economy by 2025. Reaching this stage is only possible when SMEs and MSMEs prioritise the adoption of technology. They have not generated significant employment prospects with minimal investment, failed to address regional disparities, and facilitated the growth of industries in both

DOI: 10.1201/9781003475989-30

urban and rural areas. India has already experienced rapid adoption of advanced technology by small-scale sectors. The proportion of tech adoption among SMEs and MSMEs had a substantial increase compared to the pre-pandemic rate of 29%. Half of the MSMEs have integrated WhatsApp and Video Conferencing tools into their routine company operations.

According to a 2020 Cisco study, SMEs are projected to contribute between $158 million and $216 million to the country's GDP by 2024 through their adoption of digital business practices. India currently possesses an estimated 63 million MSMEs. MSMEs, following the pandemic, have exhibited the most significant shift in behaviour towards the adoption and utilisation of technology. A significant digital transformation has occurred in various aspects of companies, including communication, marketing, payments, and recruiting. The digital environment for MSMEs has undergone significant transformations in recent years. Several novel advancements have surfaced that have the potential to influence the attributes of a prosperous digital ecosystem. These include enhanced cloud services, the adoption of blockchain technology, and the incorporation of Computer Science and Artificial Intelligence, among others (Singh and Bhambri, 2023). Recent learning experiences have primarily centred on the utilisation of technology to enhance both the sustainability and performance of companies. SMEs and MSMEs that are leveraging technology have sustainable solutions for all aspects of their ecosystem development, encompassing stakeholder engagement, marketing, supply chain management, product innovation, and more. An illustrative instance is the expansion of digital payment systems, which diminishes reliance on the physical infrastructure of banks, enabling payments to be conducted at any location and time.

26.2 LITERATURE REVIEW

The literature review will analyse previous studies and academic works about the portrayal of technology adoption in the domains of MSMEs. Technology adoption provides numerous advantages for MSMEs in the face of limited expertise and budgetary limitations at the company level. MSMEs can utilise technology adoption to actively pursue disruptive innovation (Nguyen, 2009) and improve their product or service offerings in order to create sustained value (Vrontis et al., 2022). Adopting technology in MSMEs requires a greater level of awareness because MSMEs have unique characteristics that differ from large organisations. However, it is also argued that these characteristics, particularly their flexibility, can allow them to adapt to market changes by rapidly changing their business models (Shaikh et al., 2021), especially when utilising digital technology.

During our literature analysis on the tactics and impact of technology on the survival of SMEs, we came across multiple studies that focused on the creation of creative business models, communication methods, and the use of digital marketing. Untaru and Han (2021) conducted a study investigating the impact of gender, age, education, and income level on company strategy. The researchers determined that merchants had to formulate distinct communication strategies tailored to specific client categories, based on their intended demographic profiles. This approach is expected to enhance defensive countermeasures for COVID-19 and mitigate potential losses. Thorgren and Williams (2020) analysed a sample of 456 SMEs that were facing difficulties to investigate the measures they employed to ensure their survival amidst the COVID-19 pandemic. It was discovered that the majority of SMEs postponed making investments, decreased labour expenses and other costs, and restructured the conditions of contracts and loans. Omar et al. (2020) investigated the survival tactics implemented by Malaysian SMEs in response to the Movement Control Order (MCO). Their study revealed alterations in both financial and marketing strategies. According to AlKoliby et al. (2023), digital marketing has a beneficial impact on the performance of SMEs.

In their study, Hussain et al. (2021) examined the influence of many factors, including technological (such as comparative superiority and level of preparedness for technology), organisational (such as adoption expenses and help from senior executives), and environmental (such as support from the government and the influence of competition) factors, on the adoption of Business-to-Business e-commerce by Pakistan's SMEs in the manufacturing sector. This study conducted an analysis using multiple groups of the obtained data and found that technological elements have a substantial impact on Business-to-Business e-commerce. According to Amat and Ishak (2019), the prosperity of SMEs is contingent upon their capacity to engage in innovation, secure funding, establish networks, and leverage technology. Marin Bustamante (2020) conducted a review on the impact of emerging technologies (such as blockchain, AI, mobile payments, the Internet of Things, big data and analytics, and AI) on corporate operations during the epidemic lockdowns. She said that a total of 65 studies were published between 2011 and 2020, and none of them examined the impact of new technology on the international expansion of businesses during the COVID-19 epidemic.

26.3 RESEARCH GAP

After conducting a thorough literature study, it was found that a substantial body of research has already been conducted, and there are active initiatives now in progress. These studies concentrate on a particular technology while examining various aspects that contribute to the adoption of digital technology. This narrow emphasis makes it challenging to apply the findings to other technologies. The primary objective of this study is to address the existing research void by investigating several factors that influence the adoption of a diverse range of digital technologies by MSMEs, and the issues they face (Bhambri and Khang, 2024). There is a pressing need to provide more emphasis on prioritising the adoption of ICT in MSMEs. Understanding the significance of technology in the creation of new businesses is particularly important in a country like India, where the government is strongly dedicated to promoting and nurturing an entrepreneurial culture. This study aims to rectify this inadequacy. The aim of this study is to examine the adoption of technology among entrepreneurs in rural areas, as these factors have a key role in fostering the flourishment of entrepreneurial culture in the nation.

26.4 OBJECTIVES OF THE STUDY

- To assess the level of technology use for boosting sustainable development in MSMEs of Kerala.
- To identify and analyse the obstacles and challenges encountered by MSMEs in Kerala when adopting technology.

26.5 METHODOLOGY

The current study is characterised by its descriptive nature. This study utilises a combination of primary and secondary data sources. The aggregation of secondary data involves gathering information obtained from websites, publications distributed in newspapers, and studies conducted by other research institutions. Data is primarily acquired through the utilisation of an interview schedule. The current investigation has been centred on MSMEs that are officially registered in the Kollam District of Kerala. The study consists of a cohort of 140 micro rural units. The current study has utilised a random sampling methodology to choose a sample from a population of registered MSMEs in the District Industries Centre, Kollam.

26.6 RESULTS AND DISCUSSIONS

Kerala's economic development is distinguished by a unique strategy that leverages its highly proficient workforce and increased quality of life (Bhambri and Rani, 2024a). The State's capacity to foster a highly literate populace and construct strong education systems across various levels has enabled the growth of a skilled workforce that exhibits extraordinary capability in multiple domains (Bhambri and Rani, 2024b). Owing to its restricted industrial development, the State has been unable to efficiently utilise its human capital, leading to the exodus of individuals from Kerala to various global regions, especially the Gulf countries. The rise of knowledge-based sectors has led to a notable slowdown of this pattern in recent years. The Government of Kerala has prioritised the development of MSMEs to establish a viable ecosystem for entrepreneurship. The distribution of India's MSMEs by sector is presented in Table 26.1.

TABLE 26.1
Proportional Allocation of Businesses Owned by Male and Female MSMEs

Category	Male	Female	Total
Micro	79.56	20.44	100
Small	94.74	5.26	100
Medium	97.33	2.67	100
Total	79.63	20.37	100

Source: MSME Yearly Report 2022–23

TABLE 26.2
Geographical Distribution of the Number of MSMEs in India, Categorized by State, Measured in Units of Hundreds of Thousands

State	Number
Uttar Pradesh	89.99
West Bengal	88.67
Tamil Nadu	49.48
Maharashtra	47.78
Karnataka	38.34
Bihar	34.46
Andhra Pradesh	33.87
Gujarat	33.16
Rajasthan	26.87
Madhya Pradesh	26.74
Telengana	26.05
Kerala	23.79

Source: MSME Annual Report 2022–23

Table 26.2 displays the allocation of the number of MSMEs among the leading states across the country. Uttar Pradesh and West Bengal possess the largest share of ownership among MSMEs. Kerala is positioned as the 12th state in terms of the overall estimated count of MSMEs in the nation.

The Ministry – MSME establishes "Tool Rooms & Technical Institutions" that have a vital role in delivering annual training programmes aimed at developing practical skills for unemployed youth and the industry workforce. In the fiscal year 2021–22, a total of 18 were established nationwide. These institutions have imparted training to 143,217 trainees, assisted 31,554 units, and generated gains amounting to Rs. 219.72 Crore. Among the 18, 10 of them offer technological assistance to enterprises by creating and producing tools, precision components, moulds, dies, and other related items. These technical colleges (TCs) also benefit businesses by supplying highly trained workforce in fields such as tool engineering and the manufacturing sector. These talented individuals are extremely skilled in their specific areas of expertise. There are eight technology resource centres (TRs) and technology information centres (TIs) that provide specialised support to MSMEs in various sectors. These centres offer tech services to help develop technologies, processes, and products. They also provide sessions in specific product groups, including "Forging & Foundry, Electronics, Electrical Measuring Instruments, Fragrance & Flavour, Glass, Footwear, and Sport Goods". In addition to providing design, development, and manufacturing assistance to MSMEs for intricate tools, parts, and components, certain technical centres (TCs) have also aided the strategic sectors of the country, such as Defence and Aerospace, in their product development endeavours.

26.6.1 Driving Factors for Technology Adoption by MSMEs

The current study includes a sample of 140 MSMEs chosen from the Kollam district of Kerala. The sample consists of 10 different business categories, namely: (1) Textiles/Tailoring, (2) Food Processing, (3) Construction/Electrical Products, (4) Furniture/Wood Products, (5) Plastic/Rubber Products, (6) IT-Enabled Services, (7) Printing/Photography, (8) Beauty Care/Salon Services, (9) Flour Milling, and (10) Miscellaneous.

Table 26.3 displays the distribution of sample micro, small, and medium enterprises (MSMEs) in the Kollam district of Kerala, categorised according to their specific form of business. The data shows that 25% of the total sample units are engaged in the textiles/tailoring business, while 20% are involved in the Food Processing category. Just 9% of the sample is involved in the production of Construction or Electrical Products. There are 11 units engaged in the production of plastic/rubber products. Within the service sector, 6% of the sample units are providing beauty care/salon services, while another 6% are engaged in IT-enabled services. There are now 7 units providing flour milling service, which accounts for 5% of the total. Approximately 14% of the chosen sample falls into the miscellaneous group (Table 26.3).

TABLE 26.3
Classification of Sample Micro, Small, and Medium Enterprises (MSMEs) in Kollam District, Kerala, According to Their Business Type

	Kollam	
Nature of Business	**N**	**%**
Textiles/Tailoring	35	25
Food Processing	28	20
Construction/Electrical Products	13	9
Furniture/Wood Products	5	4
Plastic/Rubber Products	11	8
IT-Enabled Services	8	6
Printing/Photography	4	3
Beauty Care/Saloon Services	9	6
Flour Milling	7	5
Miscellaneous	20	14
Total	140	100

Source: Raw Data

Among the 140 units selected, just 35 percent are owned by women. Age typically plays a significant role in affecting the expansion of entrepreneurial activity. Among the 140 MSMEs, it was noted that 42% are in the age range of 36–45, while just 24% are above the age of 45. Education plays a vital role in fostering and cultivating an entrepreneurial mindset. Twenty percent of the individuals studied up to SSLC, 29% have completed undergraduate studies, 32% have completed graduate studies, and 10% have completed postgraduate studies. Furthermore, it also indicates that the bulk of the respondents are individuals who have completed their studies at a university, with undergraduates being the second largest group. However, the percentage of responders with schooling below SSLC is quite low, accounting for only 9%. Approximately 90% of the participants were entrepreneurs who were the first in their family to start a business. The given information corresponds to Table 26.4.

According to Table 26.5, 64.39% of the MSME units were involved in social media promotion, 64.36% of the respondents received support from Online Marketing, and 57.56% of the sample used POS machines. A significant proportion of the respondents in the flour milling service industry have a low level of technology adoption. However, 72.73% of those involved in producing furniture/wood products have utilised online marketing. Additionally, approximately 85% of businesses offering beauty care services have engaged in online order capturing, social media promotion, and online marketing. Additionally, the bulk of the units did not utilise quick messaging service to enhance their business operations.

Table 26.6 indicates that 79.04% of the units have implemented technology in response to customer needs, while 71.09% of the respondents have chosen technology due to competition. The majority of respondents (83.33%) in the beauty care

TABLE 26.4
Sample Distribution Based on Selected Characteristics

Characteristics		Number	%
Gender	Male	91	65
	Female	49	35
Age	<=35	47	34
	36–45	58	42
	45+	35	24
Education	Below SSLC	12	9
	SSLC	28	20
	Plus-Two	40	29
	Degree	46	32
	PG	14	10
No. of Generations in Business	First	125	89
	Second	15	11
Total		140	100

Source: Primary Data

TABLE 26.5
MSMEs' Perception of Technology Adoption Based on Business Nature

Nature of Business	Technology Adoption				
	Online Order Capturing	Online Marketing	Social Media Promotion	POS Machine	Quick SMS Service
	%	%	%	%	%
Textiles/Tailoring	68.62	60.66	70.82	61.85	34.43
Food Processing	50.00	62.27	71.00	65.91	45.45
Construction/Electrical Products	40.00	65.00	60.00	60.00	20.00
Furniture/Wood Products	36.36	72.73	63.64	63.64	27.27
Plastic/Rubber Products	26.32	63.16	57.89	57.89	15.79
IT-Enabled Services	68.33	78.33	83.33	75.00	50.00
Printing/Photography	69.00	69.00	60.00	63.33	33.33
Beauty Care/Saloon Services	88.89	81.11	86.67	58.22	33.33
Flour Milling	33.33	38.33	30.00	9.33	11.67
Miscellaneous	32.26	50.97	60.52	60.42	35.48
Total	51.31	64.16	64.39	57.56	30.68

Source: Primary Data (Multiple Response)

TABLE 26.6
MSMEs' Perceptions on Technology Adoption Motivations Based on Business Nature

Nature of Business	Reasons for Technology Adoption					
	Competition	Customer Needs	Survival Thrive	Cost Savings	Profit Enhancement	Regulatory Needs
	%	%	%	%	%	%
Textiles/Tailoring	67.21	75.41	79.18	67.87	52.36	79.02
Food Processing	65.91	84.09	70.45	77.73	77.73	80.42
Construction/ Electrical Products	75.00	90.00	70.00	65.00	63.02	75.00
Furniture/Wood Products	63.64	90.91	63.64	65.36	72.27	64.55
Plastic/Rubber Products	84.21	89.47	63.16	62.32	63.29	36.95
IT-Enabled Services	66.67	83.33	62.32	71.67	68.33	60.36
Printing/ Photography	89.33	50	58.36	66.67	83.33	63.21
Beauty Care/Saloon Services	61.11	83.33	68.88	66.67	65.00	65.56
Flour Milling	56.67	63.25	68.88	41.67	50.00	41.67
Miscellaneous	87.1	80.65	78.43	89.43	65.16	65.15
Total	71.09	79.04	68.33	67.44	66.05	63.19

Chi-Square value=77.009, Sig. 0.165

Source: Primary Data * Multiple Response

service industry have embraced technology in response to customer demands, whereas 89.33% of respondents in the printing/photography industry have implemented technology due to competitive pressures.

The Chi-Square Test indicated that there are no significant differences across units with diverse types of businesses and motives for technology adoption since the significance level exceeds 0.05. Put simply, the differences in motives for technology adoption among various types of women entrepreneurs are not statistically significant at a significance threshold of 5%. Hence, it can be deduced that a significant proportion of the participants from various industries have embraced technology due to factors such as client demands and market competition.

26.6.2 Issues/Challenges Faced by MSMEs in Technology Adoption

The unit owners were asked to respond to the stated hurdles regarding technology adoption in MSME units and indicate the extent of their agreement. The data collected consists of the replies from the units, which were rated on a 5-point scale ranging from "Strongly Agree" to "Strongly Disagree" was converted into mean scores. These scores were then presented in Table 26.7. The table shows that the units generally agreed with the variables related to challenges, such as lack of awareness, budget constraints, resistance to change, cybersecurity concerns, and poor connectivity and infrastructure. The mean scores for these variables were significantly above 3.00, as determined by a one-sample *t*-test. Regarding the variable "Budget Constraint", most units have mean scores above 4.00, indicating a strong positive agreement. However, units in the plastic/rubber products sector show a strong disagreement, with a mean score of 2.42.

Regarding the issue of poor connectivity and infrastructure, most women entrepreneurs agree with this problem, which falls into the higher category of grouping. However, those in the beauty care service sector have good infrastructure and connectivity, which means they disagree with the variable of poor networking (with a mean score of 2.59).

Table 26.8 displays the application of principal component analysis (PCA) with assigned weights to different components of difficulties for the purpose of hypothesis testing. In the significance tests, the variable "Cyber Security Concerns" has been allocated the highest weight of 0.186, while the variable "Budget Constraint" has been assigned the least weight of 0.062.

The data in Table 26.9 indicate that there is a considerable variation in the opinions of units from different types of organisations regarding their issues. This is supported by the fact that the significance level of the F-test is less than 0.05. The MSMEs are in consensus regarding the specified variables, as indicated by the total mean score of 3.81. The units in IT-enabled services are in strong agreement with the specified variables.

26.7 SUGGESTIONS AND IMPLICATIONS

The future of sustainable social and economic development relies on establishing networks and collaborations between large corporations and the countless

TABLE 26.7
Perceptions of MSMEs Regarding the Challenges They Face in Adopting Technology, Together with a Statistical Analysis to Determine Significance

Nature of Business	Statistics	Lack of Awareness	Budget Constraints	Resistance to Change	Cyber Security Concerns	Poor Connectivity and Infrastructure
Textiles/Tailoring	Mean	4.21*	4.07^{*}_{2}	3.65*	4.08*	4.01^{*}_{2}
	SD	0.80	0.93	1.91	0.84	0.78
Food Processing	Mean	4.13*	4.04^{*}_{2}	3.52*	4.08*	3.94^{*2}
	SD	0.72	0.98	1.31	0.76	0.85
Construction/ Electrical Products	Mean	4.14*	4.14^{*}_{2}	3.16	4.05*	3.76^{*}_{2}
	SD	0.71	0.67	1.44	0.74	1.01
Furniture/Wood Products	Mean	3.94*	3.25_{12}	2.81	3.69*	4.06^{*}_{2}
	SD	0.57	1.34	1.28	0.87	0.93
Plastic/Rubber Products	Mean	4.15*	2.42^{*}_{1}	3.88*	4.15*	4.15^{*}_{2}
	SD	0.76	1.50	0.89	0.57	0.71
IT-Enabled Services	Mean	4.46*	4.29^{*}_{2}	3.92*	4.29*	3.96^{*}_{2}
	SD	0.59	0.69	0.97	0.86	0.86
Printing/ Photography	Mean	4.58*	4.33^{*}_{2}	3.92*	4.08*	4.00^{*}_{2}
	SD	0.51	0.49	1.00	1.08	0.74
Beauty Care/Saloon Services	Mean	4.15*	4.07^{*}_{2}	3.41	4.07*	2.59_{1}
	SD	0.86	1.11	1.42	0.87	1.34
Flour Milling	Mean	4.00*	4.15^{*}_{2}	3.85*	4.00*	3.60_{12}
	SD	0.92	1.18	1.31	1.12	1.31
Miscellaneous	Mean	3.93*	4.05^{*}_{2}	3.62*	3.98*	3.74^{*}_{2}
	SD	1.00	1.07	1.21	0.88	1.12
Total	Mean	4.14*	3.93*	3.58*	4.06*	3.83*
	SD	0.80	1.12	1.24	0.83	1.00
F-Value		1.575	10.030	1.910	0.689	6.511
Sig.		0.120	0.000	0.049	0.719	0.000

Source: Primary Data *significantly vary from average (3.00) as per one-sample *t*-test with test value 3.00
1 = lower subset, 2 = higher subset, 12 = includable in both subsets as per ScheffeTest

TABLE 26.8
Addressing the Challenges Faced by MSMEs through the Application of Principal Component Analysis (PCA) to Assign Weights to Different Components.

Challenges	Wt.
Lack of Awareness	0.169
Budget Constraints	0.062
Resistance to Change	0.085
Cyber Security Concerns	0.186
Poor Connectivity and Infrastructure	0.099

Source: Primary Data

TABLE 26.9
Mean Opinion Scores of MSMEs about Challenges with Statistical Significance Testing

Nature of Business	Mean	SD	F-Value	Sig.
Textiles/Tailoring	3.92	0.44	5.349	0.000
Food Processing	3.78	0.48		
Construction/Electrical Products	3.66	0.50		
Furniture/Wood Products	3.61	0.40		
Plastic/Rubber Products	3.85	0.53		
IT-Enabled Services	4.18	0.43		
Printing/Photography	4.15	0.39		
Beauty Care/Saloon Services	3.82	0.57		
Flour Milling	3.92	0.60		
Miscellaneous	3.56	0.55		
Total	3.81	0.51		

Source: Primary Data

MSMEs worldwide. These firms play a vital role in advancing innovation, creativity, and providing fair employment opportunities for everyone. Due to their inherent adaptability and expertise in local markets, MSMEs are particularly well-suited to address market-specific obstacles or fulfil increasing demands at the grassroots level. The World Trade Organisation (WTO) and the collaborative effort between the United Nations and the WTO, known as the International Trade Centre (ITC), have initiated significant endeavours to provide assistance to MSMEs operating in the technology industry. These encompass the ITC Global Trade Helpdesk and the WTO MSME platform, which are novel online resources offering immediate and pragmatic assistance to technology entrepreneurs worldwide. Here are a few of the recommendations:

1. Establish Awareness and Outreach Initiatives: Implement awareness campaigns and workshops to enlighten MSMEs regarding the advantages and utilisation of technology in their respective sectors. Engage in partnerships with government agencies, industry associations, and technology suppliers to coordinate events and initiatives that highlight successful cases and illustrate the influence of technology adoption on MSMEs.
2. Create Technology Solutions Tailored to Specific Sectors: Acknowledge the varied requirements of several industries within the MSME domain and promote the advancement of technology solutions tailored to each area. This can be accomplished through collaborations between public and private entities, wherein technology providers collaborate closely with MSMEs and industry specialists to develop tailored solutions that effectively tackle specific difficulties and meet specific needs.
3. Create Technology Incubation Centres: Establish technology incubation centres at regional levels to offer MSMEs access to technological infrastructure, training, and mentorship. These centres serve as central locations where MSMEs can actively engage in the exploration and experimentation of cutting-edge technology, get advice from specialists, and form partnerships with like-minded enterprises.
4. Facilitate Skill Enhancement: Establish partnerships with educational institutions and industry organisations to provide specialised skill development programmes and training courses tailored to the needs of MSMEs. Emphasise the development of digital literacy, technical expertise, and management proficiencies among MSME owners and staff to empower them to successfully embrace and utilise technology.
5. Enhance Digital Infrastructure: Allocate resources to enhance digital infrastructure, including internet access, power supply, and data centres, with a specific focus on rural and semi-urban regions where a significant number of MSMEs are situated. Guarantee dependable and reasonably priced availability of fast internet and sufficient electricity, all of which are essential for the efficient utilisation of technology.
6. Enhance Regulatory Compliance: Optimise and streamline regulatory compliance procedures for MSMEs pertaining to the deployment of technology. Simplify administrative obstacles and offer explicit instructions and assistance to aid MSMEs in managing legal and regulatory obligations related to the adoption and implementation of technological solutions.
7. Implement Support Mechanisms: Create specialised support mechanisms to assist MSMEs throughout their process of adopting technology. These resources encompass helplines, online forums, and technology support centres that offer technical aid, problem-solving, and guidance to MSMEs encountering difficulties or problems with technology integration.

8. Enhance the Adoption of Cybersecurity Measures: Educate MSMEs on the significance of cybersecurity and offer comprehensive guidelines and exemplary methods to safeguard their digital infrastructure. Promote the implementation of cybersecurity protocols, including data encryption, frequent backups, and employee training on cybersecurity awareness, in order to safeguard against cyber threats.

26.8 CONCLUSION

Technology adoption is crucial for the growth of MSMEs in the business world. Small enterprises constitute the vital core of numerous economies, although they encounter distinctive obstacles. In order to surmount these obstacles and achieve their maximum capabilities, MSMEs must adopt and incorporate technology to thrive in the current highly competitive business environment. MSMEs can fully realise their potential by strategically implementing management information systems, information technology, and technology-driven practices. To accomplish this, it is necessary for them to confront the difficulties associated with the adoption of technology and fully embrace the process of digital transformation. By adopting this approach, individuals or organisations can broaden their influence, improve efficiency, and provide a solid foundation for long-term development. Amid a constantly changing environment, micro, small, and medium enterprises (MSMEs) must adapt in order to maintain their competitiveness and relevance. Adopt technology today and ensure a more promising future for your MSME. India's favourable business climate, innovative visions, and initiatives such as Atmanirbhar Bharat and Make in India have expedited the expansion of the MSME sector. These advancements have created further opportunities for firms to rethink and innovate their processes. In light of the rapid rate of transition, businesses have recognised the imperative to enhance efficiency and prepare for the future through digitalisation. To achieve the goal of establishing India as a prominent global manufacturing centre, it is imperative to develop an economy that is both environmentally sustainable and capable of withstanding challenges. Digitalisation will play a crucial part in this endeavour. Proficiency in technical skills and digital literacy will be crucial for effectively harnessing digital technology as MSMEs transition to the digital realm. Organisations must address the skill deficiencies and cultivate a digitally adaptable workforce. In the digital era, the future is heavily reliant on technology. To stay competitive, efficient, productive, and relevant, it is crucial to prioritise projects that focus on acquiring new skills, improving existing skills, and adapting to new skills.

REFERENCES

AlKoliby, I. S. M., Abdullah, H. H., & Suki, N. M. (2023). Linking knowledge application, digital marketing, and manufacturing SMEs' sustainable performance: The mediating role of innovation. *Journal of Knowledge Economy*. https://doi.org/10.1007/s13132-023-01157-4

Amat, M., & Ishak, S. (2019).FaktorPSiKKIT: Pendoronginovasipembungkusandalamka langanIndustri Kecil danSederhanaberasaskanperusahaanmakanan. *Malaysian Journal of Society and Space.* https://doi.org/10.17576/geo-2019-1503-07

Bhambri, P., & Khang, A. (2024). Biosensor applications and principles of agricultural and aquacultural sectors. In A. Khang (Ed.), *Agriculture and Aquaculture Applications of Biosensors and Bioelectronics* (pp. 1–17). IGI Global. https://doi.org/10.4018/979-8-3693-2069–3.ch001

Bhambri, P., & Rani, S. (2024a). Ethical issues for climate change and mental health. In D. Samanta & M. Garg (Eds.), *Impact of Climate Change on Mental Health and Well-Being* (pp. 178–198). IGI Global. https://doi.org/10.4018/979-8-3693-2177–5.ch012

Bhambri, P., & Rani, S. (2024b). Challenges, opportunities, and the future of industrial engineering with IoT and AI. In *Integration of AI-Based Manufacturing and Industrial Engineering Systems with the Internet of Things* (pp. 1–18). CRC Press.

Hussain, A., Shahzad, A., Hassan, R., &Doski, S. A. (2021). COVID-19 impact on B2B e-commerce: A multi-group analysis of sports and surgical SME's. *Pakistan Journal of Commerce and Social Sciences (PJCSS), 15*(1), 166–195.

Marin Bustamante, D. F. (2020). *The Role of New Technologies in International Business in the Context of COVID-19: A Literature Review.* Politecnico.

Nguyen, T. H. (2009). Information technology adoption in SMEs: An integrated framework. *International Journal of Entrepreneurial Behaviour & Research, 15*(2), 162–186.

Omar, A. R. C., Ishak, S., & Jusoh, M. A. (2020). The impact of COVID-19 movement control order on SMEs' businesses and survival strategies. *Malaysian Journal of Society and Space, 16*(2), 139–150.

Rana, R., Bhambri, P., & Chhabra, Y. (2024). Evolution and the future of industrial engineering with the IoT and AI. In *Integration of AI-Based Manufacturing and Industrial Engineering Systems with the Internet of Things* (pp. 19–37). CRC Press.

Shaikh, D., Ara, A., Kumar, M., Syed, D., Ali, A., &Shaikh, M. Z. (2021).A two-decade literature review on challenges faced by SMEs in technology adoption. *Academy of Marketing Studies Journal, 25*, 3.

Singh, G., & Bhambri, P. (2023). Simulation analysis of AODV and DSDV routing protocols for secure and reliable service in mobile adhoc networks (MANETs). In *Integration of AI-Based Manufacturing and Industrial Engineering Systems with the Internet of Things* (pp. 205–216). CRC Press.

Thorgren, S., & Williams, T. A. (2020).Staying alive during an unfolding crisis: How SMEs ward off impending disaster. *Journal of Business Venturing Insights, 14*, e00187.

Untaru, E. N., & Han, H. (2021). Protective measures against COVID-19 and the business strategies of the retail enterprises: Differences in gender, age, education, and income among shoppers. *Journal of Retailing and Consumer Services, 60*, 102446.

Vrontis, D., Chaudhuri, R., & Chatterjee, S. (2022). Adoption of digital technologies by SMEs for sustainability and value creation: moderating role of entrepreneurial orientation. *Sustainability, 14*(13), 7949. https://doi.org/10.3390/su14137949

27 Ethical Considerations in Artificial Intelligence for Environmental Solutions

Striking a Balance for Sustainable Innovation

Rachna Rana and Pankaj Bhambri

27.1 INTRODUCTION

Artificial Intelligence (AI) has undergone significant improvements in modern days and has impacted diverse features of life, including communication, work, and daily living. Since AI continues to infiltrate different facets of civilization, ethical practice in its investigation and employment is becoming increasingly more important. This chapter explores the advantages and possible drawbacks of AI, the moral dilemmas raised by its use, and the efforts being made to address them. It also highlights the importance of achieving a responsible and innovative balance in AI research and use.

According to industry experts, the artificial intelligence industry is predicted to maintain an extraordinary growth rate over the next decade. It is estimated that the market will grow from USD 94.5 billion in 2021 to USD 2,967.42 billion by 2032. This suggests that the market size and demand for the AI industry will rise significantly in the upcoming years.

The advent of AI has brought numerous advantages across various sectors, simplifying complex procedures and providing inventive solutions to previously challenging issues. In healthcare, for instance, AI-powered approaches have been instrumental in identifying illnesses with unparalleled precision, leading to better patient outcomes and saving lives. The transportation industry is also set to be revolutionized with the development of personality-compelling cars, which will decrease the number of accidents caused by human being inaccuracies.

27.2 BENEFITS OF AI

One of the most important benefits of AI is its capacity to computerize repetitive as well as time-consuming development that would otherwise necessitate human

DOI: 10.1201/9781003475989-31

effort. By doing this, companies can save time and money, and their employees can focus on higher-level tasks that require innovative problem-solving techniques. This results in increased productivity and overall efficiency in the workplace.

27.2.1 Superior Effectiveness

AI can enhance effectiveness by scrutinizing large amounts of information and recognizing patterns and trends that humans may not notice. This leads to better decision-making and accurate predictions, which could be especially helpful in industries like finance, healthcare, and logistics.

27.2.2 Customization

AI can also help businesses supply more customized experiences for their customers (Bhambri et al., 2005). By analyzing a customer's purchase history and behavior, AI may suggest products or services that may be of interest to them, or personalize marketing messages to better match their preferences and needs.

27.2.3 Superior Protection

AI can improve protection in diverse fields. For instance, AI-powered systems in the automotive industry help prevent accidents by detecting and responding to potential road hazards. Similarly, the healthcare industry uses AI for patient safety by analyzing medical data and alerting healthcare professionals to potential risks or complications.

27.2.4 Artificial Intelligence

AI has brought about unprecedented innovation and opened up new opportunities for businesses and entrepreneurs. With its ability to automate creative processes such as logo creation and content writing, as well as provide 24/7 customer support via chatbots, AI has become increasingly popular.

In addition, the use of artificial intelligence for environmental sustainability is important in terms of monitoring forest damage, climate change prediction, and energy efficiency. AI-powered language models have transformed the way people interact by eliminating language barriers, facilitating cross-cultural communication, and facilitating international collaboration. The far-reaching impact of AI is immense and is expected to change many industries in the future. Synthetic Brainpower is modifying the way we live and work.

It has many benefits that make our lives easier and enable us to work better. However, there are some significant difficulties in the development and use of this technology. One problem is that AI can amplify and exacerbate existing injustices in society. These algorithms can detect biases in the data and produce unfair or discriminatory results. Another challenge is that the ability of smart devices to recognize and understand personal information can lead to privacy violations. This poses a serious threat to our progress because it can enhance surveillance tools and restrict personal freedom.

The development of autonomous weapons also raises important ethical questions regarding the use of intelligence. These weapons can undermine security by introducing new risks (Bhambri et al., 2005). Negative effects of intelligence must be carefully identified and prevented through development and responsible use. The complexity of AI technology raises many important ethical issues that require attention. One of the biggest challenges is accountability. If an AI system causes harm, who should be responsible: the AI itself, its developers, or the users who interact with it?

The impact of AI on society as a whole has major moral implications. The potential for automation to replace human jobs could lead to unemployment and income inequality. In addition, since artificial intelligence systems are not transparent, complex and difficult to understand, their use in decision-making may lead to negative consequences. These challenges highlight the need to use AI technology responsibly and efficiently.

27.3 CURRENT ENDEAVORS TO ADDRESS AI ETHICS

As AI ethics becomes more important, many groups and initiatives have emerged to address these issues. OpenAI, the AI organization, and the Machine Learning Ethics and Ethics Lab are dedicated to researching and developing best practices for advancing and using AI. Additionally, governments around the world are embracing AI by providing strategies and processes to ensure the effective use of AI. For example, the European Union has developed guidelines to create a legal framework for AI that includes accountability, transparency, and actions to address demand.

27.4 STRIKING AN ADJUST BETWEEN DEVELOPMENT AND OBLIGATION IN AI ADVANCEMENT IS VITAL

Because the digital business world is constantly changing, businesses need the freedom to grow and expand. But managing the balance between innovation and management is crucial. Latest data show that the United States and China are leading the European Union in technological development, while the United States and Europe are lagging behind. Before 2019, China's research and development investment exceeded that of the EU. The European Union allocated €31 billion for the intelligence project to be implemented within five years in order to keep pace with other countries. The recently proposed EU Artificial Intelligence Bill aims to regulate artificial intelligence and ensure the technology is developed and used effectively.

However, this will lead to more regulation for businesses operating in the EU, and some will decide to move to a less restrictive environment. One idea is to include ethical considerations in the plan. By integrating ethics into the artificial intelligence (AI) research and development process, companies can anticipate risks and reduce their likelihood of occurring in unprecedented ways.

Mueller (2021) states that it is possible that some EU businesses may decide to move outside the EU to places with fewer restrictions. Mao (2021) notes that the EU Artificial Intelligence Act supports various initiatives to promote progress in the so-called "sandboxes" that allow national authorities to test the use of new solutions at the environmental fair. However, the impact of EU legislation on intellectual property will create some issues, which will be discussed with high compliance costs, especially for small- and medium-sized enterprises (SMEs).

A report from the Center for Information Technology suggests that using high-risk AI techniques could cost SMEs an average of €400,000 (Mueller, 2021). The private sector is important for the advancement and organization of artificial intelligence, especially in the United States (Parker, 2021) and China. The US National Artificial Intelligence Strategy emphasizes the importance of companies developing AI, especially large American technology companies.

In this way, the country's cooperation with actions to open up AI assets by supporting policies that affect the advancement of AI promotes rapid growth in the AI innovation race in the United States (Parker, 2021). Companies should also accept self-management and contribute to the management of information through self-management. In addition, China's national intellectual strategy relies on the private sector, which often selects the country's "pioneering companies" and technology companies whose achievements and innovations stimulate the advancement of the country's skills beyond expectations.

It establishes a regular regime by promoting certain regimes. It can increase the "competitiveness of technology selection in the development of skills". The state-led approach to AI development thus reflects China's state funding and careful approach to private sector integration and social projects. Collaboration between academia, industry, and government is also critical in achieving this transformation. By making agreements and sharing messages, partners can agree on AI recommendations regarding ethical values and work together to create reliable plans.

27.5 WHAT IS ETHICAL AI?

Ethical AI refers to AI that follows ethical principles. These rules are based on requirements such as personal rights, protection, ethics, and management avoidance (Singh, Singh, Kaur et al., 2005). When organizations use ethical AI, they have clear plans and checklists to follow the rules without question. Ethical AI goes beyond the legal. Although the law has determined the most satisfactory measure regarding the use of artificial intelligence, it has set higher guidelines regarding respect for human values. In the 1940s, famous author Isaac Asimov developed three ethical principles he called "The Three Laws of Robotics".

Use artificial intelligence. This can be seen as the beginning of setting standards: First, the performance dictates that robots should not harm or be allowed to harm humans. When the performance begins, the robot must follow human decisions, unless such decisions are contrary to law. The third law states that the robot must take care of the health of its objects as long as it does not violate the first two important rules. In 2017, a meeting was held at the Asilomar Conference in California to examine the negative effects of artificial intelligence and find ways to solve the problem. So experts wrote a codebook of 23 standards called the Asilomar Standards for Artificial Intelligence, which set the rules for AI ethics.

27.6 SITUATIONS OF ETHICAL AI

It is ensured that the ethics of wisdom also includes confronting and responding to the various problems that arise along the way.

27.6.1 Performance vs. Interpretability

AI faces a balance between execution and explanation. Completion means that the base AI can perform tasks, while explanation means understanding how the base AI makes its choices, such as looking inside its "brain" (Singh, Bhambri et al., 2005). The current phenomenon is that the most powerful AI models are becoming increasingly difficult to obtain. They work like magic, but we cannot replicate the "trick". AI models, on the other hand, take less effort to achieve this, but this may not be true. It seems obvious, but there is less truth. As we increase the size and complexity of AI models to increase performance, AI will become more obscure or difficult to assess. The need for disclosure can lead to a lack of confidence in the findings, making integrity difficult to maintain. Understanding the relationship between AI achievements and explanations means making progress in AI roles without losing our ability to grasp how they work. Rational AI is a development method that aims to provide the necessary intelligence so that we can get accurate results by knowing how the results are produced. In this case, postdoctoral researchers are developing artificial intelligence techniques to teach intended models without compromising their accuracy.

27.6.2 Protection vs. Information Utilization

The issue of data security and use is akin to striking a balance between keeping data private and using data to drive intelligence processes. Security, on the one hand, means protecting sensitive information and ensuring that it is not used or obtained without permission. Data processing uses data to build cognitive models, make predictions, or make recommendations. Adaptation means finding a way to use data, obtaining consent, and taking steps to protect personal data, while respecting protection rights. Ethical AI must address data interests without compromising privacy. Analysts are looking for different ways to balance the protection and use of data. Some important improvements in this context include the use of artificial intelligence:

- Co-authoring
- Privacy-preserving AI program
- Development and Ethical Thinking

Finding a balance between growth and ethical thinking is crucial when creating innovative and successful ideas. While innovation involves researching and testing new ideas to achieve new changes, ethical considerations must govern the impact of these advances on people, community energy, and the environment. Generally speaking, this is a challenge with many angles and dimensions. Here are some key points.

27.6.3 Advancement vs. Natural Duty

Many studies have identified the negative impacts of AI models on the environment when compared to the amount of electricity a car emits over its expected lifespan. This suggests that progress in the development of intelligence and its natural consequences needs to be updated. Economic artificial intelligence has become an effort to reduce the perception of artificial intelligence in terms of intelligence and planning. This includes monitoring data quality rather than quantity, building small but efficient AI models, creating power-efficient AI, optimizing workflows, and improving training memory.

27.6.4 Development vs. Work Uprooting

Artificial intelligence can lead to success and improve performance. It may cause some jobs to be replaced by machines and people to lose their jobs (Singh, Singh, Bhambri et al., 2005). Although artificial intelligence can create jobs today, the changes need to be seen, and the impact on employees needs to be addressed. The plan includes AD planning programs to recall unused skills, work with AI to determine operational tasks, and provide protection to those affected by the technology.

27.6.5 Innovation vs. Deception

The issue of improving intellectual integrity and preventing fraud will be an important task. Two examples that illustrate this challenge are deep learning and chat-bots. Deep fakes are effective but controlled data that can spread misinformation; AI-powered chat-bots can be used to spread false or problematic content. Achieving the balance between development and preventing the spread of fraud must result in a strategy that keeps customers informed and follows best practices. It is important to ensure that AI is used effectively and the potential for harm is minimal.

27.7 THE FOOT LINE

Artificial intelligence has brought advancements in business, but it has also raised ethical concerns. It can be used for fraud detection and tamper protection. It is important to find a balance. Key issues include:

- Performance Comparison and Translation: AI models can be complex, making it difficult to understand how they work. Reasonable intelligence is about preserving the truth while making more reasonable decisions.
- Protection and Use of Information: It is important to have protection when using information for intellectual property purposes. Techniques such as combinatorial learning and differential security can help address this issue.

Development and Ethical Thinking: Development, ethics, and security AI refer to the inherent impacts and needs to support people affected by changing workplaces. To resolve the spoofing issue, the device needs to be installed. By solving these problems, we prepare to develop intellectual skills while making fair and responsible use.

27.7.1 Tending to the Administration Challenges Postured by AI

There are currently two intellectual property management bodies in the world: the WTO and the Organization for Economic Cooperation and Development (OECD). OECD developed the artificial intelligence model in May 2019, which was supported by 42 countries at that time.

The OECD system offers five standards for the operation of artificial intelligence and provides recommendations for governments to achieve these standards. Shortly thereafter, the G20 adopted its own human-based AI model, drawn from (essentially a simplified version of) the OECD model.

The OECD model is also supported by the European Commission, which has been developing an artificial intelligence proposal since April 2018. The EU system has a wide-ranging need, but it also makes it difficult to prepare for socio-economic changes supported by ethical procedures.

27.8 CONCLUSION

Rapid advances in artificial intelligence raise both practical and important ethical issues. For AI to benefit people and minimize harm, it is important to improve support and accountability. Through the importance of AI ethics, people and organizations can better contribute to a future that creates and disseminates innovations that support our collective values and progress. As artificial intelligence becomes increasingly integrated into our daily lives, it is important to be aware of the ethical issues surrounding its use.

People can make a difference by supporting the development of AI capabilities and continuing to understand AI ethics. Companies should prioritize accountability and responsibility in AI operations and work to develop a comprehensive and fair process. In addition, the government should enact regulations that support the development of trustworthy AI and address social impact.

This could include supporting training and career development to reduce job losses and creating social security to support those affected by technological change. Ultimately, the long-term development of intelligence is in our hands. By tackling the ethical issues of AI and working together to find effective solutions, we will help create a future that solves the problems of AI governance and thus increases the power of society. A prosperous future and shared responsibility ensure the unlimited potential of artificial intelligence is realized in a way that protects our ethics and promotes unity.

REFERENCES

Bhambri, P., Gupta, S., & Bhandari, A. (2005, March). Soft computing techniques. In National Conference on Emerging Computing Technologies (pp. 35–41).

Mao, Y. M. (2021). *The EU's Artificial Intelligence Act Could Become a Brake on Innovation.* Finextra.

Mueller, B. (2021). *AI Act Would Cost the EU Economy €31 Billion Over 5 Years, and Reduce AI Investments by Almost 20 Percent, New Report Finds.* Center for Data Innovation.

Parker, L. (2021, June 11). *The American AI Initiative: The U.S. Strategy for Leadership in Artificial Intelligence.* OECD AI Blog.

Singh, M., Bhambri, P., & Kaur, K. (2005, March). Network security. In National Conference on Future Trends in Information Technology (pp. 51–56).

Singh, M., Singh, P., Bhambri, P., & Sachdeva, R. (2005). A comparative study: Security algorithms. In Seminar on Network Security and Its Implementations (p. 14).

Singh, M., Singh, P., Kaur, K., & Bhambri, P. (2005, March). Database security. In National Conference on Future Trends in Information Technology (pp. 57–62).

Index

For Product Safety Concerns and Information please contact our EU representative GPSR@taylorandfrancis.com Taylor & Francis Verlag GmbH, Kaufingerstraße 24, 80331 München, Germany

Batch number: 10397790

Printed by Printforce, the Netherlands